Applied Survey Methods

WILEY SERIES IN SURVEY METHODOLOGY
Established in Part by WALTER A. SHEWHART AND SAMUEL S. WILKS

Editors: *Robert M. Groves, Graham Kalton, J. N. K. Rao, Norbert Schwarz, Christopher Skinner*

A complete list of the titles in this series appears at the end of this volume.

Applied Survey Methods
A Statistical Perspective

JELKE BETHLEHEM

A JOHN WILEY & SONS, INC., PUBLICATION

Cover design image copyrighted by permission of Hayo Bethlehem

Copyright © 2009 by John Wiley & Sons, Inc. All rights reserved

Published by John Wiley & Sons, Inc., Hoboken, New Jersey
Published simultaneously in Canada

No part of this publication may be reproduced, stored in a retrieval system, or transmitted in any form or by any means, electronic, mechanical, photocopying, recording, scanning, or otherwise, except as permitted under Section 107 or 108 of the 1976 United States Copyright Act, without either the prior written permission of the Publisher, or authorization through payment of the appropriate per-copy fee to the Copyright Clearance Center, Inc., 222 Rosewood Drive, Danvers, MA 01923, (978) 750-8400, fax (978) 750-4470, or on the web at www.copyright.com. Requests to the Publisher for permission should be addressed to the Permissions Department, John Wiley & Sons, Inc., 111 River Street, Hoboken, NJ 07030, (201) 748-6011, fax (201) 748-6008, or online at http://www.wiley.com/go/permission.

Limit of Liability/Disclaimer of Warranty: While the publisher and author have used their best efforts in preparing this book, they make no representations or warranties with respect to the accuracy or completeness of the contents of this book and specifically disclaim any implied warranties of merchantability or fitness for a particular purpose. No warranty may be created or extended by sales representatives or written sales materials. The advice and strategies contained herein may not be suitable for your situation. You should consult with a professional where appropriate. Neither the publisher nor author shall be liable for any loss of profit or any other commercial damages, including but not limited to special, incidental, consequential, or other damages.

For general information on our other products and services or for technical support, please contact our Customer Care Department within the United States at (800) 762-2974, outside the United States at (317) 572-3993 or fax (317) 572-4002.

Wiley also publishes its books in a variety of electronic formats. Some content that appears in print may not be available in electronic formats. For more information about Wiley products, visit our web site at www.wiley.com.

Library of Congress Cataloging-in-Publication Data:

Bethlehem, Jelke G.
  Applied survey methods : a statistical perspective / Jelke Bethlehem.
    p. cm. – (Wiley series in survey methodology)
  Includes bibliographical references and index.
  ISBN 978-0-470-37308-8 (cloth)
 1. Surveys–Statistical methods. 2. Sampling (Statistics)
3. Surveys–Methodology. 4. Estimation theory. I. Title.
  QA276.B429 2009
  001.4'33–dc22   2009001788

Printed in the United States of America
10 9 8 7 6 5 4 3 2 1

Contents

Preface ix

1. The Survey Process 1

    1.1. About Surveys, 1
    1.2. A Survey, Step-by-Step, 2
    1.3. Some History of Survey Research, 4
    1.4. This Book, 10
    1.5. Samplonia, 11
    Exercises, 13

2. Basic Concepts 15

    2.1. The Survey Objectives, 15
    2.2. The Target Population, 16
    2.3. The Sampling Frame, 20
    2.4. Sampling, 22
    2.5. Estimation, 33
    Exercises, 41

3. Questionnaire Design 43

    3.1. The Questionnaire, 43
    3.2. Factual and Nonfactual Questions, 44
    3.3. The Question Text, 45
    3.4. Answer Types, 50
    3.5. Question Order, 55
    3.6. Questionnaire Testing, 58
    Exercises, 63

4. Single Sampling Designs — 65

4.1. Simple Random Sampling, 65
4.2. Systematic Sampling, 75
4.3. Unequal Probability Sampling, 82
4.4. Systematic Sampling with Unequal Probabilities, 89
Exercises, 96

5. Composite Sampling Designs — 100

5.1. Stratified Sampling, 100
5.2. Cluster Sampling, 108
5.3. Two-Stage Sampling, 113
5.4. Two-Dimensional Sampling, 122
Exercises, 130

6. Estimators — 134

6.1. Use of Auxiliary Information, 134
6.2. A Descriptive Model, 134
6.3. The Direct Estimator, 137
6.4. The Ratio Estimator, 139
6.5. The Regression Estimator, 143
6.6. The Poststratification Estimator, 146
Exercises, 149

7. Data Collection — 153

7.1. Traditional Data Collection, 153
7.2. Computer-Assisted Interviewing, 155
7.3. Mixed-Mode Data Collection, 160
7.4. Electronic Questionnaires, 163
7.5. Data Collection with Blaise, 167
Exercises, 176

8. The Quality of the Results — 178

8.1. Errors in Surveys, 178
8.2. Detection and Correction of Errors, 181
8.3. Imputation Techniques, 185
8.4. Data Editing Strategies, 195
Exercises, 206

CONTENTS

9. The Nonresponse Problem — 209

- 9.1. Nonresponse, 209
- 9.2. Response Rates, 212
- 9.3. Models for Nonresponse, 218
- 9.4. Analysis of Nonresponse, 225
- 9.5. Nonresponse Correction Techniques, 236
- Exercises, 245

10. Weighting Adjustment — 249

- 10.1. Introduction, 249
- 10.2. Poststratification, 250
- 10.3. Linear Weighting, 253
- 10.4. Multiplicative Weighting, 260
- 10.5. Calibration Estimation, 263
- 10.6. Other Weighting Issues, 264
- 10.7. Use of Propensity Scores, 266
- 10.8. A Practical Example, 268
- Exercises, 272

11. Online Surveys — 276

- 11.1. The Popularity of Online Research, 276
- 11.2. Errors in Online Surveys, 277
- 11.3. The Theoretical Framework, 283
- 11.4. Correction by Adjustment Weighting, 288
- 11.5. Correction Using a Reference Survey, 293
- 11.6. Sampling the Non-Internet Population, 296
- 11.7. Propensity Weighting, 297
- 11.8. Simulating the Effects of Undercoverage, 299
- 11.9. Simulating the Effects of Self-Selection, 301
- 11.10. About the Use of Online Surveys, 305
- Exercises, 307

12. Analysis and Publication — 310

- 12.1. About Data Analysis, 310
- 12.2. The Analysis of Dirty Data, 312
- 12.3. Preparing a Survey Report, 317
- 12.4. Use of Graphs, 322
- Exercises, 339

13. Statistical Disclosure Control — 342

  13.1. Introduction, 342
  13.2. The Basic Disclosure Problem, 343
  13.3. The Concept of Uniqueness, 344
  13.4. Disclosure Scenarios, 347
  13.5. Models for the Disclosure Risk, 349
  13.6. Practical Disclosure Protection, 353
  Exercises, 356

References — 359

Index — 369

Preface

This is a book about surveys. It describes the whole survey process, from design to publication. It not only presents an overview of the theory from a statistical perspective, but also pays attention to practical problems. Therefore, it can be seen as a handbook for those involved in practical survey research. This includes survey researchers working in official statistics (e.g., in national statistical institutes), academics, and commercial market research.

The book is the result of many years of research in official statistics at Statistics Netherlands. Since the 1980s there have been important developments in computer technology that have had a substantial impact on the way in which surveys are carried out. These developments have reduced costs of surveys and improved the quality of survey data. However, there are also new challenges, such as increasing nonresponse rates.

The book starts by explaining what a survey is, and why it is useful. There is a historic overview describing how the first ideas have developed since 1895. Basic concepts such as target population, population parameters, variables, and samples are defined, leading to the Horvitz–Thompson estimator as the basis for estimation procedures.

The questionnaire is the measuring instrument used in a survey. Unfortunately, it is not a perfect instrument. A lot can go wrong in the process of asking and answering questions. Therefore, it is important to pay careful attention to the design of the questionnaire. The book describes rules of thumb and stresses the importance of questionnaire testing.

Taking the Horvitz–Thompson estimator as a starting point, a number of sampling designs are discussed. It begins with simple sampling designs such as simple random sampling, systematic sampling, sampling with unequal probabilities, and systematic sampling with unequal probabilities. This is followed by some more complex sampling designs that use simple designs as ingredients: stratified sampling, cluster sampling, two-stage sampling, and two-dimensional sampling (including sampling in space and time).

Several estimation procedures are described that use more information than the Horvitz–Thompson estimator. They are all based on a general descriptive model using auxiliary information to estimate population characteristics. Estimators discussed include the direct estimator, the ratio estimator, the regression estimator, and the poststratification estimator.

The book pays attention to various ways of data collection. It shows how traditional data collection using paper forms (PAPI) evolved into computer-assisted data collection (CAPI, CATI, etc.). Also, online surveys are introduced. Owing to its special nature and problems, and large popularity, online surveys are discussed separately and more extensively. Particularly, attention is paid to undercoverage and self-selection problems. It is explored whether adjustment weighting may help reduce problems. A short overview is given of the Blaise system. It is the de facto software standard (in official statistics) for computer-assisted interviewing.

A researcher carrying out a survey can be confronted with many practical problems. A taxonomy of possible errors is described. Various data editing techniques are discussed to correct detected errors. Focus is on data editing in large statistical institutes. Aspects discussed include the Felligi–Holt methodology, selective editing, automated editing, and macroediting. Also, a number of imputation techniques are described (including the effect they may have on the properties of estimators).

Nonresponse is one of the most important problems in survey research. The book pays a lot of attention to this problem. Two theoretical models are introduced to analyze the effects of nonresponse: the fixed response model and the random response model. To obtain insight into the possible effects of nonresponse, analysis of nonresponse is important. An example of such an analysis is given. Two approaches are discussed to reduce the negative effects of nonresponse: a follow-up survey among nonrespondents and the Basic Question Approach.

Weighting adjustment is the most important technique to correct a possible nonresponse bias. Several adjustment techniques are described: simple poststratification, linear weighting (as a form of generalized regression estimation), and multiplicative weighting (raking ratio estimation, iterative proportional fitting). A short overview of calibration estimation is included. It provides a general theoretical framework for adjustment weighting. Also, some attention is paid to propensity weighting.

The book shows what can go wrong if in the analysis of survey data not all aspects of the survey design and survey process are taken into account (e.g., unequal probability sampling, imputation, weighting). The survey results will be published in some kind of survey report. Checklists are provided of what should be included in such a report. The book also discusses the use of graphs in publications and how to prevent misuse.

The final chapter of the book is devoted to disclosure control. It describes the problem of prevention of disclosing sensitive information in survey data files. It shows how simple disclosure can be accomplished. It gives some theory to estimate disclosure risks. And it discusses some techniques to prevent disclosure.

The fictitious country of Samplonia is introduced in the book. Data from this country are used in many examples throughout the book. There is a small computer program *SimSam* that can be downloaded from the book website (www.applied-survey-methods.com). With this program, one can simulate samples from finite populations and show the effects of sample size, use of different estimation procedures, and nonresponse.

A demo version of the *Blaise system* can also be downloaded from the website. Small and simple surveys can be carried out with this is demo version.

The website www.applied-survey-methods.com gives an overview of some basic concepts of survey sampling. It includes some dynamic demonstrations and has some helpful tools, for example, to determine the sample size.

JELKE BETHLEHEM

CHAPTER 1

The Survey Process

1.1 ABOUT SURVEYS

We live in an information society. There is an ever-growing demand for statistical information about the economic, social, political, and cultural shape of countries. Such information will enable policy makers and others to make informed decisions for a better future. Sometimes, such statistical information can be retrieved from existing sources, for example, administrative records. More often, there is a lack of such sources. Then, a survey is a powerful instrument to collect new statistical information.

A *survey* collects information about a well-defined population. This population need not necessarily consist of persons. For example, the elements of the population can be households, farms, companies, or schools. Typically, information is collected by asking questions to the representatives of the elements in the population. To do this in a uniform and consistent way, a questionnaire is used.

One way to obtain information about a population is to collect data about all its elements. Such an investigation is called a *census* or *complete enumeration*. This approach has a number of disadvantages:

- It is very expensive. Investigating a large population involves a lot of people (e.g., interviewers) and other resources.
- It is very time-consuming. Collecting and processing a large amount of data takes time. This affects the timeliness of the results. Less timely information is less useful.
- Large investigations increase the response burden on people. As many people are more frequently asked to participate, they will experience it more and more as a burden. Therefore, people will be less and less inclined to cooperate.

A survey is a solution to many of the problems of a census. Surveys collect information on only a small part of the population. This small part is called the *sample*.

Applied Survey Methods: A Statistical Perspective, Jelke Bethlehem
Copyright © 2009 John Wiley & Sons, Inc.

In principle, the sample provides information only on the sampled elements of the population. No information will be obtained on the nonsampled elements. Still, if the sample is selected in a "clever" way, it is possible to make inference about the population as a whole. In this context, "clever" means that the sample is selected using probability sampling. A random selection procedure uses an element of chance to determine which elements are selected, and which are not. If it is clear how this selection mechanism works and it is possible to compute the probabilities of being selected in the sample, survey results allow making reliable and precise statements about the population as a whole.

At first sight, the idea of introducing an element of uncertainty in an investigation seems odd. It looks like magic that it is possible to say something about a complete population by investigating only a small randomly selected part of it. However, there is no magic about sample surveys. There is a well-founded theoretical framework underlying survey research. This framework will be described in this book.

1.2 A SURVEY, STEP-BY-STEP

Carrying out a survey is often a complex process that requires careful consideration and decision making. This section gives a global overview of the various steps in the process, the problems that may be encountered, and the decisions that have to be made. The rest of the book describes these steps in much more detail. Figure 1.1 shows the steps in the survey process.

The first step in the survey process is *survey design*. Before data collection can start, a number of important decisions have to be made. First, it has to become clear which population will be investigated (the *target population*). Consequently, this is the population to which the conclusions apply. Next, the general research questions must

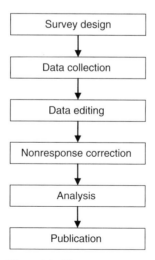

Figure 1.1 The survey process.

be translated into specification of population characteristics to be estimated. This specification determines the contents of the questionnaire. Furthermore, to select a proper sample, a sampling design must be defined, and the sample size must be determined such that the required accuracy of the results can be obtained.

The second step in the process is data collection. Traditionally, in many surveys paper questionnaires were used. They could be completed in face-to-face interviews: interviewers visited respondents, asked questions, and recorded the answers on (paper) forms. The quality of the collected data tended to be good. However, since face-to-face interviewing typically requires a large number of interviewers, who all may have to do much traveling, it was expensive and time-consuming. Therefore, telephone interviewing was often used as an alternative. The interviewers called the respondents from the survey agency, and thus no more traveling was necessary. However, telephone interviewing is not always feasible: only connected (or listed) people can be contacted, and the questionnaire should not be too long or too complicated. A mail survey was cheaper still: no interviewers at all were needed. Questionnaires were mailed to potential respondents with the request to return the completed forms to the survey agency. Although reminders could be sent, the persuasive power of the interviewers was lacking, and therefore response tended to be lower in this type of survey, and so was the quality of the collected data.

Nowadays paper questionnaires are often replaced with electronic ones. *Computer-assisted interviewing* (CAI) allows to speed up the survey process, improve the quality of the collected data, and simplify the work of the interviewers. In addition, computer-assisted interviewing comes in three forms: *computer-assisted personal interviewing* (CAPI), *computer-assisted telephone interviewing* (CATI), and *computer-assisted self-interviewing* (CASI). More and more, the Internet is used for completing survey questionnaires. This is called *computer-assisted web interviewing* (CAWI). It can be seen as a special case of CASI.

Particularly if the data are collected by means of paper questionnaire forms, the completed questionnaires have to undergo extensive treatment. To produce high-quality statistics, it is vital to remove any error. This step of the survey process is called *data editing*. Three types of errors can be distinguished: A *range error* occurs if a given answer is outside the valid domain of answers; for example, a person with an age of 348 years. A *consistency error* indicates an inconsistency in the answers to a set of questions. An age of 8 years may be valid, a marital status "married" is not uncommon, but if both answers are given by the same person, there is something definitely wrong. The third type of error is a *routing error*. This type of error occurs if interviewers or respondents fail to follow the specified branch or skip instructions; that is, the route through the questionnaire is incorrect: irrelevant questions are answered, or relevant questions are left unanswered.

Detected errors have to be corrected, but this can be very difficult if it has to be done afterward, at the survey agency. In many cases, particularly for household surveys, respondents cannot be contacted again, so other ways have to be found out to solve the problem. Sometimes, it is possible to determine a reasonable approximation of a correct value by means of an *imputation* procedure, but in other cases an incorrect value is replaced with the special code indicating the value is "unknown."

After data editing, the result is a "clean" data file, that is, a data file in which no errors can be detected any more. However, this file is not yet ready for analysis. The collected data may not be representative of the population because the sample is affected by nonresponse; that is, for some elements in the sample, the required information is not obtained. If nonrespondents behave differently with respect to the population characteristics to be investigated, the results will be biased. To correct for unequal selection probabilities and nonresponse, a *weighting adjustment* procedure is often carried out. Every record is assigned some weight. These weights are computed in such a way that the weighted sample distribution of characteristics such as gender, age, marital status, and region reflects the known distribution of these characteristics in the population.

In the case of item nonresponse, that is, answers are missing on some questions, not all questions, an *imputation* procedure can also be carried out. Using some kind of model, an estimate for a missing value is computed and substituted in the record.

Finally, a data file is obtained that is ready for analysis. The first step in the analysis will probably nearly always be tabulation of the basic characteristics. Next, a more extensive analysis will be carried out. Depending on the nature of the study, this will take the form of an exploratory analysis or an inductive analysis. An *exploratory analysis* will be carried out if there are no preset ideas, and the aim is to detect possibly existing patterns, structures, and relationships in the collected data. To make inference on the population as a whole, an *inductive analysis* can be carried out. This can take the form of estimation of population characteristics or the testing of hypotheses that have been formulated about the population.

The survey results will be published in some kind of report. On the one hand, this report must present the results of the study in a form that makes them readable for nonexperts in the field of survey research. On the other hand, the report must contain a sufficient amount of information for experts to establish whether the study was carried out properly and to assess the validity of the conclusions.

Carrying out a survey is a time-consuming and expensive way of collecting information. If done well, the reward is a data file full of valuable information. It is not unlikely that other researchers may want to use these data in additional analysis. This brings up the question of protecting the privacy of the participants in the survey. Is it possible to disseminate survey data sets without revealing sensitive information of individuals? Disclosure control techniques help establish disclosure risks and protect data sets against disclosing such sensitive information.

1.3 SOME HISTORY OF SURVEY RESEARCH

The idea of compiling statistical overviews of the state of affairs in a country is already very old. As far back as Babylonian times, censuses of agriculture were taken. This happened fairly shortly after the art of writing was invented. Ancient China counted its people to determine the revenues and the military strength of its provinces. There are also accounts of statistical overviews compiled by Egyptian rulers long before Christ. Rome regularly took a census of people and of property. The data were used to establish the political status of citizens and to assess their military and tax obligations to the state. And of course, there was numbering of the people of Israel, leading to the birth of Jesus in the small town of Bethlehem.

In the Middle Ages, censuses were rare. The most famous one was the census of England taken by the order of William the Conqueror, King of England. The compilation of this *Domesday Book* started in the year 1086 AD. The book records a wealth of information about each manor and each village in the country. There is information about more than 13,000 places, and on each county there are more than 10,000 facts. To collect all these data, the country was divided into a number of regions, and in each region, a group of commissioners was appointed from among the greater lords. Each county within a region was dealt with separately. Sessions were held in each county town. The commissioners summoned all those required to appear before them. They had prepared a standard list of questions. For example, there were questions about the owner of the manor, the number of free men and slaves, the area of woodland, pasture, and meadow, the number of mills and fishponds, to the total value, and the prospects of getting more profit. The *Domesday Book* still exists, and county data files are available on CD-ROM or the Internet.

Another interesting example can be found in the Inca Empire that existed between 1000 and 1500 AD in South America. Each Inca tribe had its own statistician, called *Quipucamayoc* (Fig. 1.2). This man kept records of, for example, the number of people, the number of houses, the number of llamas, the number of marriages, and the number of young men who could be recruited to the army. All these facts were recorded on a *quipu*, a system of knots in colored ropes. A decimal system was used for this.

Figure 1.2 The Quipucamayoc, the Inca statistician. Reprinted by permission of ThiemeMeulenhoff.

At regular intervals, couriers brought the quipus to Cusco, the capital of the kingdom, where all regional statistics were compiled into national statistics. The system of Quipucamayocs and quipus worked remarkably well. Unfortunately, the system vanished with the fall of the empire.

An early census also took place in Canada in 1666. Jean Talon, the intendant of New France, ordered an official census of the colony to measure the increase in population since the founding of Quebec in 1608. The enumeration, which recorded a total of 3215 people, included the name, age, gender, marital status, and occupation of every person. The first censuses in Europe were undertaken by the Nordic countries: The first census in Sweden–Finland took place in 1746. It had already been suggested earlier, but the initiative was rejected because "it corresponded to the attempt of King David who wanted to count his people."

The first known attempt to make statements about a population using only information about part of it was made by the English merchant John Graunt (1620–1674). In his famous tract, Graunt describes a method to estimate the population of London on the basis of partial information (Graunt, 1662). Graunt surveyed families in a sample of parishes where the registers were well kept. He found that on average there were 3 burials per year in 11 families. Assuming this ratio to be more or less constant for all parishes, and knowing the total number of burials per year in London to be about 13,000, he concluded that the total number of families was approximately 48,000. Putting the average family size at 8, he estimated the population of London to be 384,000. Although Graunt was aware of the fact that averages such as the number of burials per family varied in space and time, he did not make any provisions for this phenomenon. Lacking a proper scientific foundation for his method, John Graunt could not make any statement about the accuracy of his method.

Another survey-like method was applied more than a century later. Pierre Simon Laplace (1749–1827) realized that it was important to have some indication of the accuracy of the estimate of the French population. Laplace (1812) implemented an approach that was more or less similar to that of John Graunt. He selected 30 departments distributed over the area of France. Two criteria controlled the selection process. First, he saw to it that all types of climate were represented. In this way, he could compensate for climate effects. Second, he selected departments for which the mayors of the communes could provide accurate information. By using the central limit theorem, he proved that his estimator had a normal distribution. Unfortunately, he overlooked the fact that he used a cluster sample instead of a simple random sample, and moreover communes were selected within departments purposively, and not at random. These problems made the application of the central limit theorem at least doubtful. The work of Laplace was buried in oblivion in the course of the nineteenth century.

In the period until the late 1880s, there were many *partial investigations*. These were statistical inquiries in which only a part of human population was investigated. The selection from the population came to hand incidentally, or was made specifically for the investigation. In general, the selection mechanism was unclear and undocumented. While by that time considerable progress had already been made in the areas of probability theory and mathematical statistics, little attention was paid to applying these theoretical developments to survey sampling. Nevertheless, gradually probability theory found its way in official statistics. An important role was played by the Dutch/Belgian scientist, Lambert Adolphe Jacques Quetelet (1796–1874). He was involved in the first attempt in 1826 to establish The Netherlands Central Bureau of Statistics. In 1830, Belgium separated from The Netherlands, and Quetelet continued his work in Belgium.

Quetelet was the supervisor of statistics for Belgium (from 1830), and in this position, he developed many of the rules governing modern census taking. He also stimulated statistical activities in other countries. The Belgian census of 1846, directed by him, has been claimed to be the most influential in its time because it introduced careful analysis and critical evaluation of the data compiled. Quetelet dealt only with censuses and did not carry out any partial investigations.

According to Quetelet, many physical and moral data have a natural variability. This variability can be described by a normal distribution around a fixed, true value. He assumed the existence of something called the *true value*. He proved that this true value could be estimated by taking the mean of a number of observations. Quetelet introduced the concept of *average man* ("l'homme moyenne") as a person of which all characteristics were equal to the true value. For more information, see Quetelet (1835, 1846).

In the second half of the nineteenth century, so-called *monograph studies* or surveys became popular. They were based on Quetelet's idea of the average man (see Desrosières, 1998). According to this idea, it suffices to collect information only on typical people. Investigation of extreme people was avoided. This type of inquiry was still applied widely at the beginning of the twentieth century. It was an "officially" accepted method.

Industrial revolution was also an important era in the history of statistics. It brought about drastic and extensive changes in society, as well as in science and technology. Among many other things, urbanization started from industrialization, and also democratization and the emerging social movements at the end of the industrial revolution created new statistical demands. The rise of statistical thinking originated partly from the demands of society and partly from work and innovations of men such as Quetelet. In this period, the foundations for many principles of modern social statistics were laid. Several central statistical bureaus, statistical societies, conferences, and journals were established soon after this period.

The development of modern sampling theory started around the year 1895. In that year, Anders Kiaer (1895, 1997), the founder and first director of Statistics Norway, published his *Representative Method*. It was a partial inquiry in which a large number of persons were questioned. This selection should form a "miniature" of the population. Persons were selected arbitrarily but according to some rational scheme based on general results of previous investigations. Kiaer stressed the importance of *representativeness*. His argument was that if a sample was representative with respect to variables for which the population distribution was known, it would also be representative with respect to the other survey variables.

Kiaer was way ahead of his time with ideas about survey sampling. This becomes clear in the reactions on the paper he presented at a meeting of the International

Statistical Institute in Bern in 1895. The last sentence of a lengthy comment by the influential Bavarian statistician von Mayr almost became a catch phrase: "Il faut rester ferme et dire: pas de calcul là où l'obervation peut être faite." The Italian statistician Bodio supported von Mayr's views. The Austrian statistician Rauchberg said that further discussion of the matter was unnecessary. And the Swiss statistician Milliet demanded that such incomplete surveys should not be granted a status equal to "la statistique serieuse."

A basic problem of the representative method was that there was no way of establishing the accuracy of estimates. The method lacked a formal theory of inference. It was Bowley (1906) who made the first steps in this direction. He showed that for large samples, selected at random from the population, the estimate had an approximately normal distribution.

From this moment on, there were two methods of sample selection. The first one was Kiaer's representative method, based on purposive selection, in which representativeness played a crucial role, and for which no measure of the accuracy of the estimates could be obtained. The second was Bowley's approach, based on simple random sampling, for which an indication of the accuracy of estimates could be computed. Both methods existed side by side for a number of years. This situation lasted until 1934, when the Polish scientist Jerzy Neyman published his now famous paper (see Neyman, 1934). Neyman developed a new theory based on the concept of the confidence interval. By using random selection instead of purposive selection, there was no need any more to make prior assumptions about the population.

Neyman's contribution was not restricted to the confidence interval that he invented. By making an empirical evaluation of Italian census data, he could prove that the representative method based on purposive sampling failed to provide satisfactory estimates of population characteristics. The result of Neyman's evaluation of purposive sampling was that the method fell into disrepute in official statistics.

Random selection became an essential element of survey sampling. Although theoretically very attractive, it was not very simple to realize this in practical situations. How to randomly select a sample of thousands of persons from a population of several millions? How to generate thousands of random numbers? To avoid this problem, often systematic samples were selected. Using a list of elements in the population, a starting point and a step size were specified. By stepping through this list from the starting point, elements were selected. Provided the order of the elements is more or less arbitrary, this systematic selection resembles random selection. W.G. and L.H. Madow made the first theoretical study of the precision of systematic sampling only in 1944 (see Madow and Madow, 1944). The use of the first tables of random numbers published by Tippet (1927) also made it easier to select real random samples.

In 1943, Hansen and Hurvitz published their theory of multistage samples. According to their theory, in the first stage, primary sampling units are selected with probabilities proportional to their size. Within selected primary units, a fixed number of secondary units are selected. This proved to be a useful extension of the survey sampling theory. On the one hand, this approach guaranteed every secondary unit to have the same probability of selection in the sample, and on the other, the sampled units were distributed over the population in such a way that the fieldwork could be carried out efficiently.

The classical theory of survey sampling was more or less completed in 1952. Horvitz and Thompson (1952) developed a general theory for constructing unbiased estimates. Whatever the selection probabilities are, as long as they are known and positive, it is always possible to construct a reliable estimate. Horvitz and Thompson completed the classical theory, and the random sampling approach was almost unanimously accepted. Most of the classical books about sampling were also published by then: Cochran (1953), Deming (1950), Hansen et al. (1953), and Yates (1949).

Official statistics was not the only area where sampling was introduced. Opinion polls can be seen as a special type of sample surveys, in which attitudes or opinions of a group of people are measured on political, economic, or social topics. The history of opinion polls in the United States goes back to 1824, when two newspapers, the *Harrisburg Pennsylvanian* and the *Raleigh Star*, attempted to determine political preferences of voters before the presidential election. The early polls did not pay much attention to sampling. Therefore, it was difficult to establish the accuracy of results. Such opinion polls were often called *straw polls*. This expression goes back to rural America. Farmers would throw a handful of straws into the air to see which way the wind was blowing. In the 1820s, newspapers began doing straw polls in the streets to see how political winds blew.

It took until the 1920s before more attention was paid to sampling aspects. At that time, Archibald Crossley developed new techniques for measuring American public's radio listening habits. And George Gallup worked out new ways to assess reader interest in newspaper articles (see, for example, Linehard, 2003). The sampling technique used by Gallup was *quota sampling*. The idea was to investigate groups of people who were representative for the population. Gallup sent out hundreds of interviewers across the country. Each interviewer was given quota for different types of respondents: so many middle-class urban women, so many lower class rural men, and so on. In total, approximately 3000 interviews were carried out for a survey.

Gallup's approach was in great contrast with that of the *Literary Digest* magazine, which was at that time the leading polling organization. This magazine conducted regular "America Speaks" polls. It based its predictions on returned ballot forms that were sent to addresses obtained from telephone directories and automobile registration lists. The sample size for these polls was very large, something like 2 million people.

The presidential election of 1936 turned out to be decisive for both approaches (see Utts, 1999). Gallup correctly predicted Franklin Roosevelt to be the new President, whereas *Literary Digest* predicted that Alf Landon would beat Franklin Roosevelt. How could a prediction based on such a large sample be so wrong? The explanation was a fatal flaw in the *Literary Digest*'s sampling mechanism. The automobile registration lists and telephone directories were not representative samples. In the 1930s, cars and telephones were typically owned by the middle and upper classes. More well-to-do Americans tended to vote Republican and the less well-to-do were inclined to vote Democrat. Therefore, Republicans were overrepresented in the *Literary Digest* sample.

As a result of this historic mistake, the *Literary Digest* magazine ceased publication in 1937. And opinion researchers learned that they should rely on more scientific ways of sample selection. They also learned that the way a sample is selected is more important than the size of the sample.

1.4 THIS BOOK

This book deals with the theoretical and practical aspects of sample survey sampling. It follows the steps in the survey process described in Section 1.1.

Chapter 2 deals with various aspects related to the design of a survey. Basic concepts are introduced, such as population, population parameters, sampling, sampling frame, and estimation. It introduces the Horvitz–Thompson estimator as the basis for estimation under different sampling designs.

Chapter 3 is devoted to questionnaire designing. It shows the vital importance of properly defined questions. Its also discusses various question types, routing (branching and skipping) in the questionnaire, and testing of questionnaires.

Chapters 4 and 5 describe a number of sampling designs in more detail. Chapter 3 starts with some simple sampling designs: simple random sampling, systematic sampling, unequal probability sampling, and systematic sampling with unequal probabilities. Chapter 4 continues with composite sampling designs: stratified sampling, cluster sampling, two-stage sampling, and sampling in space and time.

Chapter 6 presents a general framework for estimation. Starting point is a linear model that explains the target variable of a survey from one or more auxiliary variables. Some well-known estimators, such as the ratio estimator, the regression estimator, and the poststratification estimator, emerge as special cases of this model.

Chapter 7 is about data collection. It compares traditional data collection with paper questionnaire forms with computer-assisted data collection. Advantages and disadvantages of various modes of data collection are discussed. To give some insight into the attractive properties of computer-assisted interviewing, a software package is described that can be seen as the de facto standard for CAI in official statistics. It is the Blaise system.

Chapter 8 is devoted to the quality aspects. Collected survey data always contain errors. This chapter presents a classification of things that can go wrong. Errors can have a serious impact on the reliability of survey results. Therefore, extensive error checking must be carried out. It is also shown that correction of errors is not always simple. Imputation is discussed as one of the error correction techniques.

Nonresponse is one of the most important problems in survey research. Nonresponse can cause survey estimates to be seriously biased. Chapter 9 describes the causes of nonresponse. It also incorporates this phenomenon in sampling theory, thereby showing what the effects of nonresponse can be. Usually, it is not possible to avoid nonresponse in surveys. This calls for techniques that attempt to correct the negative effect of nonresponse. Two approaches are discussed in this chapter: the follow-up survey and the Basic Question Approach.

Adjustment weighting is one of the most important nonresponse correction techniques. This technique assigns weights to responding elements. Overrepresented groups get a small weight and underrepresented groups get a large weight. Therefore, the weighted sample becomes more representative for the population, and the estimates based on weighted data have a smaller bias than estimates based on unweighted data. Several adjustment weighting techniques are discussed in Chapter 10. The simplest one is poststratification. Linear weighting and multiplicative weighting are techniques that can be applied when poststratification is not possible.

Chapter 11 is devoted to online surveys. They become more and more popular, because such surveys are relatively cheap and fast. Also, it is relatively simple to obtain cooperation from large groups of people. However, there are also serious methodological problems. These are discussed in this chapter.

Chapter 12 is about the analysis of survey data. Due to their special nature (unequal selection probabilities, error correction with imputation, and nonresponse correction by adjustment weighting), analysis of such data is not straightforward. Standard software for statistical analysis may not interpret these data correctly. Therefore, analysis techniques may produce wrong results. Some issues are discussed in this chapter. Also, attention is paid to the publication of survey results. In particular, the advantages and disadvantages of the use of graphs in publications are described.

The final chapter is devoted to statistical disclosure control. It is shown how large the risks of disclosing sensitive information can be. Some techniques are presented to estimate these risks. It becomes clear that it is not easy to reduce the risks without affecting the amount of information in the survey data.

1.5 SAMPLONIA

Examples will be used extensively in this book to illustrate concepts from survey theory. To keep these examples simple and clear, they are all taken from an artificial data set. The small country of Samplonia has been created, and a file with data for all inhabitants has been generated (see Fig. 1.3). Almost all examples of sampling designs and estimation procedures are based on data taken from this population file.

Samplonia is a small, independent island with a population of 1000 souls. A mountain range splits the country into the northern province of Agria and the southern province of Induston. Agria is rural province with mainly agricultural activities. The province has three districts. Wheaton is the major supplier of vegetables, potatoes, and fruits. Greenham is known for growing cattle. Newbay is a fairly new area that is still under development. Particularly, young farmers from Wheaton and Greenham attempt to start a new life here.

The other province, Induston, is for a large part an industrial area. There are four districts. Smokeley and Mudwater have a lot of industrial activity. Crowdon is a commuter district. Many of its inhabitants work in Smokeley and Mudwater. The small district of Oakdale is situated in the woods near the mountains. This is where the rich and retired live.

Figure 1.3 The country of Samplonia. Reprinted by permission of Imre Kortbeek.

Samplonia has a central population register. This register contains information such as district of residence, age, and gender for each inhabitant. Other variables that will be used are employment status (has or does not have a job) and income (in Samplonian dollars). Table 1.1 contains the population distribution. Using an

Table 1.1 The Population of Samplonia by Province and District

Province/District	Inhabitants
Agria	293
Wheaton	144
Greenham	94
Newbay	55
Induston	707
Oakdale	61
Smokeley	244
Crowdon	147
Mudwater	255
Total	1000

Table 1.2 Milk Production by Dairy Farms in Samplonia

	Mean	Standard Deviation	Minimum	Maximum
Milk production (liters per day)	723.5	251.9	10.0	1875.0
Area of grassland (hectares)	11.4	2.8	4.0	22.0
Number of cows	28.9	9.0	8.0	67.0

artificial data file has the advantage that all population data are exactly known. Therefore, it is possible to compare computed estimates with true population figures. The result of such a comparison will make clear how well an estimation procedure performs.

Some survey techniques are illustrated by using another artificial example. There are 200 dairy farms in the rural part of Samplonia. Surveys are regularly conducted with as objective estimation of the average daily milk production per farm. There is a register containing the number of cows and the total area of grassland for each farm. Table 1.2 summarizes these variables.

Included in the book is the software package SimSam. This is a program for simulating samples from finite populations. By repeating the selection of a sample and the computation of an estimate a large number of times, the distribution of the estimates can be characterized in both graphical and numerical ways. SimSam can be used to simulate samples from the population of Samplonia. It supports several of the sampling designs and estimation procedures used in this book. It is a useful tool to illustrate the behavior of various sampling strategies. Moreover, it is also possible to generate nonresponse in the samples. Thus, the effect of nonresponse on estimation procedures can be studied.

EXERCISES

1.1 The last census in The Netherlands took place in 1971. One of the reasons to stop it was the concern about a possible refusal of a substantial group of people to participate. Another was that a large amount of information could be obtained from other sources, such as population registers. Which of statements below about a census is correct?

   a. In fact, a census is a sample survey, because there are always people who refuse to cooperate.

   b. A census is not a form of statistical research because the collected data are used only for administrative purposes.

   c. A census is a complete enumeration of the population because, in principle, every member of the population is asked to provide information.

   d. The first census was carried out by John Graunt in England around 1662.

1.2 The authorities in the district of Oakdale want to know how satisfied the citizens are with the new public swimming pool. It is decided to carry out a survey. What would be the group of people to be sampled?

  a. All inhabitants of Oakdale.

  b. All adult inhabitants of Oakdale.

  c. All inhabitants of Oakdale who have visited the swimming pool in a specific week.

  d. All inhabitants of Oakdale who have an annual season ticket.

1.3 No samples were selected by national statistical offices until the year 1895. Before that data collection was mainly based on complete enumeration. Why did they not use sampling techniques?

  a. The idea of investigating just a part of the population had not yet emerged.

  b. They considered it improper to replace real data by mathematical manipulations.

  c. Probability theory had not been invented yet.

  d. National statistical offices did not yet exist.

1.4 Arthur Bowley suggested in 1906 to use random sampling to select a sample from a population. Why was this idea so important?

  a. It made it possible to introduce the "average man" ("l'homme moyenne") in statistics.

  b. It was not important because it is too difficult to select probability samples in practice.

  c. It made it possible to carry out partial investigations.

  d. It made it possible to apply probability theory to determine characteristics of estimates.

1.5 Why could Gallup provide a better prediction of the outcome of the 1936 Presidential election than the poll of the *Literary Digest* magazine?

  a. Gallup used automobile registration lists and telephone directories.

  b. Gallup used a much larger sample than *Literary Digest* magazine.

  c. Gallup used quota sampling, which resulted in a more representative sample.

  d. Gallup interviewed people only by telephone.

CHAPTER 2

Basic Concepts

2.1 THE SURVEY OBJECTIVES

The *survey design* starts by specifying the *survey objectives*. These objectives may initially be vague and formulated in terms of abstract concepts. They often take the form of obtaining the answer to a general question. Examples are

- Do people feel safe on the streets?
- Has the employment situation changed in the country?
- Make people more and different use of the Internet?

Such general questions have to be translated into a more concrete survey instrument. Several aspects have to be addressed. A number of them will be discussed in this chapter:

- The exact definition of the population that has to be investigated (the target population).
- The specification of what has to be measured (the variables) and what has to be estimated (the population characteristics).
- Where the sample is selected from (the sampling frame).
- How the sample is selected (the sampling design and the sample size).

It is important to pay careful attention to these initial steps. Wrong decisions have their impact on all subsequent phases of the survey process. In the end, it may turn out that the general survey questions have not been answered.

Surveys can serve several purposes. One purpose is to explore and describe a specific population. The information obtained must provide more insight into the behavior or attitudes of the population. Such a survey should produce estimates of

---

Applied Survey Methods: A Statistical Perspective, Jelke Bethlehem
Copyright © 2009 John Wiley & Sons, Inc.

all kinds of population characteristics. Another purpose could be to test a hypothesis about a population. Such a survey results in a statement that the hypothesis is rejected or not. Due to conditions that have to be satisfied, hypothesis testing may require a different survey design. This book focuses on descriptive surveys.

2.2 THE TARGET POPULATION

Defining the target population of the survey is one of the first steps in the survey design phase. The target population is the population that should be investigated. It is also the population to which the outcomes of the survey refer. The elements of the target population are often people, households, or companies. So, the population does not necessarily consist of persons.

Definition 2.1 The *target population U* is a finite set

$$U = \{1, 2, \ldots, N\} \quad (2.1)$$

of N elements. The quantity N is the size of the population. The numbers $1, 2, \ldots, N$ denote the sequence numbers of the elements in the target population. When the text refers to "element k," this should be understood as the element with sequence number k, where k can assume a value in the range from 1 to N.

It is important to define the target population properly. Mistakes made during this phase will affect the outcomes of the survey. Therefore, the definition of the target population requires careful consideration. It must be determined without error whether an element encountered "in the field" does or does not belong to the target population.

Take, for example, a labor force survey. What is the target population of this survey? Every inhabitant of the country above or below a certain age? What about foreigners temporarily working in the country? What about natives temporarily working abroad? What about illegal immigrants? If these questions cannot be answered unambiguously, errors can and will be made in the field. People are incorrectly excluded from or included in the survey. Conclusions drawn from the survey results may apply to a different population.

A next step in the survey design phase is to specify the variables to be measured. These variables measure characteristics of the elements in the target population. Two types of variables are distinguished: target variables and auxiliary variables.

The objective of a survey usually is to provide information about certain aspects of the population. How is the employment situation? How do people spend their holidays? What about Internet penetration? Target variables measure characteristics of the elements that contribute to answering these general survey questions. Also, these variables provide the building blocks to get insight into the behavior or attitudes of the population. For example, the target variables of a holiday survey could be the destination of a holiday trip, the length of the holiday, and the amount of money spent.

Definition 2.2 A *target variable* will be denoted by the letter Y, and its values for the elements in the target population by

$$Y_1, Y_2, \ldots, Y_N. \tag{2.2}$$

So Y_k is the value of Y for element k, where $k = 1, 2, \ldots, N$. For example, if Y represents the income of a person, Y_1 is the income of person 1, Y_2 is the income of person 2, and so on.

For reasons of simplicity, it is assumed that there is only one target variable Y in the survey. Of course, many surveys will have more than just one.

Other variables than just the target variables will usually be measured in a survey. At first sight, they may seem unrelated to the objectives of the survey. These variables are called *auxiliary variables*. They often measure background characteristics of the elements. Examples for a survey among persons could be gender, age, marital status, and region. Such auxiliary variables can be useful for improving the precision of estimates (see Chapter 6). They also play a role in correcting the negative effects of nonresponse (see Chapter 10). Furthermore, they offer possibilities for a more detailed analysis of the survey results.

Definition 2.3 An *auxiliary variable* is denoted by the letter X, and its values in the target population by

$$X_1, X_2, \ldots, X_N. \tag{2.3}$$

So X_k is the value of variable X for element k, where $k = 1, 2, \ldots, N$.

Data that have been collected in the survey must be used to obtain more insight into the behavior of the target population. This comes down to summarizing its behavior in a number of indicators. Such indicators are called population parameters.

Definition 2.4 A *population parameter* is numerical indicator, the value of which depends only on the values $Y_1, Y_2, \ldots, Y_N$ of a target variable Y.

Examples of population parameters are the mean income, the percentage of unemployed, and the yearly consumption of beer. Population parameters can also be defined for auxiliary variables. Typically, the values of population parameters for target variables are unknown. It is the objective of the survey to estimate them. Population parameters for auxiliary variables are often known. Examples of such parameters are the mean age in the population and the percentages of males and females. Therefore, these variables can be used to improve the accuracy of estimates for other variables.

Some types of population parameters often appear in surveys. They are the population total, the population mean, the population percentage, and the (adjusted) population variance.

Definition 2.5 The *population total* of target variable Y is equal to

$$Y_T = \sum_{k=1}^{N} Y_k = Y_1 + Y_2 + \cdots + Y_N. \tag{2.4}$$

So the population total is simply obtained by adding up all values of the variable in the population. Suppose, the target population consists of all households in a country, and Y is the number of computers in the household, then the population total is the total number of computers in all households in the country.

Definition 2.6 The *population mean* of target variable Y is equal to

$$\bar{Y} = \frac{1}{N} \sum_{k=1}^{N} Y_k = \frac{Y_1 + Y_2 + \cdots + Y_N}{N} = \frac{Y_T}{N}. \tag{2.5}$$

The population mean is obtained by dividing the population total by the size of the population. Suppose the target population consists of all employees of a company. Then the population mean is the mean age of the employees of the company.

A target variable Y can also be used to record whether an element has a specific property or not. Such a variables can only assume two possible values: $Y_k = 1$ if element k has the property, and $Y_k = 0$ if it does not have the property. Such a variable is called dichotomous variable or a dummy variable. It can be used to determine the percentage of elements in the population having a specific property.

Definition 2.7 If the target variable Y measures whether or not elements in the target population have a specific property, where $Y_k = 1$ if element k has the property and otherwise $Y_k = 0$, then the *population percentage* is equal to

$$P = 100\bar{Y} = \frac{100}{N} \sum_{k=1}^{N} Y_k = 100 \frac{Y_1 + Y_2 + \cdots + Y_N}{N} = 100 \frac{Y_T}{N}. \tag{2.6}$$

Since Y can only assume the values 1 and 0, its mean is equal to the fraction of 1s in the population, and therefore the percentage of 1s is obtained by multiplying the mean by 100. Examples of this type of variable are an indicator whether or not some element is employed and an indicator for having Internet access at home.

This book focuses on estimating the population mean and to a lesser extent on population percentages. It should be realized that population total and population mean differ only by a factor N. Therefore, it is easy to adapt the theory for estimating totals. Most of the time, it is just a matter of multiplying by N.

Another important population parameter is introduced here, and that is the population variance. This parameter is an indicator of the amount of variation of the values of a target variable.

Definition 2.8 The *population variance* of a target variable Y is equal to

$$\sigma^2 = \frac{1}{N} \sum_{k=1}^{N} (Y_k - \bar{Y})^2. \tag{2.7}$$

This quantity can be seen as a kind of mean distance between the individual values and their mean. This distance is the squared difference. Without taking squares, all differences would cancel out, resulting always in a mean equal to 0.

The theory of sampling from a finite population that is described in this book uses a slightly adjusted version of the population variance. It is the adjusted population variance.

Definition 2.9 The *adjusted population variance* of a target variable Y is equal to

$$S^2 = \frac{1}{N-1} \sum_{k=1}^{N} (Y_k - \bar{Y})^2. \tag{2.8}$$

The difference with the population variance is that the sum of squares is not divided by N but by $N-1$. Use of the adjusted variance is somewhat more convenient. It makes many formulas of estimators simpler. Note that for large value of N, there is hardly any difference between both variances.

The (adjusted) variance can be interpreted as an indicator for the homogeneity of the population. The variance is equal to 0 if all values of Y are equal. The variance will increase as the values of Y differ more. For example, if in a country the variance of the incomes is small, then all inhabitants will approximately have the same income. A large variance is an indicator of substantial income inequality.

Estimation of the (adjusted) population variance will often not be a goal in itself. However, this parameter is important, because the precision of other estimators depends on it.

2.3 THE SAMPLING FRAME

How to draw a sample from a population? How to select a number of people that can be considered representative? There are many examples of doing this wrongly:

- In a survey on local radio listening behavior among inhabitants of a town, people were approached in the local shopping center at Saturday afternoon. There were many people there at that time, so a lot of questionnaire forms were filled. It turned out that no one listened to the sports program broadcasted on Saturday afternoon.
- To carry out a survey on reading a free distributed magazine, a questionnaire was included in the magazine. It turned out that all respondents at least browsed through the magazine.
- If Dutch TV news programs want to know how the Dutch think about political issues, they often interview people at one particular market in the old part of Amsterdam. They go there because people often respond in an attractive, funny, and sometimes unexpected way. Unwanted responses are ignored and the remaining part is edited such that a specific impression is created.

It is clear that this is not the proper way to select a sample that correctly represents the population. The survey results would be severely biased in all examples mentioned above. To select a sample in a scientifically justified way, two ingredients are required: a sampling design based on probability sampling and a sampling frame. Several sampling designs are described in detail in Chapters 4 and 5. This section will discuss sampling frames.

A *sampling frame* is a list of all elements in the target population. For every element in the list, there must be information on how to contact that element. Such contact information can comprise of, for example, name and address, telephone number, or e-mail address. Such lists can exist on paper (a card index box for the members of a club, a telephone directory) or in a computer (a database containing a register of all companies). If such lists are not available, detailed geographical maps are sometimes used.

For selecting a sample from the total population of The Netherlands, a population register is available. In principle, it contains all permanent residents in the country. It is a decentralized system. Each municipality maintains its own register. Demographic changes related to their inhabitants are recorded. It contains information on gender, date of birth, marital status, and nationality. Periodically, all municipal information is combined into one large register, which is used by Statistics Netherlands as a sampling frame for its surveys.

Another frequently used sampling frame in The Netherlands is the Postal Delivery Points file of TNT Post, the postal service company. This is a computer file containing all addresses (of both private houses and companies) where post can be delivered. Typically, this file can be used to draw a sample of households.

The sampling frame should be an accurate representation of the population. There is a risk of drawing wrong conclusion from the survey if the sample has been selected from a sampling frame that differs from the population. Figure 2.1 shows what can go wrong.

THE SAMPLING FRAME

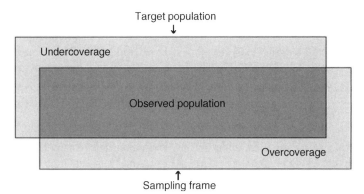

Figure 2.1 Target population and sampling frame.

The first problem is *undercoverage*. This occurs if the target population contains elements that do not have a counterpart in the sampling frame. Such elements can never be selected in the sample. An example of undercoverage is the survey where the sample is selected from a population register. Illegal immigrants are part of the population, but they are never encountered in the sampling frame. Another example is an online survey, where respondents are selected via the Internet. In this case, there will be undercoverage due to people having no Internet access. Undercoverage can have serious consequences. If the elements outside the sampling frame systematically differ from the elements in the sampling frame, estimates of population parameters may be seriously biased. A complicating factor is that it is often not very easy to detect the existence of undercoverage.

The second sampling frame problem is *overcoverage*. This refers to the situation where the sampling frame contains elements that do not belong to the target population. If such elements end up in the sample and their data are used in the analysis, estimates of population parameters may be affected. It should be rather simple to detect overcoverage in the field. This should become clear from the answers to the questions.

Another example is given to describe coverage problems. Suppose a survey is carried out among the inhabitants of a town. It is decided to collect data by means of telephone interviewing. At first sight, it might be a good idea to use the telephone directory of the town as the sampling frame. But this sampling frame can have serious coverage problems. Undercoverage occurs because many people have unlisted numbers, and some will have no phone at all. Moreover, there is a rapid increase in the use of mobile phones. In many countries, mobile phone numbers are not listed in directories. In a country like The Netherlands, only two out of three people can be found in the telephone directory. A telephone directory also suffers from overcoverage, because it contains the telephone numbers of shops, companies, and so on. Hence, it may happen that persons are contacted who do not belong to the target population. Moreover, some people may have a higher than assumed contact probability, because they can be contacted both at home and in the office.

A survey is often supposed to measure the status of a population at a specific moment of time. This is called *reference date*. The sampling frame should reflect the status at this reference date. Since the sample will be selected from the sampling frame before the reference date, this might not be the case. The sampling frame may contain elements that do not exist anymore at the reference date. People may have died or companies may have ceased to exist. These are the cases of overcoverage. It may also happen that new elements come into existence after the sample selection and before the reference date; for example, a person moves into the town or a new company is created. These are cases of undercoverage.

Suppose a survey is carried out in a town among the people of age 18 and older. The objective is to describe the situation at the reference date of May 1. The sample is selected in the design phase of the survey, say on April 1. It is a large survey, so data collection cannot be completed in 1 day. Therefore, interviews are conducted in a period of 2 weeks, starting 1 week before the reference date and ending 1 week after the reference date. Now suppose an interviewer contacts a selected person on April 29. Thereafter, it turns out that the person has moved to another town. It becomes a case of overcoverage. What counts is the difference in the situation on May 1, as the person does not belong anymore to the target population at the reference date. So, there is no problem. Since this is a case of overcoverage, it can be ignored. The situation is different if an interviewer attempts to contact a person on May 5, and this person turns out to have moved on May 2. This person belonged to the target population at the reference date, and therefore should have been interviewed. This is no coverage problem, but a case of nonresponse. The person should be tracked down and interviewed.

Problems can also occur if the units in the sampling frame are different from those in the target population. Typical is the case where one consists of addresses and the other of persons. First, the case is considered where the target population consists of persons and the sampling frame of addresses. This may happen if a telephone directory is used as a sampling frame. Suppose persons are to be selected with equal probabilities. A naïve way to do this would be to randomly select a sample of addresses and to draw one person from each selected address. At first sight, this is reasonable, but it ignores the fact that now not every person has the same selection probability: members in large families have a smaller probability of being selected than members of small families.

A second case is a survey in which households have to be selected with equal probabilities, and the sampling frame consists of persons. This can happen if the sample is selected from a population register. Now large families have a larger selection probability than smaller families, because larger families have more people in the sampling frame. In fact, the selection probability of a family is proportional to the size of the family.

2.4 SAMPLING

The basic idea of a survey is to measure the characteristics of only a sample of elements from the target population. This sample must be selected in such a way that it allows drawing conclusions that are valid for the population as a whole. Of course, a researcher could just take some elements from the population randomly. Unfortunately, people do

SAMPLING   23

not do very well in selecting samples that reflect the population. Conscious or unconscious preferences always seem to play a role. The result is a selective sample, a sample that cannot be seen as representative of the population. Consequently, the conclusions drawn from the survey data do not apply to the target population.

2.4.1 Representative Samples

To select a sample, two elements are required: a sampling frame and a selection procedure. The sampling frame is an administrative copy of the target population. This is the file used to select the sample from. Sampling frames have already been described in Section 2.3. Once a sampling frame has been found, the next step is to select a sample. Now the question is: How to select a sample? What is the good way to select a sample and what is the bad way to do it? It is often said that a sample must be *representative*, but what does it mean?

Kruskal and Mosteller, (1979a, 1979b, 1979c) present an extensive overview of what representative is supposed to mean in nonscientific literature, scientific literature excluding statistics, and in the current statistical literature. They found the following meanings for "representative sampling":

(1) *General acclaim for data*. It means not much more than a general assurance, without evidence, that the data are OK. This meaning of "representative" is typically used by the media, without explaining what it exactly means.

(2) *Absence of selective forces*. No elements or groups of elements were favored in the selection process, either consciously or unconsciously.

(3) *Miniature of the population*. The sample can be seen as a scale model of the population. The sample has the same characteristics as the population. The sample proportions are in all respects similar to population proportions.

(4) *Typical or ideal case(s)*. The sample consists of elements that are "typical" of the population. These are "representative elements." This meaning probably goes back to the idea of *l'homme moyenne* (average man) that was introduced by the Dutch/Belgian statistician Quetelet, (1835, 1846).

(5) *Coverage of the population's heterogeneity*. Variation that exists in the population must also be encountered in the sample. So, the sample should also contain atypical elements.

(6) *A vague term, to be made precise*. Initially the term is simply used without describing what it is. Later it is explained what is meant by it.

(7) *A specific sampling method has been applied*. A form of probability sampling must have been used giving equal selection probabilities to each element in the population.

(8) *As permitting good estimation*. All characteristics of the population and the variability must be found back in the sample, so that it is possible to compute reliable estimates of population parameters.

(9) *Good enough for a particular purpose*. Any sample that shows that a phenomenon thought to be very rare or absent occurs with some frequency will do.

Due to the many different meanings the term "representative" can have, it is recommended not to use it in practice unless it is made clear what is meant by it. In this book, the term "representative" is used only in one way: a sample is said to be *representative* with respect to a variable if its relative distribution in the sample is equal to its relative distribution in the population. For example, a sample is representative with respect to the variable gender, if the percentages of males and females in the sample are equal to the percentages of males and females in the population.

The foundations of survey sampling learn that samples have to be selected with some kind of probability mechanism. Intuitively, it seems a good idea to select a probability sample in which each element has the same probability of being selected. It produces samples that are "on average" representative with respect to all variables. Indeed, this is a scientifically sound way of sample selection, and probably also the one that is most frequently applied in practice.

In subsequent chapters, it will be shown that selecting samples with unequal probabilities can also be meaningful. Under specific conditions, this type of sampling can lead to even more accurate estimates of population characteristics. The remainder of this chapter limits itself to sampling with equal probabilities.

2.4.2 Randomizers

Drawing an equal probability sample requires a selection procedure that indeed gives each element in the population the same probability of selection. Elements must be selected without prejudice. Human beings, however, are not able to select such a sample. They just cannot pick a number of elements giving each element the same probability of selection. Conscious or unconscious preferences always seem to play a role. An illustration of this phenomenon is an experiment in which a sample of 413 persons were asked to pick an arbitrary number in the range from 1 up to and including 9. The results are summarized in Fig. 2.2.

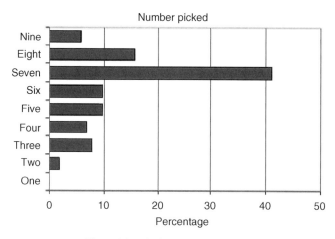

Figure 2.2 Picking a random number.

SAMPLING

If people can behave as a random number generator, each number should be mentioned with approximately the same frequency of 11%. This is definitely not the case. People seem to have a high preference for the number "seven." More than 40% mentioned it. Apparently, "seven" is more random than any other number. The numbers "one" and "two" are almost never mentioned. The conclusion is clear that people should not select a random sample. If they did, they would behave as train travelers who conclude that railroad crossings are always closed for road traffic.

Samples have to be drawn by means of an objective probability mechanism that guarantees that every element in the population has exactly the same probability of being selected. Such a mechanism will be called a randomizer (see Hemelrijk, 1968).

Definition 2.10 A *randomizer* is a machine (electronic or mechanical) with the following properties:

- It can be used repeatedly.
- It has N possible outcomes that are numbered 1, 2, ..., N, where N is known.
- It produces one of the N possible outcomes every time it is activated.
- Each time it is activated, all possible outcomes are equally probable.

The main property of a randomizer is that its outcome is unpredictable in the highest possible degree. All methods of prediction, with or without knowledge or use of past results, are equivalent.

A randomizer is a theoretical concept. The perfect randomizer does not exist in practice. There are, however, devices that come close to a randomizer. They serve the purpose of a randomizer. The proof of the pudding is in the eating: the people living in the princedom of Monaco do not pay taxes as the randomizers in the casino of Monaco provide sufficient income for the princedom.

A simple example of a randomizer is a coin. The two outcomes "heads" and "tails" are equally probable. Another example of a randomizer is a dice. Each of the numbers 1–6 has the same probability, provided the dice is "fair" (Fig. 2.3).

A coin can be used to draw a sample only if the population consists of two elements. A dice can be used only for a population of six elements. This is not very realistic. Target populations are usually much larger than that. Suppose a sample of size 50 is to

Figure 2.3 Example of a randomizer: dice. Reprinted by permission of Imre Kortbeek.

Figure 2.4 A 20-sided dice.

be selected from a population of size 750. In this case, a coin and a dice cannot be used. What can be used is a 20-sided dice; see Fig. 2.4.

Such a dice contains the numbers from 1 to 20. If 10 is subtracted from the outcomes of 10 and higher, and 10 is interpreted as 0, then the dice contains twice the numbers from 0 to 9. Three throws of such a dice produce a three-digit number in the range from 0 to 999. If the outcome 0 and all outcomes over 750 are ignored, a sequence number of an element is obtained. This element is selected in the sample. By repeating this procedure 50 times, a sample of size 50 is obtained.

The use of a 20-sided dice guarantees that all elements in the population have the same probability of being selected. Unfortunately, selecting a sample in this manner is a lengthy process. One way to reduce the work of selecting a sample to some extent is to use a table of random numbers. The first table of random numbers was published by Tippet (1927). Such a table contains numbers that have been obtained with a randomizer. To select a sample, an arbitrary starting point must be chosen in the table. Starting from that point, as many numbers as needed are taken. Figure 2.5 contains a fragment of such a table.

Suppose a sample is to be drawn from the numbers from 1 to 750. To that end, groups of three digits are taken from the table. This produces three-digit numbers. Values 000 and over 750 are ignored. For example, the table is processed row-wise.

```
06966 75356 46464 15180 23367 31416 36083 38160 44008 26146
62536 89638 84821 38178 50736 43399 83761 76306 73190 70916
65271 44898 09655 67118 28879 96698 82099 03184 76955 40133
07572 02571 94154 81909 58844 64524 32589 87196 02715 56356
30320 70670 75538 94204 57243 26340 15414 52496 01390 78802

94830 56343 45319 85736 71418 47124 11027 15995 68274 45056
17838 77075 43361 69690 40430 74734 66769 26999 58469 75469
82789 17393 52499 87798 09954 02758 41015 87161 52600 94263
64429 42371 14248 93327 86923 12453 46224 85187 66357 14125
76370 72909 63535 42073 26337 96565 38496 28701 52074 21346
```

Figure 2.5 A table with random numbers.

SAMPLING

Of each group of five digits, only the first three are used. This produces the following sequence of numbers:

069 753 464 151 233 314 360 381 440 261 625 896 848 381 ...

If irrelevant numbers (753, 896, 848) are removed, the series reduces to

069 464 151 233 314 360 381 440 261 625 381 ...

These are the sequence numbers of the elements that have been selected in the sample. For larger samples from much larger populations, this manual process of sample selection remains cumbersome and time-consuming. It seems obvious to let a computer draw a sample. However, it should be realized that a computer is deterministic machine, and therefore is not able to generate random numbers. The way out is to use a *pseudorandomizer*. Most computers or computer programming languages have algorithms that produce pseudorandom numbers. Such an algorithm generates a deterministic sequence of numbers. For all practical purposes, these numbers cannot be distinguished from real random numbers. Most implementations of pseudorandom number generators use an algorithm like

$$u_{i+1} = f(u_i). \qquad (2.9)$$

The next pseudorandom number u_{i+1} is computed by using the previous number u_i as the argument in some function f. When the pseudorandomizer is used for the first time, it must be given some initial value u_0. This initial value is sometimes also called a *seed value*. The initial value is used to compute the first random number u_1. This first value is used to compute the second value u_2, and so on. So, if the same seed value is used, the same sequence of random numbers will be generated.

Many pseudorandomizers in computers generate values u_i in the interval [0, 1). Possibly, the value 0 may be produced, but never the value 1. The values of this basic randomizer can easily be transformed into other random numbers. To draw a random sample from a population of size N, integer random numbers from the range 1–N are needed. A value u_i from the interval [0, 1) can be transformed in such an integer by applying the formula

$$1 + [u_i \times N], \qquad (2.10)$$

where the square brackets denote rounding downward (truncation) to the nearest integer. Multiplying u_i by N produces a value in the interval [0, N). Truncation gives an integer in the range from 0 to N–1. Finally, adding 1 results in an integer in the range from 1 to N.

An example illustrates this approach. The district of Mudwater in Samplonia has 255 inhabitants. Random integers in the range from 1 to 255 are needed to select a sample of persons from this district. A hand calculator is used to generate a value from [0, 1). Let the result be 0.638. Multiplication by 255 gives 162.69 and truncation results in 162. Adding 1 gives the random integer 163.

2.4.3 Samples With and Without Replacement

Carrying out a survey means that a sample is selected from some population U (as described in Section 2.2). Information is obtained only about the sampled elements. This information must be used to say something about the population as whole. Reliable conclusions can only be drawn if the sample is selected by means of a probability sample, where every element in the population has a nonzero probability of being selected. Researchers can control the characteristics of the sampling mechanism (sample size, selection probabilities), but they have no control over which elements are ultimately selected.

Samples that have been selected by means of a probability mechanism are called *probability samples*. Such samples allow not only good estimates of population parameters but also quantification of the precision of the estimates. Usually, indicators such as the variance, the standard error, or the confidence interval are used for this purpose. Only probability samples are considered in this book. For convenience, they will just be called samples here.

Definition 2.11 A *sample a* from a target population $U = \{1, 2, \ldots, N\}$ is a sequence of indicators,

$$a = (a_1, a_2, \ldots, a_N). \tag{2.11}$$

The value of the indicator a_k (for $k = 1, 2, \ldots, N$) is equal to the number of times element k is selected in the sample.

Definition 2.12 The *sample size n* of the sample a from the population U is equal to

$$n = \sum_{k=1}^{N} a_k. \tag{2.12}$$

Here only sample selection mechanisms will be considered where each possible sample has the same size. Note, however, that there are probability mechanisms where not every sample has the same size. Each possible sample may have a different size.

The sample selection mechanism is formally defined in the sampling design.

Definition 2.13 A *sampling design p* assigns to every possible sample a from the population U a probability $p(a)$ of being selected, where $0 \leq p(a) \leq 1$ and

$$\sum_{a} p(a) = 1. \tag{2.13}$$

SAMPLING

Summation is over every possible sample a from U. The set of all samples a from U that have a nonzero probability $p(a)$ of being selected under the sampling design p is defined by

$$A = \{a|p(a)>0\}. \tag{2.14}$$

Every sampling design can be characterized by a set of first-order, second-order, and higher order *inclusion expectations*. The first-order inclusion expectations are needed to construct estimators of population parameters. The second-order inclusion expectations are needed to compute the precision of estimators.

Definition 2.14 The *first-order inclusion expectation* π_k of element k is defined by

$$\pi_k = E(a_k) = \sum_{a \in A_p} a_k p(a), \tag{2.15}$$

for $k = 1, 2, \ldots, N$. E denotes the expected value of the random variable a_k. So the first-order inclusion expectation π_k of element k is equal to its expected frequency of appearance in one sample.

For example, a sample of size 2 is selected from a population of size 6. A six-sided dice is thrown for both the first and the second sample element. In total, 36 different outcomes are possible: (1, 1), (1, 2), ..., (6, 6). Each sample has a probability 1/36 of being selected. There is only one sample possible in which 5 appears twice: (5, 5). There are 10 samples in which 5 appears once: (1, 5), (2, 5), ..., (5, 1). All other samples do not contain 5. Therefore, the first-order inclusion expectation of element 5 is equal to $\pi_5 = (1 \times 2 + 10 \times 1 + 25 \times 0)/36 = 0.333$.

Definition 2.15 The *second-order inclusion expectation* π_{kl} of two elements k and l is defined by

$$\pi_{kl} = E(a_k a_l) = \sum_{a \in A} a_k a_l p(a), \tag{2.16}$$

for $k = 1, 2, \ldots, N$ and $l = 1, 2, \ldots, N$. So the second-order inclusion expectation π_{kl} of the two elements k and l is equal to the expected value of the product of the sample frequencies of the these two elements.

Coming back to the example of the sample of size 2 from a population of size 6, there are only two samples containing both elements 3 and 5: (3, 5) and (5, 3). For all other samples, either $a_3 = 0$ or $a_5 = 0$. Therefore, the second-order inclusion probability π_{35} of elements 3 and 5 is equal to 2/36.

If a dice is thrown a number of times, it is quite possible that a certain number appears more than once. The same applies to any other randomizer. If a sequence of

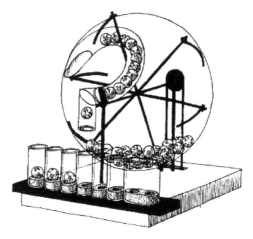

Figure 2.6 A lotto machine: a sample without replacement. Reprinted by permission of Imre Kortbeek.

random numbers is generated, numbers may occur more than once. The consequence would be that an element is selected more than once in the sample. This is not very meaningful. It would mean repeating the measurements for this element. If the questions are answered a second time, the answers will not be different. Therefore, *sampling without replacement* is preferred. This is a way of sampling in which each element can appear at most only once in a sample.

A lotto machine is a good example of sampling without replacement. A selected ball is not replaced in the population. Therefore, it cannot be selected for a second time (Fig. 2.6).

The procedure for selecting a sample without replacement is straightforward. A sequence of random numbers in the range from 1 to N is selected using some randomizer. If a number is generated that has already been selected previously, it is simply ignored. The process is continued until the sample size has been reached.

If a sample is selected without replacement, there are only two possibilities for each element k in the population: the element is selected or the element is not selected. Consequently, the indicator a_k in the sample a can assume only the two possible values 1 and 0. The inclusion expectation π_k of element k is now equal to the probability that element k is selected in the sample. Therefore, the first-order inclusion expectation is called the *first-order inclusion probability* in this situation. Likewise, π_{kl} is called the *second-order inclusion probability* of elements k and l.

Suppose a population consists of four elements: $U = \{1, 2, 3, 4\}$. The possible samples of size 2 without replacement are as follows:

(1, 2)	(1, 3)	(1, 4)
(2, 1)	(2, 3)	(2, 4)
(3, 1)	(3, 2)	(3, 4)
(4, 1)	(4, 2)	(4, 3)

SAMPLING   **31**

Figure 2.7 A roulette: a sample with replacement. Reprinted by permission of Imre Kortbeek.

There are some situations in practice where sampling with replacement is preferred. This is the case for samples selected with unequal probabilities. This sampling design is described in Section 4.3. A sample with replacement can also be represented by a sequence of indicators $a = (a_1, a_2, \ldots, a_N)$. But now every indicator a_k can assume the values $0, 1, 2, \ldots, n$.

A roulette wheel is a good example of sampling with replacement. At every turn, each of the possible numbers can again be produced with the same probability. It is possible (but not very likely) that in a sequence of 10 turns, the number 7 is produced 10 times (Fig. 2.7).

A simple example illustrates sampling with replacement. Suppose, the population consists of four elements: $U = \{1, 2, 3, 4\}$. Then the possible samples of size 2 are as follows:

(1, 1)	(1, 2)	(1, 3)	(1, 4)
(2, 1)	(2, 2)	(2, 3)	(2, 4)
(3, 1)	(3, 2)	(3, 3)	(3, 4)
(4, 1)	(4, 2)	(4, 3)	(4, 4)

So a sampling design describes which samples a from U are possible. The set of all possible samples is denoted by A. The sampling design also fixes the probability $p(a)$ of realization of each possible sample a in A. A number of different sampling designs will be treated in Chapters 4 and 5.

The sampling design described which samples are possible and what their probabilities are. The sampling design does not describe how it must be implemented in practice. The practical implementation is described in the sample selection scheme.

Definition 2.16 The *sample selection scheme* describes a practical algorithm for selecting elements from the population with some randomizer. This algorithm may produce samples only in the set A. The probability of a sample being selected must be equal to the probability $p(a)$ as specified in the sampling design.

Definition 2.17 The *first-order selection probability*

$$p_k^{(i)} \tag{2.17}$$

is equal to the probability that ith draw of the sample selection scheme results in element k, where $k = 1, 2, \ldots, N$ and $i = 1, 2, \ldots$.

Suppose a sample of size 2 is to be selected from a population of size 6. Furthermore, the following sample selection scheme is used:

- Make six lottery tickets, numbered 1–6.
- Arbitrarily select one ticket and write down the number.
- Arbitrarily select one more ticket from the remaining five tickets and write down the number.

According to this selection scheme, the first-order selection probability $p_4^{(1)}$ to select element in the first draw is equal to 1/6. The first-order selection probability $p_4^{(2)}$ to select element 4 in the second draw is less simple to compute. Element 4 can only be selected in the second draw if it has not been selected in the first draw. Therefore,

$$p_4^{(2)} = \frac{1}{6} \times 0 + \frac{5}{6} \times \frac{1}{5} = \frac{1}{6}.$$

Definition 2.18 The *second-order selection probability*

$$p_{kl}^{(i,j)} \tag{2.18}$$

is equal to the probability that in the ith draw element k is selected and in the jth draw element l, for $k, l = 1, 2, \ldots, N$ and $i, j = 1, 2, \ldots$.

Going back to the example of selection of a sample of size 2 from a population of size 6, the probability of selecting element 3 in the first draw, and element 5 in the second draw, is equal to

$$p_{35}^{(1,2)} = \frac{1}{6} \times \frac{1}{5} = \frac{1}{30}.$$

If a sample is selected with replacement, then the selection mechanism is the same for each consecutive draw. Therefore,

$$p_{kl}^{(i,j)} = \begin{cases} p_k p_l, & \text{if } i \neq j, \\ 0, & \text{if } k \neq l \text{ and } i = j, \\ p_k, & \text{if } k = l \text{ and } i = j. \end{cases} \tag{2.19}$$

ESTIMATION

For sampling designs with a fixed sample size n, the following relations hold:

$$\pi_k = \sum_{i=1}^{n} p_k^{(i)}, \tag{2.20}$$

$$\pi_{kl} = \sum_{i=1}^{n} \sum_{j=1}^{n} p_{kl}^{(i,j)}. \tag{2.21}$$

For sampling designs without replacement and with a fixed sample size n, the following relations hold:

$$\sum_{k=1}^{N} \pi_k = n, \tag{2.22}$$

$$\sum_{l=1}^{N} \pi_{kl} = n\pi_k, \tag{2.23}$$

$$\sum_{k=1}^{N} \sum_{l \neq k}^{N} \pi_{kl} = n(n-1). \tag{2.24}$$

2.5 ESTIMATION

Estimation of population parameters has to be based on the information that is collected in the sample. Values of variables can be measured only for those elements k in the population for which the indicator a_k has a nonzero value.

Definition 2.19 Suppose a sample of size n has been selected with a sampling design. The *sample values* of the target variable Y are denoted by

$$y_1, y_2, \ldots, y_n. \tag{2.25}$$

Each y_i represents one of the values $Y_1, Y_2, \ldots, Y_N$ of the target variable. If a sample is selected without replacement, all y_i will be obtained from different elements in the population. If the sample is selected with replacement, some y_i may be equal because they are obtained from the same population element that has been selected more than once.

A similar notation is used if auxiliary variables have also been measured in the sample. So the measured values of the auxiliary variable X are denoted by

$$x_1, x_2, \ldots, x_n. \tag{2.26}$$

The sample values are to be used to say something about the values of population parameters. Such statements take the form of *estimates* of the population parameters. The recipe (algorithm) to compute such an estimate is called an *estimator*.

2.5.1 Estimators

The values of the target variable Y can only be measured for the sampled elements. Furthermore, if an auxiliary variable X is included in the survey, its sample values also become available. This auxiliary information can be used to improve estimates. This only works if more information about the auxiliary variables is available than just the sample values. Typically, it turns out to be very useful to have the population distribution auxiliary variables. Such improved estimators are treated in detail in Chapter 6.

The first step in defining estimators is the definition of a statistic. Estimators are statistics with special properties.

Definition 2.20 A *statistic* is a real-valued function t that with respect to the target variable Y only depends on its sample values $y_1, y_2, \ldots, y_n$. Possibly, t may depend on the sample values $x_1, x_2, \ldots, x_n$ of an auxiliary variable X and of some population parameter for X.

The sample mean and the sample variance are introduced as examples of statistics.

Definition 2.21 The *sample mean* of a target variable Y is equal to

$$\bar{y} = \frac{1}{n} \sum_{i=1}^{n} y_i. \tag{2.27}$$

The sample mean is simply obtained by summing the sample values and dividing the result by the sample size.

Definition 2.22 The *sample variance* of a target variable Y is equal to

$$s^2 = \frac{1}{n-1} \sum_{i=1}^{n} (y_i - \bar{y})^2. \tag{2.28}$$

This statistic is obtained by subtracting the sample mean from each sample value, by summing the squared differences, and by dividing the result by $n-1$.

Note that the definitions of the sample mean and sample variance are similar to those of the population mean and (adjusted) population variance. The only difference is that for the population quantities all values in the population are used whereas for the sample quantities only the sample values are used.

Definition 2.23 An *estimator t* for a population parameter θ is a statistic t that is used for estimating the population parameter θ. Given the population U and the sampling design p, the value of the statistic t only depends on the sample a. Therefore, the notation $t = t(a)$ is sometimes used.

The performance of an estimator is determined to a large extent by the sampling design. The combination of a sampling design p and an estimator t is called the *sampling strategy* (p, t).

2.5.2 Properties of Estimators

Application of an estimator to the sample data provides a value. This value is an estimate for a population value θ. An estimator performs well if it produces estimates that are close to the true value of the population parameter. Four quantities play an important role in judging the performance of estimators: expected value, bias, variance, and mean square error.

Definition 2.24 The *expected value* of an estimator t for a population parameter θ is under the sampling strategy (p, t) equal to

$$E(t) = \sum_{a \in A} t(a) p(a). \qquad (2.29)$$

So the expected value of an estimator is obtained by multiplying its value for each possible sample a by the probability $p(a)$ of its selection and summing the results. The expected value can be seen as a weighted average of its possible values, where the weights are the selection probabilities. The expected value is a kind of center value around which its possible values vary.

The first condition that an estimator must satisfy is that its expected value is equal to the value of the population parameter to be estimated.

Definition 2.25 An estimator t for a population parameter θ under the sampling strategy (p, t) is called an *unbiased estimator* if $E(t) = \theta$.

Repeated use of an unbiased estimator results in estimates that are on average equal to the value of the population parameter. There will be no systematic under- or overestimation.

Definition 2.26 The *bias* of an estimator t for a population parameter θ under the sampling strategy (p, t) is equal to

$$B(t) = E(t) - \theta. \tag{2.30}$$

The bias of an unbiased estimator is equal to 0.

Definition 2.27 The *variance* of an estimator t for a population parameter θ under the sampling strategy (p, t) is equal to

$$V(t) = E(t - E(t))^2 = \sum_{a \in A} (t(a) - E(t))^2 p(a). \tag{2.31}$$

The variance is equal to the expected value of the squared difference between the estimators and the expected value of the estimator. The variance is an indicator of the amount of variation in the possible outcomes of the estimator. The variance is small if all possible values are close to each other. The variance is large if there are large differences in the possible outcomes.

The unit of measurement of the variance is the square of the unit of measurement of the target variable itself. For example, if Y measures income in euros, the variance is measured in squared euros. This makes interpretation of the value of the variance somewhat cumbersome. Moreover, its values tend to be large numbers. To simplify things, a different quantity is often used: the standard error.

Definition 2.28 The *standard error* of an estimator t for a population parameter θ under the sampling strategy (p, t) is equal to

$$S(t) = \sqrt{V(t)}. \tag{2.32}$$

So the standard error is simply equal to the square root of the variance. An estimator is called *precise* of its variance (or standard error is small). An estimator is said to perform well if it is unbiased and precise. If these conditions are satisfied, the possible values of the estimator will always be close to the true, but unknown, value of the population parameter. There is a theoretical quantity that measures both aspects of an estimator simultaneously, and that is the mean square error.

Definition 2.29 The *mean square error* of an estimator t for a population parameter θ under the sampling strategy (p, t) is equal to

$$M(t) = E(t - \theta)^2. \tag{2.33}$$

So the mean square error is equal to the expected value of the squared difference between the estimator and the value of the population parameter. A small mean square

ESTIMATION

error indicates that all possible values of the estimator will be close to the value to be estimated. By working out expression 2.33, it can be shown that it can be written as

$$M(t) = V(t) + B^2(t) = S^2(t) + B^2(t). \tag{2.34}$$

It can be concluded from this expression that a small mean square implies both a small bias and a small variance. So, an estimator with a small mean square error is a good estimator. Note that for unbiased estimators, the bias component in (2.34) vanishes. Consequently, the mean square error is then equal to the variance.

2.5.3 The Confidence Interval

The value of the variance of standard is not easy to interpret in practical situation. What does a specific value mean? When is it large and when is it small? The Polish statistician Jerzy Neyman (1934) invented a more meaningful indicator, the confidence interval, for the precision of estimators.

Definition 2.30 The $100 \times (1-\alpha)\%$ *confidence interval* is determined by a lower bound and an upper bound that have been computed by using the available sample information such that the probability that the interval covers the unknown population value is at least equal to predetermined large probability $1-\alpha$. The quantity $1-\alpha$ is called the *confidence level*.

The use of the confidence interval is based on the *central limit theorem*. According to this theorem, many (unbiased) estimators have approximately a normal distribution with as expected value the population parameter θ to be estimated and as variance the variance $V(t)$ of the estimator. The approximation works better as the sample size increases.

Often, the value of α is set to 0.05. This implies the confidence level is equal to 0.95. This can be interpreted as follows: if the sample selection and the subsequent computation of the estimator are repeated a large number of times, on average in 95 out of the 100 cases the confidence interval will contain the true population value. To say it differently: in 5% of the cases, the statement that the confidence interval contains the true population parameter is wrong. So in 1 out of 20 cases, the researcher will draw the wrong conclusion.

If the estimator has an approximate normal distribution, the 95% confidence interval is equal to

$$(t - 1.96 \times S(t), t + 1.96 \times S(t)). \tag{2.35}$$

Researchers are free in their choice of the confidence level. If they want to draw a more reliable conclusion, they can choose a smaller value of α. For example, a value of $\alpha = 0.01$ could be considered. The corresponding 99% confidence interval is then equal to

$$(t - 2.58 \times S(t), t + 2.58 \times S(t)). \tag{2.36}$$

A price has to be paid for a more reliable conclusion. This price is that the confidence interval is wider. In fact, there is always a compromise between reliability and precision. Either a more reliable conclusion can be drawn about a less precise estimate, or a less reliable conclusion is drawn about a more precise estimate.

The confidence interval can only be computed if the value $S(t)$ of the standard error is known. This is generally not the case, because it requires knowledge of all possible values of the estimator. The distribution of the estimator can only be computed if all values of Y in the population are known. If all these values were known, there was no need to carry out the survey. However, the confidence interval can be estimated. First, the variance $V(t)$ is estimated using the available data. This estimator is indicated by $v(t)$. Next, an estimator $s(t)$ for $S(t)$ is obtained by taking the square root of $v(t)$. This estimator $s(t)$ is used to compute an estimated confidence interval:

$$(t-1.96 \times s(t), t+1.96 \times s(t)). \tag{2.37}$$

2.5.4 The Horvitz–Thompson Estimator

The accuracy of the conclusions of a survey is to a large extent based on the choice of the sampling design and the estimator. All kinds of combinations are possible, but one combination may lead to more precise estimates than another combination. On the one hand, the advantages of a well-chosen sampling design can be undone by a badly chosen estimator. On the other hand, a badly chosen sampling design can be compensated for by using an effective estimator.

The properties of a sampling design can only be investigated in combination with an estimator. Such an estimator must be unbiased, or approximately unbiased. Furthermore, often the additional condition must be satisfied that the estimator must be simple to compute, for example, as a linear combination of the sample values of the target variable. Fortunately, it is always possible to construct an estimator with these properties. This estimator was first developed by Horvitz and Thompson (1952) in their seminal paper. Therefore, it is called the *Horvitz–Thompson estimator*. This estimator can be used for all sampling designs that will be discussed in Chapters 4 and 5. Note that the Horvitz–Thompson estimator does not use any auxiliary information. Chapter 6 describes a number of estimators that do use auxiliary variables.

Definition 2.31 Let $a = (a_1, a_2, \ldots, a_N)$ be a sample from a population U. Suppose a sampling design p has been used with first-order inclusion expectations π_k ($k = 1, 2, \ldots, N$) and second-order inclusion expectations π_{kl} ($k, l = 1, 2, \ldots, N$). The *Horvitz–Thompson estimator* for the population mean of the target variable Y is now defined by

$$\bar{y}_{HT} = \frac{1}{N} \sum_{k=1}^{N} a_k \frac{Y_k}{\pi_k}. \tag{2.38}$$

ESTIMATION

The Horvitz–Thompson estimator is an unbiased estimator provided that $\pi_k > 0$ for all k. Since by definition $\pi_k = E(a_k)$, the expected value of this estimator is equal to

$$E(\bar{y}_{\text{HT}}) = \frac{1}{N}\sum_{k=1}^{N} E(a_k)\frac{Y_k}{\pi_k} = \frac{1}{N}\sum_{k=1}^{N} \pi_k \frac{Y_k}{\pi_k} = \bar{Y}. \qquad (2.39)$$

If all $\pi_k >$ are positive, the variance of the Horvitz–Thompson estimator is equal to

$$V(\bar{y}_{\text{HT}}) = \frac{1}{N^2}\sum_{k=1}^{N}\sum_{l=1}^{N}(\pi_{kl} - \pi_k \pi_l)\frac{Y_k}{\pi_k}\frac{Y_l}{\pi_l}. \qquad (2.40)$$

For samples without replacement, a_k can only assume two values: $a_k = 0$ or $a_k = 1$. Therefore, $a_k = a_k^2$, and consequently $\pi_{kk} = E(a_k a_k) = E(a_k) = \pi_k$. The variance can now be written as

$$V(\bar{y}_{\text{HT}}) = \frac{1}{N^2}\sum_{k=1}^{N}(1-\pi_k)\frac{Y_k^2}{\pi_k} + \frac{1}{N^2}\sum_{k=1}^{N}\sum_{\substack{l=1\\l\neq k}}^{N}(\pi_{kl}-\pi_k\pi_l)\frac{Y_k}{\pi_k}\frac{Y_l}{\pi_l}. \qquad (2.41)$$

Expression (2.41) was derived by Horvitz and Thompson (1952). For sampling designs producing samples of a fixed size n, the variance in (2.40) can be rewritten as

$$V(\bar{y}_{\text{HT}}) = \frac{1}{2N^2}\sum_{k=1}^{N}\sum_{l=1}^{N}(\pi_k\pi_l - \pi_{kl})\left(\frac{Y_k}{\pi_k} - \frac{Y_l}{\pi_l}\right)^2. \qquad (2.42)$$

From expression (2.42), it becomes clear that the variance of the Horvitz–Thompson estimator is smaller if the values Y_k of the target variable and the inclusion expectations π_k are more proportional. If this is the case, the ratios Y_k/π_k are almost constant, and therefore the quadratic term in (2.42) will be small. In the ideal case of exact proportionality, the variance will even be equal to zero. This will not happen in practice. Taking the inclusion probabilities exactly proportional to the values of the target variables would mean that these values are known. The survey was carried out because these values were unknown.

The important message conveyed by the Horvitz–Thompson estimator is that it is not self-evident to draw elements in the sample with equal probabilities. Equal probability sampling may be simple to implement, but other sampling designs may result in more precise estimators.

Section 2.4 introduced the concept of selection probabilities for samples with replacement. The Horvitz–Thompson estimator can be adapted for this type of sampling. To keep things simple, it is assumed that selection probabilities remain the same for each consecutive draw. This implies that the probability of selecting element k in the ith draw of the sample selection scheme is equal to

$$p_k^{(i)} = p_k. \qquad (2.43)$$

The probability to select element k in the ith draw and element l in the jth draw (where $i \neq j$) is equal to

$$p_{kl}^{(i,j)} = p_k p_l. \qquad (2.44)$$

So, subsequent draws are independent. It follows from expressions 2.22 and 2.24 that

$$\pi_k = np_k \qquad (2.45)$$

and

$$\pi_{kl} = n(n-1)p_k p_l \qquad (2.46)$$

for $k \neq l$.

Substitution of (2.45) in expression (2.38) of the Horvitz–Thompson estimator leads to the following expression for the estimator in case of with replacement sampling:

$$\bar{y}_{HT} = \frac{1}{N}\sum_{k=1}^{N} a_k \frac{Y_k}{\pi_k} = \frac{1}{Nn}\sum_{k=1}^{N} a_k \frac{Y_k}{p_k}. \qquad (2.47)$$

Substitution of (2.45) and (2.46) in expression (2.42) gives the following expression for the variance of estimator (2.47):

$$V(\bar{y}_{HT}) = \frac{1}{2nN^2}\sum_{k=1}^{N}\sum_{l=1}^{N} p_k p_l \left(\frac{Y_k}{p_k} - \frac{Y_l}{p_l}\right)^2. \qquad (2.48)$$

Taking into account that subsequent draws are independent, the expression for the variance can be written in a much simpler form:

$$V(\bar{y}_{HT}) = \frac{1}{n}\sum_{k=1}^{N} p_k \left(\frac{Y_k}{Np_k} - \bar{Y}\right)^2. \qquad (2.49)$$

The above theory introduced the Horvitz–Thompson estimator for estimating the population mean of a quantitative variable. This estimator can also be used to estimate population totals. The estimator for the population total Y_T of Y is equal to

$$y_{HT} = N\bar{y}_{HT}. \qquad (2.50)$$

The variance of this estimator is equal to

$$V(y_{HT}) = N^2 V(\bar{y}_{HT}). \qquad (2.51)$$

If the target variable measures whether or not elements have a specific property (with values 1 and 0), then

$$p_{HT} = 100\bar{y}_{HT} \qquad (2.52)$$

EXERCISES

is an unbiased estimator for the population percentage P of elements with that property. The variance of this estimator is equal to

$$V(p_{HT}) = 10,000 V(\bar{y}_{HT}). \qquad (2.53)$$

The theory of the Horvitz–Thompson estimator will be applied to a number of different sampling designs in Chapters 4 and 5.

EXERCISES

2.1 The population variance σ^2 and the adjusted population variance S^2 are related. The adjusted population variance can be computed by multiplying population variance by
  a. $N/(N-1)$;
  b. $(N-1)/N$;
  c. $(N-n)/(N-1)$;
  d. $(N-n)/N$.

2.2 Undercoverage occurs in a sampling frame if it contains elements that
  a. do not belong to the target population and do not appear in the sampling frame;
  b. do belong to the target population and do not appear in the sampling frame;
  c. do not belong to the target population and do appear in the sampling frame;
  d. do belong to the target population and do appear in the sampling frame.

2.3 The local authorities of a town want to know more about the living conditions of single households. To that end, a sample frame is constructed from an address register. Due to a programming error, the sampling frame consists only of households with at least two persons. What is wrong with this sampling frame?
  a. There is both undercoverage and overcoverage.
  b. There is only undercoverage.
  c. There is only overcoverage.
  d. There is no coverage problem.

2.4 A large company regularly distributes newsletters among its employees. The management wants to know whether the employees really read the newsletter. Therefore, a questionnaire form is included in the next release of the newsletter. Questions included in the questionnaire are whether people are aware of the newsletter, and whether they read the newsletter. Employees are invited to complete the questionnaire and to send it back to the management. Explain why this is, or is not, a good way to obtain an estimate of the readership of the newsletter.

2.5 The second-order inclusion probability is equal to
  a. The unconditional probability that both the elements are selected in the sample.
  b. The conditional probability that an element is selected in the sample, given that another element is selected in the sample.
  c. The unconditional probability that an element is selected in the sample twice.
  d. The conditional probability that an element is selected in the sample a second time, given that it has already been selected in the sample.

2.6 Prove that the following relationship holds for sampling designs for samples of a fixed size n:

$$\pi_1 + \pi_2 \cdots + \pi_N = n.$$

2.7 The Horvitz–Thompson estimator is a random variable because
  a. The inclusion expectations $\pi_1, \pi_2, \ldots, \pi_N$ are random variables.
  b. Both the inclusion expectations $\pi_1, \pi_2, \ldots, \pi_N$ and the values $Y_1, Y_2, \ldots, Y_N$ of the target variable are random variables.
  c. Both the selection indicators $a_1, a_2, \ldots, a_N$ and the values $Y_1, Y_2, \ldots, Y_N$ of the target variable are random variables.
  d. The selection indicators $a_1, a_2, \ldots, a_N$ are random variables.

2.8 The Horvitz–Thompson estimator is an unbiased estimator provided
  a. The underlying distribution is the normal distribution.
  b. The standard deviation of the values of the target variable is known.
  c. All first-order inclusion expectations are positive.
  d. All second-order inclusion expectations are positive.

CHAPTER 3

Questionnaire Design

3.1 THE QUESTIONNAIRE

The survey process starts with the formulation of the general research question. The objective of the survey is to answer this question. To that end, the research question must be translated into a set of variables (both target and auxiliary variables) that are to be measured in the survey. The values of these variables will be used to estimate a set of relevant population parameters. Together, the values of these parameters should provide sufficient insight. Questions must be defined to obtain the values of the variables. Together, these questions make up the questionnaire.

So, the questionnaire is a measuring instrument. However, it is not a perfect measuring instrument. A measuring scale can be used for determining someone's length, and the weight of a person can be determined by a weighing scale. These physical measuring devices are generally very accurate. The situation is different for a questionnaire. It only indirectly measures someone's behavior or attitude. Schwarz et al. (2008) describe the tasks involved in answering a survey question. First, respondents need to understand the question. They have to determine the information they are asked to provide. Next, they need to retrieve the relevant information from their memory. In the case of a nonfactual question (e.g., an opinion question), they will not have this information readily available. Instead, they have to form an opinion on the spot with whatever information comes to mind. In the case of a factual question (e.g., a question about behavior), they have to retrieve from their memory information about events in the proper time period. Then they have to translate the relevant information in a format fit for answering the questions. Finally, respondents may hesitate to give this answer. If the question is about a sensitive topic, they may refuse to give an answer, and if an answer is socially undesirable, they may change their answer. All this complicates the use of a questionnaire as a measuring instrument.

A lot can go wrong in the process of asking and answering questions. Problems in the survey questionnaires will affect the collected data, and consequently, also the

---

Applied Survey Methods: A Statistical Perspective, Jelke Bethlehem
Copyright © 2009 John Wiley & Sons, Inc.

survey results. It is of utmost importance to carefully design and test the survey questionnaire. It is sometimes said that questionnaire design is an art and not a skill. Nevertheless, long years of experience have led to a number of useful rules. A number of these rules are described in this chapter. They deal with question texts, question types, and the structure of the questionnaire. Also, some attention is paid to testing a questionnaire.

3.2 FACTUAL AND NONFACTUAL QUESTIONS

Kalton and Schuman (1982) distinguish factual and nonfactual questions. Factual questions are asked to obtain information about facts and behavior. There is always an individual true value. This true value could also be determined, at least in theory, by some other means than asking a question to the respondent. Examples of factual questions are "What is your regular hourly rate of pay on this job," "Do you own or rent your place of residence," and "Do you have an Internet connection in your home?"

The fact to be measured by a factual question must be precisely defined. It has been shown that even a small difference in the question text may lead to a substantially different answer. As an example, a question about the number of rooms in the household can cause substantial problems if it is not clear what constitutes a room and what not. Should a kitchen, a bathroom, a hall, and a landing be included?

Nonfactual questions ask about attitudes and opinions. An opinion usually reflects views on a specific topic, such as voting behavior in the next elections. An attitude is a more general concept, reflecting views about a wider, often more complex issue. With opinions and attitudes, there is no such thing as a true value. They measure a subjective state of the respondent that cannot be observed by another means. The attitude only exists in the mind of the respondent.

There are various theories explaining how respondents determine their answer to an opinion question. One such theory is the *online processing model* described by Lodge et al. (1995). According to this theory, people maintain an overall impression of ideas, events, and persons. Every time they are confronted with new information, this summary view is updated spontaneously. When they have to answer an opinion question, their response is determined by this overall impression. The online processing model should typically be applicable to opinions about politicians and political parties.

There are situations in which people do not have formed an opinion about a specific issue. They only start to think about it when confronted with the question. According to the *memory-based model* of Zaller (1992), people collect all kinds of information from the media and their contacts with other people. Much of this information is stored in memory without paying attention to it. When respondents have to answer an opinion question, they may recall some of the relevant information stored in memory. Owing to the limitations of the human memory, only part of the information is used. This is the information that immediately comes to mind when the question is asked. This is often information that has been stored only recently in memory. Therefore, the memory-based model is able to explain why people seem to be unstable in their opinions. Their answer may easily be determined by the way the issue was recently covered in the media.

3.3 THE QUESTION TEXT

The question text is the most important part of the question. This is what the respondents respond to. If they do not understand the question, they will not give the right answer, or they will give no answer at all. Some rules of thumb are presented here that may help avoid the most obvious mistakes. Examples are given of question texts not following these rules.

- *Use Familiar Wording.* The question text must use words that are familiar to those who have to answer them. Particularly, questionnaire designers must be careful not to use jargon that is familiar to themselves but not to the respondents. Economists may understand a question such as

```
Do you think that food prices are increasing at the same rate as
a year ago, at a faster rate, or at a slower rate?
```

This question asks about the rate at which prices rise, but a less knowledgeable person may easily interpret the question as asking whether prices decreased, have stayed the same, or increased. Unnecessary and possibly unfamiliar abbreviation must be avoided. Do not expect respondents to be able to answer questions about, for example, caloric content of food, disk capacity (in megabytes) of their computer, or the bandwidth (in Mbps) of their Internet connection.

Indefinite words like "usually," "regularly," "frequently," "often," "recently," and "rarely" must be avoided if there is no additional text explaining what they mean. How regular is regularly? How frequent is frequently? These words do not have the same meaning for every respondent. One respondent may interpret "regularly" as every day, while it could mean once a month to another respondent. Here is an example of such a question:

```
Have you been to the cinema recently?
```

What does "recently" mean? It could mean the last week or the last month. The question can be improved by specifying the time period:

```
Have you been to the cinema in the last week?
```

Even this question text could cause some confusion. Does "last week" mean the past 7 days or maybe the period since last Sunday?

- *Avoid Ambiguous Questions.* If the question text is such that different respondents may interpret the question differently, their answers will not be comparable. For example, if a question asks about income, it must be clear whether it is about weekly, monthly, or annual income. It must also be clear whether the respondents should specify their income before or after tax has been deducted. Vague wording may also lead to interpretation problems. A respondent confronted with the question

```
Are you satisfied with the recreational facilities in your
neighborhood?
```

may wonder about what recreational facilities exactly are. Is this a question about parks and swimming pools? Do recreational facilities also include libraries, theaters, cinemas, playgrounds, dance studios, and community centers? What will respondents have in their mind when they answer this question? It is better to describe in the question text what is meant by recreational facilities.

- *Avoid Long Question Texts.* The question text should be as short as possible. A respondent attempting to comprehend a long question may leave out part of the text and thus change the meaning of the question. Long texts may also cause respondent fatigue, resulting in a decreased motivation to continue. Of course, the question text should not be so short that it becomes ambiguous. Here is an example of a question that may be too long:

```
During the past 7 days, were you employed for wages or other
remuneration, were you self-employed in a household enterprise,
were you engaged in both types of activities simultaneously, or
were you engaged in neither activity?
```

Some indication of the length and difficulty of a question text can be obtained by counting the total number of syllables and the average number of syllables per word. Table 3.1 gives examples of indicators for three questions. The first question is simple and short. The second one is also short, but it is much more complex. The third question is very long and has an intermediate complexity.

If a question text appears to be too long, it might be considered to split into two or more shorter questions. It should be noted that some research shows that longer question text sometimes lead to better answers. According to Kalton and Schuman (1982), longer text may work better for open questions about threatening topics.

- *Avoid Recall Questions as much as Possible.* Questions requiring recall of events that have happened in the past are a source of errors. The reason is that people

THE QUESTION TEXT

Table 3.1 Indicators for the Length and Complexity of a Question

Questions	Words	Syllables	Syllables Per Word
Have you been to the cinema in the last week?	9	12	1.3
Are you satisfied with the recreational facilities in your neighborhood?	10	21	2.1
During the past 7 days, were you employed for wages or other remuneration, were you self-employed in a household enterprise, were you engaged in both types of activities simultaneously, or were you engaged in neither activity?	38	66	1.7

make memory errors. They tend to forget events, particularly when they happened a long time ago. Recall errors are more severe as the length of the reference period is longer. Important events, more interesting events, and more frequently happening events will be remembered better than other events. For example, the question

> How many times did you contact your family doctor in the past 2 years?

is a simple question to ask but difficult to answer for many people. Recall errors may even occur for shorter periods. In the 1981 Health Survey of Statistics Netherlands, respondents had to report contacts with their family doctor over the past 3 months. Memory effects were investigated by Sikkel (1983). It turned out that the percentage of not-reported contacts increased linearly with time. The longer ago an event took place, the more likely it is that it would be forgotten. The percentage of unreported events for this question increased on average by almost 4% per week. Over the total period of 3 months, about one quarter of the contacts with the family doctor were not reported.

Recall questions may also suffer from *telescoping*. This occurs if respondents report events as having occurred either earlier or later than they actually did. As a result, an event is incorrectly reported within the reference period, or incorrectly excluded from the reference period. Bradburn et al. (2004) note that telescoping more often leads to overstating than to understating a number of events. Particularly, for short reference periods, telescoping may lead to substantial errors in estimates.

- *Avoid Leading Questions.* A leading question is a question that is not asked in a neutral way but leads the respondents in the direction of a specific answer. For example, the question

> Do you agree with the majority of people that the quality of the health care in the country is falling?

contains a reference to the "majority of people." It suggests that it is socially undesirable to not agree. A question can also become leading by including the opinion of experts in questions text, such as

```
Most doctors say that cigarette smoking cause lung cancer. Do you
agree?
```

Questionnaire designers should watch out for loaded words that have a tendency of being attached to extreme situations:

```
What should be done about murderous terrorists who threaten the
freedom of good citizens and the safety of our children?
```

Particularly, adjectives such as "murderous" and "good" increase a specific loading of the question.

Opinion questions may address topics about which respondents may not have yet made up their mind. They may even lack sufficient information for a balanced judgment. Questionnaire designers may sometime provide additional information in the question text. Such information should be objective and neutral and should not influence respondents in a specific direction. Saris (1997) performed an experiment to show the dangers of making changes in the question text. He measured the opinion of the Dutch about increasing the power of the European Parliament. Respondents were randomly assigned one of these two questions:

Question 1	Question 2
An increase of the powers of the European Parliament will be at the expense of the national parliament.	Many problems cross national borders. For example, 50% of the acid rain in The Netherlands comes from other countries.
Do you think the powers of the European Parliament should be increased?	Do you think the powers of the European Parliament should be increased?

In case respondents were offered the question on the left, 33% answered "yes" and 42% answered "no." In case respondents were offered the question on the right, 53% answered "yes" and only 23% answered "no." These substantial differences are not surprising, as the explanatory text on the left stresses a negative aspect and the text on the right stresses a positive aspect.

- *Avoid Asking Things Respondents Don't Know*. A question text can be very simple, and completely unambiguous, but still it can be impossible to answer it. This may happen if the respondents are asked for facts that they do not know. The following is an example:

```
How many hours did you listen to your local radio station in the
past 6 months?
```

Respondents do not keep record of all kinds of simple things happening in their lives. So, they can only make a guess. This guess need not necessarily be an accurate one. Answering this question is even more complicated by using a relatively long reference period.

- *Avoid Sensitive Questions*. Sensitive questions should be avoided as much as possible. Sensitive questions address topics that respondents may see as embarrassing. Such questions may result in inaccurate answers. Respondents may refuse to provide information on topics such as income or health. Respondents may also avoid giving an answer that is socially undesirable. Instead, they may provide a response that is socially more acceptable.

Sensitive questions can be asked in such a way that the likelihood of response is increased and a more honest response is facilitated. The first option is to include the question in a series of less sensitive questions about the same topic. Another option is to make it clear in the question text that the behavior or attitude is not so unusual. Bradburn et al. (2004) give the following example:

```
Even the calmest parents sometimes get angry at their children. Did
your children do anything in the past 7 days to make you angry?
```

A similar effect can be obtained by referring in question text to experts that may find the behavior not so unusual:

```
Many doctors now believe that moderate drinking of liquor helps
reduce the likelihood of heart attacks and strokes. Have you drunk
any liquor in the past month?
```

A question asking about numerical quantities (such as income) can be experienced as threatening if an exact value must be supplied. This can be avoided by letting the respondent select a range of values.

- *Avoid Double Questions (or Double-Barreled Questions)*. A question must ask one thing at a time. If more than one thing is asked in a question, it is unclear what the answer means. For example, the question

```
Do you think that people should eat less and exercise more?
```

actually consists of two questions: "Do you think that people should eat less?" and "Do you think that people should exercise more?" Suppose, someone thinks that people should not eat less but should exercise more, what answer must be given: yes or no? The solution to this problem is simple: the question must be split into two questions, each asking one thing at a time.

- *Avoid Negative Questions.* Questions must not be asked in the negative, as this is more difficult to understand for respondents. Respondents may be confused by a question such as

```
Are you against a ban on smoking?
```

Even more problematic are double-negative questions. They are a source of serious problems.

```
Would you rather not use a nonmedicated shampoo?
```

Negative questions can usually be rephrased such that negative effect is removed. For example, "are you against ..." can be replaced by "are you in favor ...."

- *Avoid Hypothetical Questions.* It is difficult for people to answer questions about imaginary situations, as they relate to circumstances they have never experienced. At best, the answer is guesswork and a total lie at worst. Here is an example of a hypothetical question:

```
If you were the president of the country, how would you stop crime?
```

Hypothetical questions are often asked to get more insight into attitudes and opinions about certain issues. However, little is known about processes in the respondent's mind that lead to an answer to such a question. So, one may wonder whether hypothetical questions really measure what a researcher wants to measure.

3.4 ANSWER TYPES

Only the text of the question has been discussed until now. Another import aspect of a survey question is the way in which the question must be answered. Several answer types are possible. Advantages and disadvantages of a number of such answer types are discussed.

An *open question* is a simple question to ask. It allows respondents to answer the question completely in their own words. An open question is typically used in situations where respondents should be able to express themselves freely. Open questions often invoke spontaneous answers. Open questions also have disadvantages. The possibility always exists that a respondent overlooks a certain answer. Consider the following question from a readership survey:

```
Which weekly magazines have you read in the past 2 weeks ?
.............................................................
```

Research in The Netherlands has shown that if this question is offered to respondents as an open question, typically television guides are overlooked. If a list is presented containing all weekly magazines, including television guides, much more people answer that they have read TV guides.

Asking an open question may also lead to vague answers. Consider the following question:

```
What do you consider the most important aspect of your job?
.............................................................
```

To many respondents it will be unclear what kind of answer is expected. They will probably answer something like "salary." What do they mean if they say this? It is important to get a high salary, or a regular salary, or maybe both?

Processing the answers to open questions is cumbersome, particularly if the answer is written down on a paper form. Entering such answers in the computer takes effort, and even more if the written text is not very well readable. Answers to open questions also take more disk space than answers to other types of questions. Furthermore, analyzing answers to open questions is not very straightforward. It is often done manually because there is no intelligent software that can do this automatically.

Considering the potential problems mentioned above, open questions should be avoided wherever possible. However, there are situations where there is no alternative. An example is a question asking for the occupation of the respondent. A list containing all possible occupations would be very long. It could easily have thousands of entries. Moreover, respondents with the same occupation may give very different descriptions. All this makes it impossible to let respondents locate their occupation in the list. The only way out is to ask for occupation by means of an open question. Extensive, time-consuming automatic and/or manual coding procedures must be implemented to find the proper classification code matching the description.

A *closed question* is used to measure qualitative variables. There is a list of possible answers corresponding to the categories of the variable. Respondents have to pick one possibility from the list. Of course, this requires the list to contain all possible answers:

```
What is your present martial status?
Never married ................................................  1
Married ......................................................  2
Divorced .....................................................  3
Separated ....................................................  4
Widowed ......................................................  5
```

There will be problem if respondents cannot find their answer. One way to avoid such a problem is to add a category "other," possibly also offering the option to enter the answer. An example is the question below for listeners to a local radio station:

```
Which type of programs do you listen the most on your
local radio station??
Music ........................................................  1
News and current affairs .....................................  2
Sports .......................................................  3
Culture ......................................................  4
Other ........................................................  5

If other, please specify: .........................................
```

If the list with answer options is long, it will be difficult for the respondent to find the proper answer, particularly in telephone surveys, where the interviewer has to read out all options. By the time the interviewer has reached the end of the list, the respondent has already forgotten the first options in the list. Use of show cards may help in face-to-face interviews. A show card contains the complete list of possible answers to a question. Such a card can be handed over to respondents who then can pick their answer from the list.

Only one answer is allowed for a closed question. Therefore, radio buttons are used to implement such a question in an electronic questionnaire. Indeed, only one option can be selected by clicking on it. Clicking on another option would deselect the current selected option. See Fig. 3.1 for an example.

If the list of options is long, the order of the options in the list matters. If the interviewer reads out the options, the first options in the list tend to be forgotten. In case

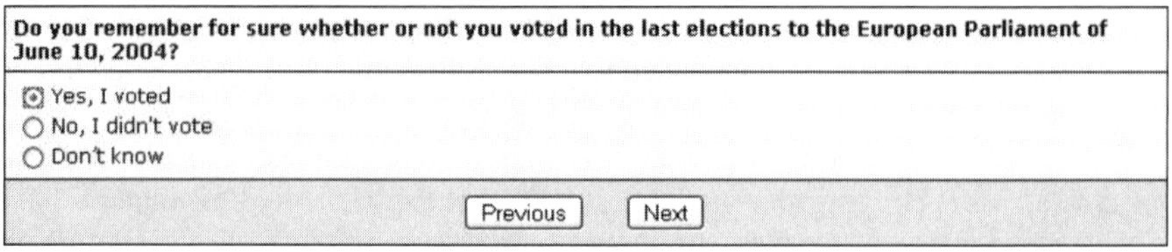

```
Do you remember for sure whether or not you voted in the last elections to the European Parliament of
June 10, 2004?
⊙ Yes, I voted
○ No, I didn't vote
○ Don't know
                              [ Previous ]  [ Next ]
```

Figure 3.1 The implementation of a closed question.

ANSWER TYPES

Figure 3.2 The implementation of a closed question with many possible answers.

of a self-completion questionnaire, the respondents have to read the list themselves. This leads to preference for the first options in the list.

Figure 3.2 show how a long list of possible answers could be implemented in an electronic questionnaire. Only five items in this list are visible. To see other options, respondents have to scroll through the list. Experiments have shown that the options initially visible tend to be selected more than the other options in the list.

Sometimes a question cannot be answered because respondents simply do not know the answer. Such respondents should have the possibility to indicate this on the questionnaire form. Forcing them to make up an answer will reduce the reliability of the data. It has always been a matter of debate how to deal with the *don't-know* option. One way to deal with *don't know* is to offer it as one of the options in a closed question:

```
Do you remember for sure whether or not you voted in the last
elections to the European Parliament of June 10, 2004?
Yes, I voted ..................................................... 1
No, I didn't vote ................................................ 2
Don't know ....................................................... 3
```

Particularly for self-completion questionnaire, this tends to lead to *satisficing* (see Krosnick, 1991). Respondents seek the easiest way to answer a question by simply selecting the *don't-know* option. Such behavior can be avoided in CAPI or CATI surveys. Interviewers are trained to assist respondents to give a real answer and to avoid *don't know* as much as possible. The option is not explicitly offered but is implicitly available. Only if respondents indicate that they really do not know the answer, the interviewer records this response as *don't know*.

Another way to avoid satisficing is to introduce a *filter question*. This question asks whether respondents have an opinion about a certain issue. And only if they say they have an opinion, they are asked to specify their opinion in a subsequent question.

The closed questions discussed until now allowed for exactly only one answer to be given. All answer options have to be mutually exclusive and exhaustive. So respondents can always find one and only one option referring to their situation. Sometimes, however, there are closed questions in which respondents must have the possibility to select more than one option. Figure 3.3 gives an example.

The question asks for modes of transport to work. Respondents may use more than one means of transport for their journey to work, so more answers must be possible.

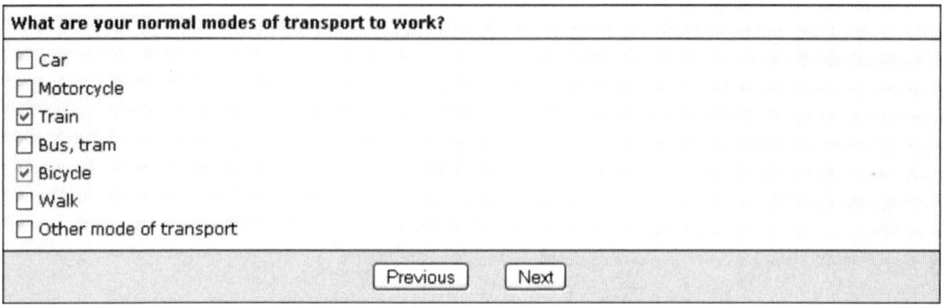

```
┌─────────────────────────────────────────────────────────────┐
│ What are your normal modes of transport to work?            │
├─────────────────────────────────────────────────────────────┤
│ ☐ Car                                                       │
│ ☐ Motorcycle                                                │
│ ☑ Train                                                     │
│ ☐ Bus, tram                                                 │
│ ☑ Bicycle                                                   │
│ ☐ Walk                                                      │
│ ☐ Other mode of transport                                   │
│                                                             │
│                           [ Previous ]   [ Next ]           │
└─────────────────────────────────────────────────────────────┘
```

Figure 3.3 A check-all-that-apply question.

Therefore, they can check every option applying to them. Figure 3.3 shows the answer of a respondent who first takes his bicycle to go to the railway station, where he takes the train.

A closed question with more than one answer is sometimes called a *check-all-that-apply question*. Often square *check boxes* are used to indicate that more than one answer can be given (see Fig. 3.3). Dillman et al. (1998) have shown that such a question may lead to problems if the list of options is very long. Respondents tend to stop after they have checked a few answers and do not look at the rest of the list anymore. Too few options are checked. Figure 3.4 shows a different format for a check-all-that-apply question. Each check box has been replaced by two radio buttons, one for "yes" and the other for "no." This approach forces respondents to do something for each option. They have to check either "yes" or "no." So, they have to go down the list option by option and give an explicit answer for each option. This approach leads to more options that apply. This approach has the disadvantage that it takes more time to answer the question.

Another frequently occurring type of question is a *numerical question*. The answer to such a question is simply a number. Examples are questions about age, income, or prices. In most household survey questionnaires, there is a question about the number of members in the household:

```
┌─────────────────────────────────────────────────────────────┐
│ How many people are there in your household?         _ _    │
└─────────────────────────────────────────────────────────────┘
```

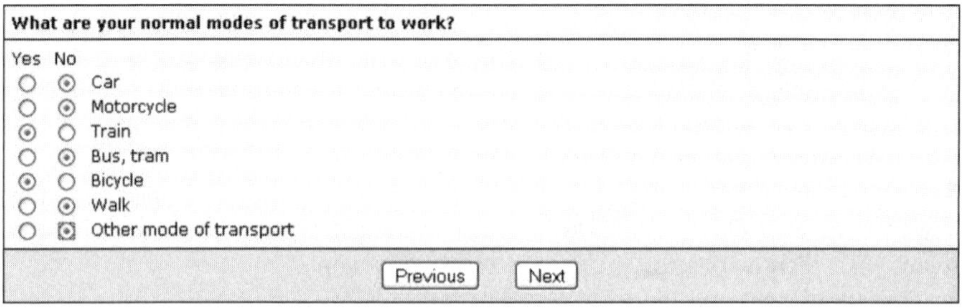

```
┌─────────────────────────────────────────────────────────────┐
│ What are your normal modes of transport to work?            │
├─────────────────────────────────────────────────────────────┤
│ Yes  No                                                     │
│  ○   ⊙   Car                                                │
│  ○   ⊙   Motorcycle                                         │
│  ⊙   ○   Train                                              │
│  ○   ⊙   Bus, tram                                          │
│  ⊙   ○   Bicycle                                            │
│  ○   ⊙   Walk                                               │
│  ○   ⊙   Other mode of transport                            │
│                                                             │
│                           [ Previous ]   [ Next ]           │
└─────────────────────────────────────────────────────────────┘
```

Figure 3.4 A check-all-that-apply question with radio buttons.

The two separate dashes give a visual clue to the respondent as to how many digits are (at most) expected. Numerical questions in electronic questionnaires may have a lower and an upper bound built in for the answer. This ensures that entered numbers are always within a valid range.

It should be noted that respondents in many situations are not able to give exact answers to numerical questions because they simply do not know the answer. An example is the following question:

```
How many hours did you listen to your local radio station in the
past 7 days?
                                                              _ _ _
```

An alternative may be to ask a closed question with a number of intervals as options:

```
How many hours did you listen to your local radio station in the
past 7 days?
0  -  1 hours ................................................  1
1  -  2 hours ................................................  2
2  -  5 hours ................................................  3
5  - 10 hours ................................................  4
More than 10 hours ...........................................  5
```

A special type of question is a *date question*. Many surveys ask respondents to specify dates, for example, date of birth, date of purchase of a product, or date of retirement:

```
What is your date of birth?          _ _    _ _    _ _
                                     day   month   year
```

Of course, a date can be asked by means of an open question, but if used in interviewing software, dedicated date questions offer much more control, and thus few errors will be made in entering a date.

3.5 QUESTION ORDER

Once all questions have been defined, they have to be included in the questionnaire in the proper order. The first aspect is grouping of questions. It is advised to keep questions about the same topic close together. This will make answering question easier for respondents and therefore will improve the quality of the collected data.

The second aspect is the potential *learning effect*. An issue addressed early in the questionnaire may make respondents think about it. This may affect answers to later questions. This phenomenon played a role in a Dutch housing demand survey. People turned out to be much more satisfied with their housing conditions if this question was asked early in the questionnaire. The questionnaire contained a number of questions about the presence of all kind of facilities in and around the house (Do you have a bath? Do you have a garden? Do you have a central heating system?). As a consequence, several people realized that they lacked these facilities and therefore became less and less satisfied about their housing conditions.

Question order can affect the results in two ways. One is that mentioning something (an idea, an issue, a brand) in one question can make people think of it while they answer a later question, when they might not have thought of it if it had not been previously mentioned. In some cases, this problem may be reduced by randomizing the order of related questions. Separating related questions by unrelated ones might also reduce this problem, though neither technique will completely eliminate it.

Tiemeijer (2008) also mentions an example where the answers to specific questions were affected by a previous question. The *Eurobarometer* (www.europa.eu/public_opinion) is an opinion survey in all member states of the European Union held since 1973. The European Commission uses this survey to monitor the evolution of public opinion in the member states. This may help in making policy decision. The following question was asked in 2007:

```
Taking everything into consideration, would you say that the coun-
try has on balance benefited or not from being a member of the
European Union?
```

It turned out that 69% of the respondents were of the opinion that the country had benefited from the EU. A similar question was included at the same time in a Dutch opinion poll (Peil.nl). However, the question was preceded by another question that asked respondents to select the most important disadvantages of being a member of the EU. Among the items in the list were the fast extension of the EU, the possibility of Turkey becoming a member state, the introduction of the Euro, the waste of money by the European Commission, the loss of identity of the member states, the lack of democratic rights of citizens, the veto rights of member states, and the possible interference of the European Commission with national issues. As a result, only 43% of the respondents considered membership of the EU beneficial.

The third aspect of the order of the questions is that a specific question order can encourage people to complete the survey questionnaire. Ideally, the early questions in a survey should be easy and pleasant to answer. Such questions encourage respondents to continue the survey. Whenever possible, difficult or sensitive questions should be asked near the end of the questionnaire. If these questions cause respondents to quit, at least many other questions have been answered.

Another aspect of the order of questions is *routing*. Usually, not every question is relevant for every respondent. For example, a labor force survey questionnaires will

QUESTION ORDER

```
1. What is your gender?
   Male ......................................................... 1
   Female ....................................................... 2

2. What is your age (in years)?                               ...

Interviewer: Ask questions below only of persons of
age 15 and older.

3. Do you have a paid job?
   Yes .......................................................... 1 → 4
   No ........................................................... 2 STOP

4. What is your occupation?.........................................

5. What is the distance from your home to your work?
   Lees than 5 km ............................................... 1
   Between 5 and 10 km .......................................... 2
   Between 10 and 20 km ......................................... 3
   More than 20 km .............................................. 4

6. What is your mode of transport to work?
   (more than one answer allowed))
   Walking....................................................... 1
   Bicycle ...................................................... 2
   Motorcycler .................................................. 3
   Car .......................................................... 4
   Bus, tram .................................................... 5
   Train ........................................................ 6
   Other mode of transport ...................................... 7
```

Figure 3.5 A simple questionnaire with route instructions.

contain questions for both employed and unemployed people. For the employed, there may be questions about working conditions, and for unemployed there may be questions about looking for work. Irrelevant questions may irritate people, possibly resulting in refusal to continue. Moreover, they may not be able to answer questions not relating to their situation. Finally, it takes more time to complete a questionnaire if irrelevant questions also have to be answered. To avoid all these problems, *route instruction* should be included in the questionnaire. Figure 3.5 contains an example of a simple questionnaire with route instructions.

There are two types of route instructions. The first type is that of a branch instruction attached to an answer option of a closed question. Question 3 has such instructions. If respondents answer "yes," they are instructed to jump to question 4 and continue from there. If the answer to question 3 is "no," they are finished with the questionnaire. Sometimes a route instruction does not depend on just an answer to a closed question. It may happen that the decision to jump to another question depends on the answer to several other questions, or on the answer to another type of question. In this case, a route instruction may take the form of an instruction to the interviewer. This is a text placed between questions. Figure 3.5 contains such an instruction between questions 2 and 3.

It was already mentioned that route instructions not only see to it that only relevant questions are asked, but also reduce the number of questions asked, so that the interview takes less time. However, it should be remarked that many and complex route instructions increase the burden for the interviewer. There may be an extra source of possible errors.

3.6 QUESTIONNAIRE TESTING

Before a questionnaire can be used to collect data, it must be tested. Errors in the questionnaire may cause wrong questions to be asked and right questions to be skipped. Also, errors in the questions themselves may lead to errors in answers. Every researcher will agree that testing is important, but this does not always happen in practice. Often there is no time to carry out a proper testing procedure. An overview of some aspects of questionnaire testing is given here. More information can be found, for example, in Converse and Presser (1986).

Questionnaire testing usually comes down to trying it out in practice. There are two approaches to do this. One is to imitate a normal interview situation. Interviewers make contact with respondents and interview them, as in a real survey situation. The respondents do not know that it is just a test, and therefore they behave like they are appearing in a normal interview. If they know it were just a test, they could very well behave differently. Another way to test a questionnaire is to inform respondents that they are part of a test. This has the advantage that the interviewers can ask the respondents whether they have understood the questions, whether things were unclear to them, and why they gave specific answers.

A number of aspects of a questionnaire should be tested. Maybe the most important aspect is the *validity* of the question. Does the question measure what the researcher wants to measure? It is not simple to establish question validity in practice. A first step may be to determine the meaning of the question. It is important that the researcher and the respondent interpret the question exactly in the same way. There are ample examples in the questionnaire design literature about small and large misunderstandings. Converse and Presser (1986) mention a question about "heavy traffic in the neighborhood," where the researcher meant "trucks" and respondents thought that the question was about "drugs." Another question asks about "family planning," where the researcher meant "birth control" and respondents did interpret this as "saving money for vacations."

The above examples make it clear how important validity testing is. Research has shown that often respondents interpret questions differently from what the researcher intended. Also, if respondents do not understand the question, they change the meaning of the question in such a way that they can answer it.

Another aspect of questionnaire testing is to check whether questions offer sufficient variation in answer possibilities. A survey question is not very interesting for analysis purposes if all respondents give the same answer. It must be possible to explore and compare the distribution of the answers to a question for several subgroups of the population.

It should be noted that there are situations where a very skew answer distribution may be interesting. For example, DeFuentes-Merillas et al. (1998) investigate addiction to scratch cards in The Netherlands. It turned out that only 0.24% of the adult population was addicted. Although this was a small percentage, it was important to have more information about the size of the group.

The meaning of a question may be clear, and it may also allow sufficient variation in answers, but this still does not mean that it can always be answered. Some questions are easy to ask but difficult to answer. A question such as

```
How many kilograms of coffee did you consume in the last year in
your household?
```

is clear, but very hard to answer, because respondents simply do not know the answer or can only determine the answer with great effort. Likewise, asking for the net yearly income is not as simple as it looks. Researchers should realize they may get only an approximate answer.

Many people are reluctant to participate in surveys. Even if they cooperate, they may not be very enthusiastic or motivated to answer the questions. Researchers should realize this may have an effect on the quality of the answers given. The more interested respondents are, the better their answers will be. One aspect of questionnaire testing is to determine how interesting questions are for respondents. The number of uninteresting questions should be as small as possible.

Another important aspect is the length of the questionnaire. The longer the questionnaire, the larger the risk of problems. *Questionnaire fatigue* may cause respondents to stop answering questions before the end of the questionnaire is reached. A rule sometimes suggested in The Netherlands is that an interview should not last longer than a class in school (50 min). However, it should be noted that this also partly depends on the mode of interviewing. For example, telephone interviews should take less time than face-to-face interviews.

Up until now, testing was aimed at individual questions. However, the structure of the questionnaire as a whole also has to be tested. Each respondent follows a specific route through the questionnaire. The topics encountered en route must have a meaningful order for all respondents. One way the researcher can check this is by reading aloud the questions (instead of silent reading). While listening to this story, unnatural turns will become apparent.

To keep the respondent interested, and to avoid questionnaire fatigue, it is recommended to start the questionnaire with interesting questions. Uninteresting and sensitive questions (gender, age, income) should come at the end of the questionnaire. This way potential problems can be postponed until the end.

It should be noted that sometimes the structure of the questionnaire requires uninteresting questions, such as gender to be asked early in the questionnaire. This may happen when they are used as filter questions. The answer of such a question determines the route through the rest of the questionnaire. For example, if a questionnaire contains separate parts for male and female respondents, first gender of the respondent must be determined.

The growing potential of computer hardware and software has made it possible to develop very large and complex electronic questionnaires. It is not uncommon for electronic questionnaires to have thousands of questions. To help respondents avoid answering all these questions, routing structures and filter questions see to it that only relevant questions are asked and irrelevant questions are skipped. Owing to the increasing size and complexity of electronic questionnaires, it has become increasingly difficult for developers, users, and managers to keep control of the content and structure of questionnaires. It takes a substantial amount of knowledge and experience to understand such questionnaires. It has become difficult to comprehend electronic questionnaires in their entirety and to understand the process that leads to responses for each of the questions as they ultimately appear in data files.

A number of concrete problems have arisen in survey agencies due to the lack of insight into complex electronic questionnaires:

- It has become very hard to test electronic questionnaires. It is no simple matter to test whether every possible person one might encounter in the field will answer the questions correctly in the correct order.
- Creating textual documentation of an electronic questionnaire has become an enormous task. It is usually a manual task and is therefore error-prone. There is no guarantee that handmade documentation exactly describes the real questionnaire. Making documentation by hand is, of course, also very time-consuming.
- There are always managers in survey organizations who have to approve questionnaires going into the field. In the earlier days of paper questionnaires, they could base their judgment on the paper questionnaire. However, for modern electronic questionnaire instruments, they have nothing to put their signature on. The printout of the questionnaire specification in the authoring language of the CAI system is usually not very readable for the nonexpert.
- Interviewers carrying out a survey with a paper questionnaires could use the paper questionnaire to get some idea of where they are in the questionnaire, of what the next question is about, and of how close they are to the end. If they have an electronic questionnaire, they lack such an overview. Therefore, they often ask for a paper document describing the global content and structure of the questionnaire, which they can use as a tool together with the electronic questionnaire.

All these problems raise the question of the feasibility of a flexible tool capable of representing content and logic of an electronic questionnaire in a human-readable way. Such a tool should not only provide a useful documentation but also help analyze the questionnaire and report possible sources of problems. Bethlehem and Hundepool (2004) have shown that there is a need of software capable of displaying the various routes through the questionnaire in the form of a flow chart. Figure 3.6 shows an example of such a flowchart.

Jabine (1985) describes flowcharts as a tool to design survey questionnaires. Particularly, flowcharts seem to be useful in the early stages of questionnaire

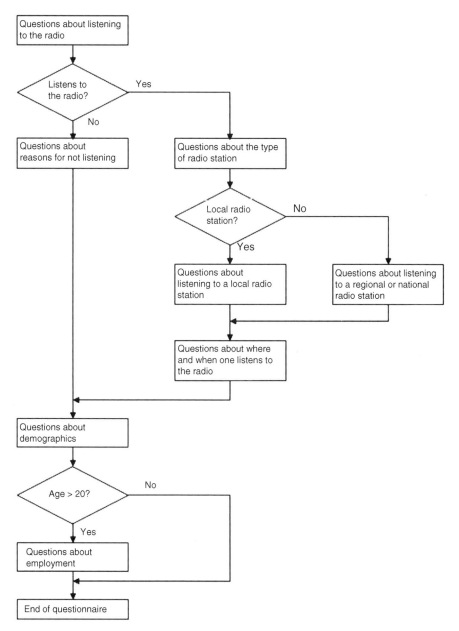

Figure 3.6 Flowchart of the global route structure of a questionnaire.

development. Sirken (1972) used flowcharts to effectively explore alternative structures and sequences for subquestionnaires. He also found that more detailed flowcharts, for example, of the structure of subquestionnaires, can be equally effective. Another example is the *QD* system developed by Katz et al. (1999).

Flowcharts can also be a useful tool in the documentation of electronic questionnaires. Their strong point is that they can give a clear idea of the routing structure. However, they also have the weak point that the amount of textual information that can be displayed about the questionnaire object is limited. Therefore, a flowchart can be a very important component of questionnaire documentation, but it will not be the only component.

There have been a number of initiatives for automatically producing survey documentation, but they pay little or no attention to documentation of survey data collection instruments. They focus on postsurvey data documentation and not on providing tools to assist in the development and analysis of the operation of the collection instrument. The *TADEQ* project was set up to develop a tool documenting these instruments. See Bethlehem and Hundepool (2004) for more information.

The last aspect that can be taken into account when developing a questionnaire is the general well-being of the respondents. Nowadays surveys are conducted over a wide range of topics, including sensitive topics such as use of alcohol and drugs, homosexual relationships, marriage and divorce, maltreatment of children, mental problems, depression, suicide, physical and mental handicaps, and religious experiences. Although the principle of informed consent should be applied to respondents, one may wonder whether respondents feel as happy after the survey interview as before the interview if sensitive issues such as the ones mentioned are addressed in the survey.

Testing a survey questionnaire may proceed in two phases. Converse and Presser (1986) suggest that the first phase should consist of 25–75 interviews. Focus is on testing closed questions. The answer options must be clear and meaningful. All respondents must be able to find the proper answer. If the answer options do not cover all possibilities, there must be a way out by having the special option "other, please specify . . . ."

To collect the experiences of the interviewers in the first phase, Converse and Presser (1986) suggest letting them complete a small survey with the following questions:

- Did any of the questions seem to make the respondents uncomfortable?
- Did you have to repeat any questions?
- Did the respondents misinterpret any questions?
- Which questions were the most difficult or awkward for you to ask? Have you come to dislike any questions? Why?
- Did any of the sections in the questionnaire seem to drag?
- Were there any sections in the questionnaire in which you felt that the respondents would have liked the opportunity to say more?

The first phase of questionnaire testing is a thorough search for essential errors. The second phase should be seen as a final rehearsal. The focus is not anymore on repairing substantial errors or on trying out a completely different approach. This is just the finishing touch. The questionnaire is tested in a real interview situation with real respondents. The respondents do not know that they are participating in a test. The number of respondents in the second phase is also 25–75. This is the phase to consult external experts about the questionnaires.

EXERCISES

3.1 Indicate what is wrong with the following question texts:
- In the past 2 years, how often did you go the cinema?
- Are you against a ban on smoking in restaurants?
- Did you ever visit a coffee shop in Amsterdam and did you buy soft drugs there?
- Should the mayor spend even more tax money trying to keep the streets in town in shape?

3.2 A survey intends to explore how frequently households are bothered by companies trying to sell products or services by telephone. One of the questions is "How often have you been called lately by insurance companies attempting to sell you a private pension insurance?" Give at least three reasons why this is not a good question.

3.3 A researcher wants to get more insight into the statistical software packages that are used for data analysis by commercial and noncommercial statistical research agencies. He considers two types of questions for this: an open question and a check-all-that-apply question. Give advantages and disadvantages of both types of questions.

3.4 The question "Do you regularly download music from the Internet" may cause problems because it contains the word "regularly." Describe what can go wrong.

3.5 An opinion poll is designed to measure the opinion of the people in The Netherlands about building new nuclear power stations. The table below contains two ways to ask this question. Explain which question would you prefer?

To be able to satisfy the future need for energy in The Netherlands, some politicians think it is necessary to build a new nuclear power station within the next 10 years. What is your opinion?	To be able to satisfy the future need for energy in The Netherlands, some politicians think it is necessary to build a new nuclear power station within the next 10 years. What is your opinion?
○ Agree ○ Do not agree	○ Agree ○ Do not agree ○ Don't know

3.6 Improve the questionnaire below by including route instructions.

10. Have you drunk any alcoholic beverages in the last week?	○ Yes ○ No
11. Have you drunk any wine in the last week?	○ Yes ○ No
12. How many glasses of wine did you drink last week?	...
13. Have you smoked any cigarettes last week?	○ Yes ○ No

3.7 A researcher wants to include a question in a readership survey questionnaire about weekly magazines. He wants to know which weekly magazines are read in the sampled households. He has the choice to format this question as an open question or as a closed question (with a list of magazines). Give at least two reasons why he should prefer a closed question.

3.8 Kalton et al. (1978) describe an experiment with wording of a question in a survey on public views on transport in a town. The question is: "Do you think that giving buses priority at traffic signals would increase or decrease traffic congestion?" Half the sample was also offered the neutral middle option "or would it make no difference." The results are summarized in the table below.

	Without Neutral Option	With Neutral Option
Increases congestion	33%	25%
Decreases congestion	20%	12%
Makes no difference	47%	63%

Note that even if the neutral option was not offered, still 47% gave that answer. Explain the differences in the answers for the two approaches.

CHAPTER 4

Single Sampling Designs

4.1 SIMPLE RANDOM SAMPLING

A simple random sample is closest to what comes into the mind of many people when they think about random sampling. It is similar to a lottery. It is also one of the simplest ways to select a random sample. The basic property is that each element in the target has the same probability of being selected in the sample (Fig. 4.1).

A simple random sample can be selected with and without replacement. Only sampling without replacement is considered in this section. This is more efficient than sampling with replacement. Sampling without replacement guarantees that elements cannot be selected more than once in the same sample. All sample elements will be different.

4.1.1 Sample Selection Schemes

There are several techniques to implement selection of a simple random sample without replacement. For small samples, the following manual sample selection schemes can be used:

- Use a 20-sided dice to create numbers in the range from 1 to at least the population size N. If, for example, the population consists of 341 elements, throw the dice three times for each number. The three dice throws produce three digits that together form a number in the range from 0 to 999. If such a number is in the range from 1 to 341, and this number has not already been drawn, it denotes the sequence number of the next element in the sample. This process is repeated until the sample size has been reached.
- Use a table with random numbers and use the same procedure as for the 20-sided dice. Take sufficient digits to form a number in the proper range. For a population of size $N = 341$, this would mean three digits. Numbers outside the valid range and already generated numbers are ignored.

Applied Survey Methods: A Statistical Perspective, Jelke Bethlehem
Copyright © 2009 John Wiley & Sons, Inc.

Figure 4.1 A simple random sample. Reprinted by permission of Imre Kortbeek.

- Use a hand calculator with a random number generator. These random number generators often produce values u in the interval [0, 1). Such values can be transformed into numbers in the range from 1 to N with the formula $1 + [u \times N]$. The square brackets indicate that the value is rounded downward to the nearest integer. If a number reappears, it is ignored.
- The dice and the random number table can also be used to construct values between 0 and 1. Simply take a number of digits and see it is a fraction by putting "0." in front of it. For example, 169971 becomes 0.169971. Then the hand calculator can be applied to form numbers between 1 and N.

Note that in fact all sample selection schemes described above select samples with replacement. By ignoring multiple numbers, the sample becomes a sample without replacement.

A more efficient way to select a simple random sample is to implement a computer algorithm for it. Recipe 4.1 describes such an algorithm.

Recipe 4.1 Selecting a Simple Random Sample Without Replacement

Ingredients	Population size N
	Sample size n
	Random values u from the interval [0, 1)
Step 1	Fill a vector v of length N with the numbers from 1 to N: $v[1] = 1$, $v[2] = 2, \ldots, v[N] = N$
Step 2	Set the counter i to 1
Step 3	Draw a random value u from [0, 1)
Step 4	Compute the sequence number $k = [i + (N - i + 1) \times u]$. The square brackets denote rounding down to the nearest integer
Step 5	Exchange the values of elements $v[i]$ and $v[k]$
Step 6	If i is smaller than n, increase the value of i with 1 and go back to step 3
Step 7	If the value of i is equal to n, sample selection is ready. The first n elements $v[1], v[2], \ldots, v[n]$ of the vector v contain the sequence numbers of the sampled elements

4.1.2 Estimation of a Population Mean

A simple random sample without replacement of size n from a population of size N assigns the same first-order inclusion probability to each element, so $\pi_k = \pi_l$ for each k and l (with $k \neq l$). According to expression (2.22), the sum of all N inclusion probabilities is equal to n. Consequently, the first-order inclusion probability of element k is equal to

$$\pi_k = \frac{n}{N} \tag{4.1}$$

for $k = 1, 2, \ldots, N$. Application of the same theorem to the second-order inclusion probabilities leads to

$$\pi_{kl} = \frac{n(n-1)}{N(N-1)}. \tag{4.2}$$

Note that the quantity

$$f = \frac{n}{N}, \tag{4.3}$$

obtained by dividing the sample size n by the population N size, is also called the *sampling fraction f*.

An unbiased estimator for the population mean can be found by substituting the first-order inclusion probabilities (4.1) in definition (2.38) of the Horvitz–Thompson estimator. Then the estimator turns out to be equal to

$$\bar{y} = \frac{1}{n}\sum_{i=1}^{n} y_i = \frac{y_1 + y_2 + \cdots + y_n}{n}. \tag{4.4}$$

For a simple random sample, the sample mean is an unbiased estimator of the population mean. This is an example of what is sometimes called the *analogy principle*: an estimator for a population characteristic is obtained by computing the same quantity for just the sample data. The analogy principle often (but not always) leads to unbiased estimators.

The variance of this estimator can be determined by substituting the inclusion probabilities (4.1) and (4.2) in formula (2.41) for the variance of the Horvitz–Thompson estimator. This leads to the expression

$$V(\bar{y}) = \frac{1-f}{n} S^2 = \frac{1-f}{n} \frac{1}{N-1} \sum_{k=1}^{N} (Y_k - \bar{Y})^2. \tag{4.5}$$

The quantity f is the sampling fraction n/N. The factor $1-f$ is also sometimes called the *finite population correction*. S^2 is the (adjusted) population variance. See also definition (2.9). Two interesting conclusions can be drawn from formula (4.5).

- Since $(1-f)/n$ can be rewritten as $(1/n - 1/N)$, the variance becomes smaller as the sample size increases. This means selecting a larger sample increases the precision of estimators.
- Since the population size N is usually much larger than the sample size n, the quantity $(1-f)/n$ is approximately equal to $1/n$. It implies that the variance of the estimator does not depend on the size of population. As long as the population variance remains the same, it does not matter for the precision whether a sample is selected from the Dutch population (16 million people) or the Chinese population (1300 million people). This may sound counterintuitive to some people. Maybe the metaphor of tasting soup helps: just tasting one spoonful of soup is sufficient to judge its quality. It does not matter whether this spoonful came from a small pan of soup or a large bathtub full of soup as long as the soup was properly stirred.

To be able to compute the precision of an estimate, the value of the variance (4.5) is required. Unfortunately, this value depends on the unknown population variance S^2. Estimating the population variance using the sample data solves this problem. Also, here the analogy principle applies. It can be shown that *sample variance* s^2, defined by

$$s^2 = \frac{1}{n-1} \sum_{i=1}^{n} (y_i - \bar{y})^2, \tag{4.6}$$

is an unbiased estimator of the population variance S^2. Therefore,

$$v(\bar{y}) = \frac{1-f}{n} s^2 \tag{4.7}$$

is an unbiased estimator of the variance of the sample mean. An (estimated) confidence interval can be computed using this estimated variance. See also Section 2.5.

Another way to investigate the precision of an estimator is to simulate the process of sample selection a large number of times. The working population of Samplonia is used as an example. This population consists of 341 persons. The target variable is the monthly net income. The objective is to estimate the mean income in the population. Then seeing what range of values the mean income in the sample can assume, the sample selection process has been repeated 1000 times. For each sample, the mean income has been computed. All values have been summarized in a histogram. The result is shown in Fig. 4.2. The graph on the left is based on samples of size 20, and the graph on the right is for samples of size 40. The true population value to be estimated (1234) is indicated by means of a vertical line.

In both simulations, the estimates are symmetrically distributed around the population value. This indicates that both estimators are unbiased. Outcomes closer

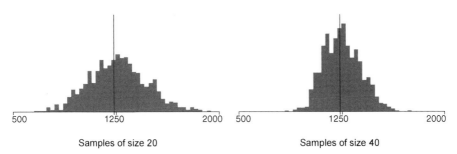

Figure 4.2 Distribution of the mean income in samples from the working population of Samplonia.

to the population value have a higher frequency than values further away. The spread of the values is larger for samples of size 20, which means that estimator has a larger variance in this case.

Simulation experiments can also illustrate how confidence intervals work. Figure 4.3 contains a graphical presentation of confidence intervals. The graph on the left shows 30 confidence intervals, constructed using samples of size 20. A 95% confidence interval has been estimated for each sample. Each confidence interval is represented by a horizontal line. The vertical line indicates the population value to be estimated. Almost all intervals contain the population value. Only in one case this value is outside the interval. Hence, in 1 out of 30 cases, the statement that confidence interval contains the population is wrong. This is 3%, which is very close to the theoretical value of 5%, corresponding to a confidence level of 95%.

The graph on the right in Fig. 4.3 contains 95% confidence intervals for samples of size 40. The horizontal lines are shorter than those for samples of size 20, implying the width of the intervals is smaller and thus that estimators are more precise. Note that also here only 1 out of 30 intervals does not contain the population value. The sample size does not affect the confidence level, but it does affect the precision.

Table 4.1 presents a practical example of the computations that have to be carried out to compute a confidence interval. A sample of 20 persons has been selected from the working population of Samplonia. Income has been recorded for each selected person.

The second column ("element") contains the sequence numbers of the selected elements. All numbers are in the range 1–341, and they are all different. The column

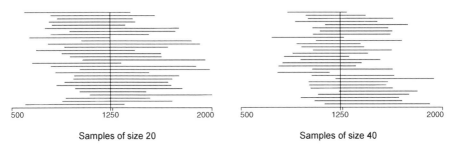

Figure 4.3 Simulation of confidence intervals.

Table 4.1 Computations for a 95% Confidence Interval

Number	Element	Value	(Value − Mean)	(Value − Mean)^2
1	73	2439	1061	1125721
2	103	2956	1578	2490084
3	296	944	−433	187489
4	44	515	−862	743044
5	303	4464	3086	9523396
6	256	531	−846	715716
7	214	951	−426	181476
8	74	627	−750	562500
9	62	158	−1219	1485961
10	166	2289	911	829921
11	169	481	−896	802816
12	210	3493	2115	4473225
13	164	193	−1184	1401856
14	115	1588	210	44100
15	289	1002	−375	140625
16	118	961	−416	173056
17	85	1873	495	245025
18	96	527	−850	722500
19	188	619	−758	574564
20	104	955	−422	178084
Total		27566		26601160
Mean		1378		1400061

Estimate: 27566/20 = 1378, sample variance: 26601160/19 = 1400061, estimated variance of the estimator: (1 − 20/341) × 1400061/20 = 65897, estimated standard error of the estimator: $\sqrt{65897} = 257$, lower bound of the confidence interval: 1379 − 1.96 × 257 = 875, and upper bound of the confidence interval: 1379 + 1.96 × 257 = 1883.

"value" contains the incomes of the sampled elements. The sum of these incomes is equal to 27,566. The mean income in the sample (1378) is obtained by dividing the sample sum (27,566) by the sample size (20). To compute the sample variance, the mean income is subtracted from each sample income and the results are squared. Summation of all these results produces a value 26,601,160. The sample variance 1,400,061 is obtained dividing by 19 (the sample size minus 1). Application of formula (4.7) gives 65,897 as the value of estimated variance of the estimator. The estimated standard error is obtained by taking the square root. This results in a value of 257. Now the 95% confidence interval can be computed by applying formula (2.37). The resulting interval has a lower bound of 875 and an upper bound of 1883. So, one can conclude that (with 95% confidence) the population mean of the income in the working population of Samplonia will be between 875 and 1883.

If a statement is required with a higher level of confidence level, value of α must be changed. For example, $\alpha = 0.01$ results in a confidence level of 99%. In this case, the value of 1.96 in expression (2.37) and in Table 4.1 must be replaced by 2.58. The 99% confidence interval becomes (716, 2042). This is a wider interval than the 95% confidence interval. This is the price that has to be paid for a higher confidence level.

Note that the population mean of the incomes is equal to 1234. Both confidence intervals indeed contain this value. So these intervals lead to correct statements about the population value.

4.1.3 Estimation of a Population Percentage

Population percentages are probably estimated more often than population means. Typical examples are the percentage of people voting for a presidential candidate, the percentage of households having an Internet connection, and the unemployment percentage.

The theory for estimating percentages does not essentially differ from the theory of estimating means. In fact, percentages are just population means multiplied by 100 where the target variable Y assumes only the value 1 (if the element has the specific property) or 0 (if the element does not have the property). Because of this restriction, formulas become even much simpler.

If Y only assumes the values 1 and 0, the population mean $\bar{Y}$ is equal to the proportion of elements having a specific property. The population percentage P is therefore equal to

$$P = 100\bar{Y}. \qquad (4.8)$$

Estimation of a population percentage comes down to first estimate the population mean. The sample mean is an unbiased estimator for this quantity. Multiplication of the sample mean by 100 produces the sample percentage. This estimator is denoted by

$$p = 100\bar{y}. \qquad (4.9)$$

Since the sample mean is an unbiased estimator for the population mean, the sample percentage is an unbiased estimator of population percentage.

The variance of this estimator can be found by working out the term S^2 in variance formula (4.5) for a population in which a percentage P of the elements has a specific property and a percentage $100 - P$ does not have this property. This results in the simple formula

$$V(p) = \frac{1-f}{n} \frac{N}{N-1} P(100-P). \qquad (4.10)$$

This variance can be estimated using the sample data. Again, the analogy principle applies. If p denotes the sample percentage, then

$$v(p) = \frac{1-f}{n-1} p(100-p) \qquad (4.11)$$

is an unbiased estimator of the variance (4.10). The estimated variance is used in practical situations to obtain an (estimated) confidence interval. See also Section 2.5.

The computations to be carried out for obtaining a confidence interval are illustrated in a numerical example. Suppose the objective of the survey is to estimate the percentage of employed people in the total population of Samplonia. From the population (of size $N = 1000$), a simple of size $n = 100$ is drawn. It turns out that 30 people in the sample are employed. So, the estimate of the employment percentage in the total population is 30%.

The sampling fraction is $f = 100/1000 = 0.1$. Substitution of $n = 100$, $p = 30$, and $f = 0.1$ in formula (4.11) produces a variance of the sample percentage $v(p) = 19.09$. The estimated standard error $s(p)$ is obtained by taking the square root. The result is $s(p) = 4.37$.

The margin M of the 95% confidence interval is estimated by $m = 1.96 \times s(p) = 1.96 \times 4.37 = 8.6$. To obtain the lower bound of the interval, the margin is subtracted from the estimate, resulting in 21.4%. Likewise, the upper bound is obtained by adding the margin to the estimate, and this gives 38.6%. So, the 95% confidence interval is equal to (21.4, 36.6). Note that since the population percentage is equal to $P = 34.1\%$, this statement is correct.

4.1.4 Determining the Sample Size

A decision to be made in the survey design phase is the size of the sample to be selected. This is an important decision. If, on the one hand, the sample is larger than what is really necessary, a lot of time and money may be wasted. And if, on the other hand, the sample is too small, the required precision will not be achieved, making the survey results less useful.

It is not so simple to determine the sample size, since it depends on a number of different factors. It has already been shown that there is a relationship between precision of estimators and the sample size: the larger the sample is, the more precise the estimators will be. Therefore, the question about the sample size can only be answered if it is clear how precise the estimators must be. Once the precision has been specified, the sample size can be computed. A very high precision nearly always requires a large sample. However, a large survey will also be costly and time-consuming. Therefore, the sample size will in practical situations always be a compromise between costs and precision.

Some formulas will be given here for the size of a simple random without replacement. The first situation to be considered is that for estimating population percentages. Then the case of estimating population means will be described.

Starting point is that the researcher gives some indication of how large the *margin of error* at most may be. The margin is defined as distance between the estimate and the lower or upper bound of the confidence interval. Formulas are given for the sample size that is at least required to achieve this margin of error. In the case of a 95% confidence interval, the margin of error is equal to

$$1.96 \times S(p). \tag{4.12}$$

For a 99% confidence interval, the value of 1.96 must be replaced by 2.58.

Suppose M is the maximum value of the margin of error the survey researcher wants to accept. This mean that the actual margin of error must not exceed M. Rewriting this condition leads to

$$S(p) \leq \frac{M}{1.96}. \qquad (4.13)$$

The variance of the estimator for a population percentage can be found in expression (4.10). Substituting in inequality (4.13) leads to the condition

$$\sqrt{\frac{1-f}{n}\frac{N}{N-1}P(100-P)} \leq \frac{M}{1.96}. \qquad (4.14)$$

The lower bound for the sample size can now be computed by solving n from this equality. However, there is a problem because expression contains an unknown quantity, and that is population percentage P. There are two ways to solve this problem:

- There is a rough indication of the value of P. Maybe there was a previous survey in which this quantity was estimated or maybe a subject matter expert provided an educated guess. Such an indication can be substituted in expression (4.14), after which it can be solved.
- Nothing at all is known about the value of P. Now $P(100-P)$ is a quadratic function that assumes its minimum value 0 in the interval [0, 100] for $P = 0$ and $P = 100$. Exactly in the middle, for $P = 50$, the function assumes its maximum value. This implies that the upper bound for the variance can be computed by filling in the value $P = 50$. So the worst case for the variance is obtained for this value of P. For any other value of P, the variance is smaller. If the value is determined so that the worst-case variance is not exceeded, then the true variance will certainly be smaller. It should be noted that for values of P between, say, 30% and 70%, the true variance will not differ much from the maximum variance.

Solving n from inequality (4.14) leads to a lower bound of n equal to

$$n \geq \frac{1}{(N-1/N)(M/1.96)^2(1/P(100-P)) + 1/N}. \qquad (4.15)$$

A simple approximation can be obtained if the population size N is very large. Then $(N-1)/N$ can be approximated by 1 and $1/N$ can be ignored, reducing (4.15) to

$$n \geq \left(\frac{1.96}{M}\right)^2 P(100-P). \qquad (4.16)$$

An example illustrates the use of this expression. Suppose an opinion poll has predicted that 38% of the voters will support a certain party. A new survey will be conducted to measure the current support for that party. No dramatic changes are expected. Therefore, it is not unreasonable to fill in a value of 38 for P in expression (4.16). Furthermore, the margin of error should not exceed $M = 3\%$. Substitution in expression (4.16) results in

$$n \geq \left(\frac{1.96}{3}\right)^2 38 \times 62 = 1005.6.$$

So, the sample size must be at least equal to 1006. The confidence level is 95%. For a confidence level of 99%, the value of 1.96 must be replaced by 2.58, leading to a minimum sample size of 1689.

Expression (4.13) is also the starting point for the computation of the sample size if the objective of the survey is to estimate the mean of a quantitative variable. However, there is no simple expression for the standard error available. Expression (4.13) can be rewritten as

$$\sqrt{\left(\frac{1}{n} - \frac{1}{N}\right) S^2} \leq \frac{M}{1.96}, \qquad (4.17)$$

in which S^2 is the adjusted population variance. The problem is that usually this variance is unknown. Sometimes a rough estimated can be made using data from a previous survey, or may be some indication can be obtained from a test survey. In these situations, the approximate value can be substituted in expression (4.17). Rewriting the inequality leads to

$$n \geq \frac{1}{(M/1.96S)^2 + 1/N}. \qquad (4.18)$$

The quantity $1/N$ can be ignored for large values of N. This produces the somewhat simpler expression

$$n \geq \left(\frac{1.96S}{M}\right)^2. \qquad (4.19)$$

If no information at all is available about the value of S, the following rules of thumb may help to determine the sample size for estimating the mean of the target variable Y:

- The values of Y have a more or less *normal* distribution over an interval of known length L. This implies that L will be approximately equal to $6 \times S$. Hence, a value of $L/6$ can be substituted for S.
- The values of Y have a more or less *homogeneous* distribution over an interval of length L. Then S will be roughly equal to $0.3 \times L$.
- The values of Y have a more or less *exponential* distribution over an interval of length L. There are many small values and only a few large values. Then S will be roughly equal to $0.4 \times L$.
- The values of Y are distributed over an interval of known length, but *nothing at all* is known about the form of the distribution. In the worst case, half of the mass of the distribution concentrates at the lower bound of the interval and the other half at the upper bound. Then S will be roughly equal to $0.5 \times L$.

4.2 SYSTEMATIC SAMPLING

A systematic sample is also an equal probability sample, and it is also selected without replacement. However, a different sample selection procedure is followed, and therefore the statistical properties of the estimator differ from those for simple random sampling.

Systematic sampling is typically convenient if samples have to be selected by hand. One could think of a sampling frame consisting of a file of cards with names and addresses or telephone numbers. Systematic sampling is also useful if a sample (of phone numbers) has to be selected from a telephone directory in book format.

Systematic sampling was used in The Netherlands in the previous century to select samples of people from a population register. Each municipality had its own register. There was a card with personal data for each inhabitant. All these cards were stored in a large number of drawers. There were no sequence numbers on the cards. To select, say, person 3163, one had to count 3163 cards from the start to reach the specific card. This makes selecting a simple random sample a cumbersome and time-consuming affair. The process was simplified by drawing systematic samples (Fig. 4.4).

The basic principle of systematic sampling is that a random starting point is selected in the sampling frame. This is the first element in the sample. From there, subsequent elements are selected by repeatedly jumping forward a fixed number of elements. The process continues until the end of the sampling frame is reached.

4.2.1 Sample Selection Schemes

Systematic sampling is often used as a kind of approximation of simple random sampling, but in fact it is a totally different sampling design. The random number generator is only used to select the first element in the sample. This first element immediately determines the rest of the sample.

A first, simple sample selection schema assumes the population size N to be a multiple of the sample size n. The *step length F* is now defined by

$$F = \frac{N}{n}. \tag{4.20}$$

Figure 4.4 A systematic sample. Reprinted by permission of Imre Kortbeek.

This is an integer number. Next, the *starting point* b is determined by drawing a random number in the range from 1 to F. This can be done by hand with a 20-sided dice, a table of random numbers, or a calculator with a random number generator. This starting point is the sequence number of the first element in the sample. The rest of the sample is obtained by selecting each next Fth element. In other words, the sample consists of the elements with sequence numbers

$$b, b+F, b+2F, b+3F, \ldots, b+(n-1)F. \tag{4.21}$$

The starting point b can only assume F different values $(1, 2, \ldots, F)$. Therefore, only F different samples are possible. This is far less than in the case of simple random sampling. For example, a systematic sample will never contain two elements that are adjacent in the sampling frame. Systematic sampling excluded many samples that are possible for simple random sampling. This affects the properties of estimators.

Suppose a systematic sample of size 1000 has to be drawn from a population of size 19,000. The step length is equal to $F = 19{,}000/1000 = 19$. The starting point is a random number from the integers 1 up to and including 19. Suppose the number 5 is produced. Then the elements with sequence numbers 5, 24, 43, ... will be selected in the sample. Note that only 19 different samples are possible.

This sample selection scheme assumes the population size to be a multiple of the sample size. If this is not the case, a different sampling scheme must be used. For example, the above sampling scheme could be adapted by rounding down N/n to the nearest integer. It turns out that then sample size depends on the value of the starting point b. An example illustrates this phenomenon. Suppose, a systematic sample of size $n = 3$ must be selected from the population of $N = 7$ districts in Samplonia. The step length is $F = [N/n] = [7/2] = 2$. So, there are two possible starting points: $b = 1$ and $b = 2$. If $b = 1$, the sample consists of elements 1, 3, 5, and 7 and for $b = 2$, the sample is equal to elements 2, 4, and 6. In the first case, the sample size is 4, and in the second case it is 3. It can be shown that the sample sizes differ by at most 1.

The problem of variable sample size can be avoided by using a different, more general, sample selection scheme. Application requires a random number generator that produces values in the interval $[0, 1)$. Such generators are often available on computers and hand calculators. Even a 20-sided dice or a table of random number can be used by generating a series of digits and putting "0." in front of it. A random value from the interval $[0, 1)$ will be denoted by u.

To select a systematic sample of size n from a population of size N, the interval $(0, N]$ is divided into N intervals of length 1:

$$(0, 1], (1, 2], \ldots, (N-1, N].$$

The step length F is now equal to the real-valued number $F = N/n$. A starting point b is defined by selecting a random value from the interval $(0, F]$. This value is obtained by selecting a random value u from $[0, 1)$, subtracting this value from 1 (producing a value in the interval $(0, 1]$), and multiplying the result by F (producing a value in the

SYSTEMATIC SAMPLING

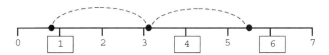

Figure 4.5 A systematic sample of size 3 from a population of size 7.

interval $(0, F]$). In short, $b = (1 - u) \times F$. Next, the values

$$b, b+F, b+2F, \ldots, b+(n-1)F$$

are determined. Each of these values will be part of one of the intervals $(0, 1], (1, 2], \ldots, (N-1, N]$. Is a value contained in the interval $(k-1, k]$, for a certain value k, then element k is selected in the sample.

Selecting a systematic sample is illustrated in a graphical way in Fig. 4.5. For a sample of size 3 from a population of size 7, the step length is $F = 7/3 = 2.333$. The starting point is a random value from the interval $(0, 2.333]$. Say, the value 0.800 is produced. This leads to the values 0.800, $0.800 + 2.333 = 3.133$, and $0.800 + 4.667 = 5.467$. Therefore, the selected sequence numbers are 1, 4, and 6.

The procedure for selecting a systematic sample is summarized in Recipe 4.2.

4.2.2 Inclusion Probabilities

In case of the first sample selection scheme (N is a multiple of n), there are F possible samples. Only one such sample will select a specific element k. Therefore, the first-order inclusion probability of element k is equal to

$$\pi_k = \frac{1}{F} = \frac{n}{N}. \tag{4.22}$$

In case of the second sampling scheme, all possible values in the interval $(0, F]$ are possible, but only the values in a subinterval of length 1 will cause a specific element k to be selected in the sample. Therefore, the first-order inclusion is also equal to n/N.

The first-order inclusion probabilities for systematic sampling are identical to those for simple random sampling. However, there are differences for the second-order

Recipe 4.2 Selecting a Systematic Sample

Ingredients	Population size N
	Sample size n
	Generator of random values u from $[0, 1)$
Step 1	Compute the step length $F = N/n$
Step 2	Select a random value u from $[0, 1)$
Step 3	Compute the starting point: $b = (1 - u) \times F$
Step 4	Compute sequence number k by rounding b upward to the nearest integer
Step 5	Select element k in the sample
Step 6	If the sample size has not been reached yet, add an amount F to b and go back to step 4

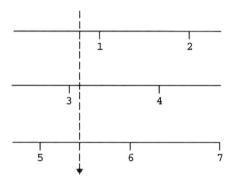

Figure 4.6 Showing the second-order inclusion probabilities.

inclusion probabilities. These probabilities are the same for each pair of elements if simple random sampling is applied. They are not the same for each pair in systematic sampling. For example, it is possible to select the pair of elements 1 and 4 in Fig. 4.5. This implies that $\pi_{14} > 0$. It is not possible to select a sample that contains the pair of elements 1 and 2. So $\pi_{12} = 0$.

There are no simple formulas for the second-order inclusion probabilities. They depend on the "distance" between elements in the sampling frame and the value of the step length. Figure 4.6 shows this graphically. The line segment of Fig. 4.6 has been split into $n = 3$ subsegments of length $F = 2.333$. These subsegments have been drawn below each other. The starting point of the sample is a random point in the first segment. The sample is selected by drawing a vertical (dashed) line through the starting point. The second-order inclusion probability of two elements is determined by the amount of overlap of their corresponding intervals. For example, the overlap for elements 2 and 5 is equal to 0.333, which means that $\pi_{25} = 0.333/F = 0.333/2.333 = 0.143$.

4.2.3 Estimation

The first-order inclusion probabilities of a systematic sample are identical to those of a simple random sample. Consequently, the Horvitz–Thompson estimator for the population mean of Y is also the same. So, if a systematic sample of size n is selected from a population of size N, then the sample mean

$$\bar{y}_S = \frac{1}{n}\sum_{i=1}^{n} y_i \qquad (4.23)$$

is an unbiased estimator of the population mean of Y. The variance of estimator (4.23) is determined to a high degree by the order of the elements in the sampling frame. There is no simple expression for the second-order inclusion probabilities, and therefore the variance expression of the Horvitz–Thompson estimator cannot be used. To get some more insight into the properties of the estimator (4.23), its variance can be written in a different form

$$V(\bar{y}_S) = \sigma^2 - E(\sigma_b^2), \qquad (4.24)$$

SYSTEMATIC SAMPLING

in which σ^2 is the unadjusted population variance (with denominator N), and

$$\sigma_b^2 = \frac{1}{n}\sum_{i=1}^{n}(Y_{bi}-\bar{Y})^2, \qquad (4.25)$$

where $Y_{b1}, Y_{b2}, \ldots, Y_{bn}$ denote the values of Y in the sample with starting point b. So, the variance of the estimator is obtained by subtracting the mean sample variance from the population variance.

Expression (4.24) shows that the variance can be reduced by ordering the elements in the sampling frame in such a way that all possible sample variances are as large as possible. In other words, the possible samples must be as heterogeneous as possible with respect to the values of the target variable. The smallest variance is obtained if each sample variance is equal to the population variance σ^2. Then the variance of the estimator is 0. The variance of the estimator obtains its maximum value if all samples are so homogeneous that each sample variance is 0. Then the variance of the estimator is equal to the population variance. Note that the variance of the estimator does not depend on the sample size. A larger sample will not necessarily lead to a more precise estimator.

4.2.4 Estimation of the Variance

It is not possible to estimate the variance of the estimator with the Horvitz–Thompson approach. The reason is that this approach requires estimation of quantities $Y_k - Y_l$ or $Y_k \times Y_l$. All these quantities can never be estimated because the sampling design excludes these quantities for a large number of combinations of Y_k and Y_l. It is, of course, possible to compute to sample variance, but this variance need not necessarily be a good indicator of the variance of the estimator. If it is not unreasonable to assume that the order of the elements in the sampling frame is completely arbitrary, the theory for simple random sampling can be applied. This implies that

$$v(\bar{y}_S) = \frac{1-f}{n}s^2 \qquad (4.26)$$

will be a reasonably good estimator for the variance of estimator (4.23).

There is an approach that can produce an unbiased estimator for the variance. This requires more than one sample to be selected. Suppose the planned sample size is n. Instead of selecting one sample of size n, r independent samples of size m are selected, where $n = r \times m$. Let

$$\bar{y}^{(j)} \qquad (4.27)$$

be the sample mean of the jth sample. This is an unbiased estimator for the population mean. The r estimators are combined into one new estimator

$$\bar{y}^{(\cdot)} = \frac{1}{r}\sum_{j=1}^{r}\bar{y}^{(j)}. \qquad (4.28)$$

This is also an unbiased estimator of the population mean. The variance of this estimator is equal to

$$V\left(\bar{y}^{(.)}\right) = \frac{1}{r^2} \sum_{j=1}^{r} V\left(\bar{y}^{(j)}\right). \tag{4.29}$$

This variance can be estimated unbiasedly by

$$v\left(\bar{y}^{(.)}\right) = \frac{1}{r(r-1)} \sum_{j=1}^{r} \left(\bar{y}^{(j)} - \bar{y}^{(.)}\right)^2. \tag{4.30}$$

This estimator is in fact based on only r observations and therefore will not be very precise.

In many practical situations, systematic samples will be treated as if they were simple random samples. One should realize that variance estimator (4.26) will be only meaningful if the order of the elements in the sampling frame is arbitrary with respect to the values of the target variable. If one suspects that this may not be the case, this estimator should not be used. For example, if each sample is much more homogeneous than the population as a whole, a optimistic impression of the precision of the estimates is obtained.

It will have by now become clear that much depends on the structure of the sampling frame. If the order of elements is random, then a systematic sample can be treated as a simple random sample. Much more precise estimators can be obtained with a specific order of the elements. For example, if the values of the target variable in the sampling frame show a linear trend (i.e., there is linear relationship between value and sequence number), the true variance of the estimator will be much smaller than its estimate based on the particular sample. Serious problems occur if there is a cyclic pattern in the values of Y, and the step length is equal to the length of the cycle. The true variance of the estimator will be much larger than its estimate based on one sample. Such a problem can be avoided if the length of the cyclic pattern is known, and the step length is taken to be half of this periodicity.

4.2.5 An Example

Some aspects of systematic sampling are now illustrated in a somewhat pronounced example. Objective is to estimate the mean number of passengers in a public transport bus between the two Samplonian towns of Crowdon and Mudwater. A sample of hours is drawn from a period of eight working days. The numbers of passengers in the bus is counted in each selected hour. The population data (which, of course, are unknown) are reproduced in Table 4.2.

Each day is divided into eight 2 hour periods. This 2 hour period is the sampling unit. There are in total $8 \times 8 = 64$ such periods. The mean number of passengers (27.6) must be estimated on the basis of a sample of eight periods.

Samples can be selected in various ways. Three different ways are shown here. The first one is to select a simple random sample of size $n=8$ periods. This gives an unbiased estimator with a variance equal to 21.9.

SYSTEMATIC SAMPLING

Table 4.2 Numbers of Passengers in the Bus Between Crowdon and Mudwater

	Day 1	Day 2	Day 3	Day 4	Day 5	Day 6	Day 7	Day 8	Mean
7–9 h	48	51	50	49	52	51	49	50	56.6
9–11 h	31	30	29	31	29	29	30	31	30.4
11–13 h	20	22	19	19	20	18	21	21	20.5
13–15 h	10	9	11	10	10	10	11	9	10.6
15–17 h	39	41	40	41	41	39	40	39	39.5
17–19 h	42	38	39	39	38	42	39	43	40.0
19–21 h	20	19	21	20	21	21	20	18	19.8
21–23 h	10	10	11	11	9	10	10	9	9.4
Mean	28.6	28.0	26.9	26.3	27.1	28.4	28.3	27.3	27.6

A second way is to select a systematic sample. One way to do this is to combine all rows into one large row consisting of 64 elements. A systematic sample is drawn from this row. The step length is equal to $F = 64/8 = 8$. Hence, the start point will always be in the first row, and $F = 8$ implies that all selected elements will be in the same column. Eight different samples are only possible, corresponding to the eight columns in the table. The characteristics of each sample are summarized in Table 4.3.

Confidence intervals have been computed under the assumption of a simple random sample, so using the sample variance s^2. All confidence intervals contain the population value to be estimated (27.6). The widths of all these intervals are surprisingly wide considering how close the sample means are to the true value. This is caused by the inappropriate assumption of simple random sampling. The true variance of the estimator for a systematic sample is 0.616. So the true margin of the confidence is obtained by taking the square root and multiplying the result by 1.96. This gives a margin of 1.5. Unfortunately, it is not possible to compute this margin with the data from a sample.

Ordering of elements row-by-row allows for systematic samples, leading to very precise estimators. If simple random sampling is assumed, computed confidence intervals are much wider than they should be. The wrong impression is created that estimators are not so precise.

Table 4.3 Systematic Samples (Row-Wise)

Starting Point b	Mean	σ_b^2	s^2	Confidence Interval
1	28.6	185.2	211.7	(19.2, 38.1)
2	28.0	190.0	217.1	(18.4, 37.6)
3	26.9	194.1	221.8	(17.2, 36.5)
4	26.3	195.2	223.1	(16.6, 35.9)
5	27.1	196.1	224.1	(17.4, 36.8)
6	28.4	207.0	236.6	(18.4, 38.3)
7	28.3	200.9	229.6	(18.4, 38.1)
8	27.3	203.4	232.5	(17.4, 37.1)

Table 4.4 Systematic Samples (Column-Wise)

Starting Point b	Mean	σ_b^2	s^2	Confidence Interval
1	50.6	1.2	1.4	(49.9, 51.4)
2	30.4	1.5	1.7	(29.5, 31.2)
3	20.5	1.3	1.4	(19.7, 21.3)
4	10.6	1.5	1.7	(9.8, 11.5)
5	39.5	1.5	1.7	(38.7, 40.3)
6	40.0	2.5	2.9	(38.9, 40.1)
7	19.8	1.2	1.4	(19.0, 20.5)
8	9.4	0.7	0.8	(8.8, 10.0)

Another way to draw a random sample is to combine the columns (instead of rows) into one large column of 64 elements and to draw a systematic sample from this column. Again, the step length is $F = 64/8 = 8$. A starting point is randomly chosen in the first column. A step length of 8 now means that a complete row will be selected in the sample. Eight different samples are possible corresponding to the eight different rows. The characteristics of the samples are summarized in Table 4.4.

Again, confidence intervals have been computed under the assumption of simple random sampling and using the sample variance s^2. All confidence intervals are very small and not any of them contains the population value to be estimated (27.6). Again, the assumption leads to misleading results. The true variance of the estimator based on systematic sampling is 195.7. The true margin of the confidence interval is obtained by taking the square root and multiplying the result by 1.96, which gives 27.4. So the true margin is much larger. As mentioned before, the true margin cannot be computed using the sample data.

The ordering of the elements is now such that systematic sampling leads to estimators that are not very precise. Assuming simple random sampling produces estimates that create a wrong impression of very high precision.

4.3 UNEQUAL PROBABILITY SAMPLING

Up until now, sampling designs have been discussed in which all elements have the same probability of being selected. In the first years of the development of survey sampling methodology, one assumed that this was only meaningful way to draw samples. A fundamental change took place in the early 1950s when Horvitz and Thompson (1952) showed in their seminal paper that samples can be selected with unequal probabilities as long as these selection probabilities are known and estimation procedures correct for these unequal probabilities (Fig. 4.7).

4.3.1 Drawing Samples with Unequal Probabilities

Selecting elements with unequal probabilities is more cumbersome, but there are also advantages. With the proper choice of selection probabilities, estimators are much

UNEQUAL PROBABILITY SAMPLING 83

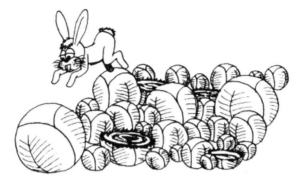

Figure 4.7 Sample selection with unequal selection probabilities. Reprinted by permission of Imre Kortbeek.

more precise. Horvitz and Thompson (1952) have shown that the variance is smaller as the selection probabilities of the elements are more proportional to values of the target variable. The variance is even 0 if the probabilities are exactly proportional to the values of the target variable. This ideal situation will never occur in practice. It would mean the values of the target variable could be computed from the selection probabilities, making a survey superfluous.

Drawing a sample with unequal probabilities is realized in practice by looking for an auxiliary variable that has a strong correlation with the target variable. All values of the auxiliary variable must be positive. A concrete example is a survey about shoplifting. The target population consists of shops, and the target variables are the number of thefts and the total value of thefts in a certain period. Shops are drawn in the sample with probabilities proportional to their floor size. The underlying assumption is that there will be more shoplifting in large shops than in small shops. Large shops have larger selection probabilities than small shops. So, there will be a lot of information about shoplifting in the sample. Of course, this is not a representative sample. Large shops are overrepresented, and small shops are underrepresented. So a correction is necessary. Horvitz–Thompson estimator just does that. Values for large shops are divided by larger selection probabilities, so their influence is reduced. The opposite effect is obtained for values of small shops.

It turns out to be not so simple to make a sample selection scheme for a without replacement unequal probability sample. A way out is to draw a with replacement sample. Let $p_1, p_2, \ldots, p_N$ be the selection probabilities. It has been described in Section 2.5 that the variance of Horvitz–Thompson estimator for the population mean is equal to

$$V(\bar{y}_{\mathrm{HT}}) = \frac{1}{n}\sum_{k=1}^{N} p_k \left(\frac{Y_k}{Np_k} - \bar{Y}\right)^2. \quad (4.31)$$

This variance is small if all quadratic terms in this formula are small. And the quadratic term is small if

$$p_k \approx cY_k, \quad (4.32)$$

where

$$c = \frac{1}{N\bar{Y}}. \tag{4.33}$$

The selection probabilities cannot, in practice, be chosen such that condition (4.32) is exactly satisfied, as this implies that the values of target variable are known. If these values are known, no survey is necessary. Still, attempts can be made to define the selection probabilities such that condition (4.32) is satisfied to some degree. To that end, one tries to find an auxiliary variable X that is more or less proportional to target variable Y. The values of X in the population must be known, and all these values must be strictly positive. This implies that

$$p_k = \frac{X_k}{X_T} = \frac{X_k}{N\bar{X}}. \tag{4.34}$$

If X_k and Y_k are approximately proportional, then $X_k \approx cY_k$ for a certain constant c. Consequently,

$$p_k = \frac{X_k}{N\bar{X}} \approx \frac{cY_k}{Nc\bar{Y}} = \frac{Y_k}{N\bar{Y}}. \tag{4.35}$$

So the more the X_k and Y_k are proportional, the smaller the variance of the estimator for the population will be.

4.3.2 Sample Selection Schemes

It is not easy to make a sample selection scheme to select an unequal probability sample without replacement. There is such a sampling scheme for systematic unequal probability sampling. This is described in Section 4.4.

Some without replacement unequal probability sampling schemes have been proposed in the literature. An overview is given by Chaudhuri and en Vos (1988). All these sampling schemes have their problems: they are so complex that they can only be used for small samples, or they lead to negative estimates of variances. To keep things simple, only with replacement sampling schemes are discussed here. Of course, with replacement sampling is less efficient compared to without replacement sampling. If the sample is selected with replacement, elements can be drawn more than once in the same sample. This reduces the amount of information that becomes available. However, if the population is much larger than the sample, differences are ignorable.

Two sample selection schemes are discussed here for selecting a sample with replacement and with unequal probabilities. There are called the *cumulative scheme* and the *Lahiri scheme*.

The *cumulative scheme* is summarized in Recipe 4.3. First, the subtotals $T_1, T_2, \ldots, T_N$ have to be computed, where

$$T_k = \sum_{i=1}^{k} X_i. \tag{4.36}$$

UNEQUAL PROBABILITY SAMPLING

Recipe 4.3 Selecting an Unequal Probability Sample with the Cumulative Scheme

Ingredients	Population size N,
	Population values $X_1, X_2, \ldots, X_N$ of an auxiliary variable X
	(all values must be positive)
	Sample size n
	Generator of random values u from $[0, 1)$
Step 1	Compute subtotals $T_k = X_1 + X_2 + \cdots + X_k$ for $k = 1, 2, \ldots, N$. $T_0 = 0$
Step 2	Draw a random value u from $[0, 1)$
Step 3	Compute $t = (1 - u) \times T_N$
Step 4	Determine the sequence number k for which $T_{k-1} < t \leq T_k$
Step 5	Select the element with sequence number k in the sample
Step 6	If the sample size has not been reached yet, go back to step 2

It follows that $T_N = X_T$. By definition $T_0 = 0$. To select an element, a random value t is drawn from the interval $(0, T_N]$. This is done by drawing a random value u from the interval $[0, 1)$ and computing $t = (1 - u) \times T_N$. This value t will lie between two subtotals. If t lies in the interval $(T_{k-1}, T_k]$ for a certain value k, then element k is selected in the sample.

The probability p_k of selecting element k is the probability that the value t turns out to lie in the interval $(T_{k-1}, T_k]$. This probability is equal to the length of this interval (X_k) divided by the length of the interval $(0, T_N]$, and this is equal to $p_k = X_k/X_T$.

The cumulative scheme has the disadvantage that first all subtotals $T_1, T_2, \ldots, T_N$ must be computed. The Lahiri scheme avoids this. This scheme was developed by Lahiri (1951). It requires the knowledge of an upper bound X_{max} of the values of the auxiliary variable in the population. So, the condition $X_k \leq X_{max}$ must be satisfied for $k = 1, 2, \ldots, N$.

Selection of an element starts by drawing a candidate. This candidate is random number k selected with equal probabilities from the range 1–N. Then a random value x is drawn from the interval $(0, X_{max}]$. Candidate k is selected in the sample if $x < X_k$. Otherwise, candidate k is not selected, and a fresh attempt is made by selecting a new candidate k and value x. This process is continued until the sample size has been reached (Recipe 4.4).

At first sight, it is not obvious that the Lahiri scheme uses probabilities that are proportional to the values of the auxiliary variable. To show this, first the probability P_{rej} is computed that an attempt (k,x) results in the rejection of a candidate k. This probability is equal to

$$P_{rej} = \sum_{i=1}^{N} P(k = i \text{ and } x > X_i) = \sum_{i=1}^{N} P(x > X_i | k = i) P(k = i)$$

$$= \sum_{i=1}^{N} \left(1 - \frac{X_i}{X_{max}}\right) \frac{1}{N} = 1 - \frac{\bar{X}}{X_{max}}.$$

(4.37)

Recipe 4.4 Selecting an Unequal Probability Sample with the Lahiri Scheme

Ingredients	Population size N
	Population values $X_1, X_2, \ldots, X_N$ of an auxiliary variable X (only for candidate elements)
	An upper bound $X_{\max}$ with $X_k \leq X_{\max}$
	Sample size n
	Generator of random values u from $[0, 1)$
Step 1	Draw a random value u_1 from $[0, 1)$
Step 2	Compute sequence number $k = 1 + [N \times u_1]$. Square brackets indicate rounding downward to the nearest integer
Step 3	Draw a random value u_2 from $[0, 1)$
Step 4	Compute $x = (1 - u_2) \times X_{\max}$
Step 5	If $x \leq X_k$, select element k in the sample
Step 6	If the sample size has not been reached yet, go back to step 1

The probability P_i that attempt (k,x) results in accepting element i is equal to

$$P_i = P(k = i \text{ and } x \leq X_i) = \frac{1}{N} \frac{X_i}{X_{\max}}. \quad (4.38)$$

The probability that for a next sample element ultimately element i is selected is equal to

$$P_i + P_{\text{rej}} P_i + P_{\text{rej}}^2 P_i + \cdots = P_i \left(1 + P_{\text{rej}} + P_{\text{rej}}^2 + \cdots \right) = \frac{P_i}{1 - P_{\text{rej}}}$$

$$= \left(\frac{1}{N} \frac{X_i}{X_{\max}}\right) \Big/ \left(\frac{\bar{X}}{X_{\max}}\right) = \frac{X_i}{X_T}. \quad (4.39)$$

Indeed, the selection probability of element i is proportional to X_i. Note that it is not required that the value $X_{\max}$ be assumed by one or more elements. For example, if people are selected proportional to their age, and the maximum age in the population is not known, a value of $X_{\max} = 200$ is safe upper bound. However, the closer the $X_{\max}$ is to the true maximum of $X_1, X_2, \ldots, X_N$, the smaller the number of rejected elements will be. A value of $X_{\max}$ far way from the real maximum will require more attempts until an element is selected.

4.3.3 Estimation

For a sample selected with replacement and with selection probabilities p_k, the Horvitz–Thompson estimator takes the form

$$\bar{y}_{\text{UP}} = \frac{1}{Nn} \sum_{k=1}^{N} a_k \frac{Y_k}{p_k}. \quad (4.40)$$

The subscript UP indicates *unequal probability* sampling. If the selection probabilities p_k must be proportional to the values X_k of the auxiliary variable X, and at the

UNEQUAL PROBABILITY SAMPLING

same time the condition must be satisfied that the selection probabilities must add up to 1, it follows that

$$p_k = \frac{X_k}{N\bar{X}}. \tag{4.41}$$

Substituting (4.41) in (4.40) leads to

$$\bar{y}_{UP} = \frac{\bar{X}}{n} \sum_{k=1}^{N} a_k \frac{Y_k}{X_k}. \tag{4.42}$$

Estimator (4.42) can be written in a different way. Suppose a new variable $Z = Y/X$ is defined. The values of Z in the population are denoted by $Z_1, Z_2, \ldots, Z_N$, where $Z_k = Y_k/X_k$, for $k = 1, 2, \ldots, N$. Note that it is assumed that $X_k > 0$ for all k, and hence the value of Z_k is always defined.

The sample provided values $y_1, y_2, \ldots, y_n$ of Y and values $x_1, x_2, \ldots, x_n$ of X. These values can be used to compute the sample values $z_1, z_2, \ldots, z_n$ of Z, where $z_i = y_i/x_i$, for $i = 1, 2, \ldots, n$. Estimator (4.42) can now be written as

$$\bar{y}_{UP} = \bar{X}\bar{z}, \tag{4.43}$$

in which

$$\bar{z} = \frac{1}{n} \sum_{i=1}^{n} z_i. \tag{4.44}$$

This estimator is equal to the product of two means: population mean of the auxiliary variable X and the sample mean of the Z.

Suppose X is a variable assuming the value 1 for each element in the population. So, X is a constant. Then, expression (4.42) reduces to the simple sample mean. This is correct because it comes down to simple random sampling with replacement and with equal probabilities.

Expression (2.49) contains the variance of the Horvitz–Thompson estimator in case of sampling with replacement. Substitution of (4.41) leads to

$$V(\bar{y}_{UP}) = \frac{\bar{X}}{Nn} \sum_{k=1}^{N} X_k \left(\frac{Y_k}{X_k} - \frac{\bar{Y}}{\bar{X}} \right)^2. \tag{4.45}$$

Suppose, the values Y_k and X_k are proportional, so $Y_k = cX_k$. Hence, $Y_k/X_k = c$, and also the ratio of the mean of Y and the mean of X is equal to c, resulting in a variance equal to 0. This ideal situation will not happen in practice. However, even if Y and X are only approximately proportional, this will reduce the variance.

In practice, variance (4.45) must be estimated using the sample data. The unbiased estimator for the variance is equal to

$$v(\bar{y}_{UP}) = \frac{\bar{X}^2}{n(n-1)} \sum_{i=1}^{n} (z_i - \bar{z})^2. \tag{4.46}$$

This estimator can be used to construct a confidence interval.

4.3.4 An Example

The effects of unequal probability sampling are shown in a simulation experiment. This time, the target population consists of 200 dairy farms in the rural part of Samplonia. Objective of the survey is to estimate the average daily milk production per farm. Several different sampling designs can be considered. Of course, the most straightforward way is a simple random sample. Unequal probability sampling is also possible because two auxiliary variables are available: the number of cows per farm and the area of grassland per farm.

It is not unreasonable to assume a relationship between milk production and the number of cows (more cows will produce more milk), or between milk production and the area of grassland (more grass means more cows and thus more milk). Therefore, one could consider drawing farms with probabilities proportional to the number of cows or the area of grass.

The upper left graph in Fig. 4.8 shows the distribution of the estimator based on 600 simple random samples of size 50 (without replacement and with equal probabilities). The standard error of the estimator is 30.9.

The upper right graph contains the distribution of the estimator in case of sampling farms with probabilities proportional to the area of grassland. The variation of the possible outcomes is less than in the case of simple random sampling. The standard error is reduced from 30.0 to 25.8. Apparently, there is a certain relationship between the target variable and the auxiliary variable.

The lower left graph in Fig. 4.8 contains the distribution of the estimator in case of sampling farms with probabilities proportional to the number of cows per farm. The variation of the possible outcomes is even much less. The standard error is reduced to

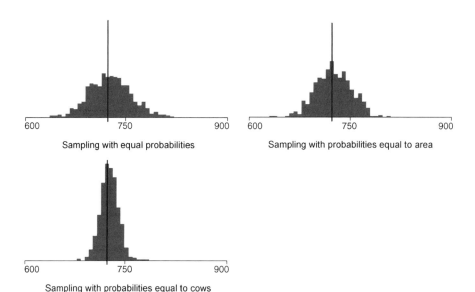

Figure 4.8 Simulation of unequal probability sampling.

13.8. This is caused by a strong relationship between the milk production per farm and the number of cows per farm, which is, of course, not surprising.

4.4 SYSTEMATIC SAMPLING WITH UNEQUAL PROBABILITIES

Manually drawing an unequal probability sample from a large sampling frame can be cumbersome and time-consuming. To avoid these problems, a systematic sampling can be selected with unequal probabilities. In fact, the same arguments apply to this sampling design as were discussed for systematic sampling with equal probabilities (see Section 4.2).

It should also be noted that it is problematic to draw an unequal probability sample without replacement. Chaudhuri and en Vos (1988) present an overview of various sample selection schemes. They all seem to have some kind of disadvantage of preventing application in a real survey. Some are so complex that they can only be applied for very small surveys, and others may produce negative variance estimates.

Systematic sampling with unequal probabilities does implement a form of without replacement sampling. This type of sampling design can be seen as a kind of crossing between systematic sampling (Section 4.2) and unequal probability sampling (Section 4.3). It combines the advantages of both types of sampling designs. By systematically working through the sampling frame an unequal probability sample without replacement is obtained (Fig. 4.9).

Systematic sampling with unequal probabilities has the potential of producing very precise estimates. This depends on the availability of an auxiliary variable that has a strong correlation with the target variable. Taking the inclusion probabilities (approximately) proportional to the values of this auxiliary variable will result in a considerable variance reduction.

Of course, the disadvantages of systematic sampling should be also taken into account. A special structure in the order of elements in the sampling frame may produce wrong estimates of variances and therefore wrong statements about the population characteristics.

Because elements are drawn without replacement, the theory can be described in terms of inclusion probabilities (instead of selection probabilities). Horvitz and Thompson (1952) have shown that precise estimators can be obtained by taking the inclusion probabilities approximately equal to the values of the target variable. This is accomplished in practice by taking the inclusion probabilities proportional to

Figure 4.9 Systematic sample selection with unequal probabilities. Reprinted by permission of Imre Kortbeek.

the values of an auxiliary variable having a strong relationship with the target variable. Suppose such a variable X is available, with values $X_1, X_2, \ldots, X_N$. It was shown in chapter 2 that the sum of the inclusion probabilities is always equal to the sample size n. Therefore, the inclusion probability of element k must be equal to

$$\pi_k = n \frac{X_k}{X_T} = n \frac{X_k}{N\bar{X}}, \qquad (4.47)$$

for $k = 1, 2, \ldots, N$.

4.4.1 A Sample Selection Scheme

The sample selection scheme for a systematic unequal probability sample is a generalized version of the scheme for a simple systematic sample (with equal probabilities) that was described in Section 4.2. First, the subtotals $T_1, T_2, \ldots, T_N$ have to be computed, where

$$T_k = \sum_{i=1}^{k} X_i. \qquad (4.48)$$

Consequently, $T_N = X_T$. Furthermore, by definition $T_0 = 0$.

To draw a sample of size n from a population of size N, the line segment $(0, T_N)$ is divided into N intervals. Each interval corresponds to an element in the population. The first interval corresponds to element 1 and has length X_1, the second interval corresponds to element 2 and has length X_2, and so on. So there are N intervals

$$(T_0, T_1], (T_1, T_2], \ldots, (T_{N-1}, T_N]. \qquad (4.49)$$

The *step length* is defined by $F = T_N/n = X_T/n$. This is a real-valued quantity. The *starting point* b is defined by drawing a random value from the interval $(0, F]$. This value is obtained by taking a random value u from the interval $[0, 1)$, subtracting it from 1 and multiplying the result by F. So b is equal to $b = (1 - u) \times F$. Next, the values

$$t_1 = b, t_2 = b + F, t_3 = b + 2F, \ldots, t_n = b + (n-1)F \qquad (4.50)$$

are computed. Each value t_i will belong to one of the intervals $(T_{k-1}, T_k]$. So, for each t_i the sequence number k is determined for which

$$T_{k-1} < t_i \leq T_k, \qquad (4.51)$$

and the corresponding element k is selected in the sample. (Recipe 4.5)

An example of systematic unequal probability sampling illustrates the above theory. A sample of size $n = 3$ is selected from the population of $N = 7$ districts in Samplonia. Inclusion probabilities are taken proportional to the number of inhabitants. The required data are shown in Table 4.5.

The step length is equal to $T_N/n = 1000/3 = 333.333$. So, the starting point is drawn from the interval $(0, 333.333]$. Suppose, this results in the value $b = 112.234$. The t-values are now equal to $t_1 = 112.234$, $t_2 = 112.234 + 333.333 = 445.567$ and $t_3 = 112.234 + 2 \times 333.333 = 778.901$. The first value lies between T_0 and T_1, so

SYSTEMATIC SAMPLING WITH UNEQUAL PROBABILITIES

Recipe 4.5 Selecting a Systematic Sample with Unequal Probabilities

Ingredients	Population size N
	Population values $X_1, X_2, \ldots, X_N$ of an auxiliary variable X ($X_k > 0$ for all k)
	Sample size n
	Generator of random values u from $[0, 1)$
Step 1	Compute the subtotals $T_k = X_1 + X_2 + \cdots + X_k$ for $k = 1, 2, \ldots, N$. $T_0 = 0$
Step 2	Compute the step length $F = T_N/n$
Step 3	Check for each element k whether $X_k \geq F$. If this is the case, select element k in the sample, reduce the sample size n by 1, remove element k from the population, and return to step 1
Step 4	Draw a random value u from $[0, 1)$
Step 5	Compute the starting point $b = (1 - u) \times F$
Step 6	Determine the sequence number k from which $T_{k-1} < b \leq T_k$
Step 7	Select the element with sequence number k in the sample
Step 8	If the sample size has not been reached yet, add an amount F to b and go back to step 6

district Wheaton is selected in the sample. The second value lies between T_4 and T_5. This means district Smokeley is added to the sample. Finally, the third value lies between T_6 and T_7, which means that Mudwater is included in the sample. See also Fig. 4.10.

To obtain more insight into the second-order inclusion probabilities, the selection process can be displayed in a different way. The line segment $(0, T_N]$ as shown in Fig. 4.10 is divided into n subsegments of length F. These subsegments are drawn below each other (Fig. 4.11). The starting point is a random value in the first subsegment. The sample is obtained by drawing a vertical line through

Table 4.5 Numbers of Inhabitants in the Districts of Samplonia

k	District	Population X_k	Subtotal T_k
1	Wheaton	144	144
2	Greenham	94	238
3	Newbay	55	293
4	Oakdale	61	354
5	Smokeley	244	598
6	Crowdon	147	745
7	Mudwater	255	1000

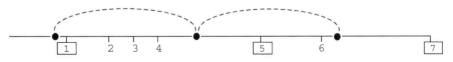

Figure 4.10 A systematic unequal probability sample of size 3 from a population of size 7.

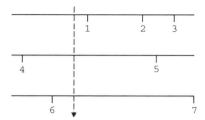

Figure 4.11 A systematic unequal probability sample of size 3.

the starting point. If the line transects interval $(T_{k-1}, T_k]$, element k is selected in the sample.

The first-order inclusion probability of an element is equal to the probability of transecting its interval. This probability is equal to $X_k/F = nX_k/X_T$ for element k. Indeed, its inclusion probability is proportional to its value of the auxiliary variable.

The second-order inclusion probabilities depend on the order of the elements and also on the values of the auxiliary variables. For example, it is clear from Fig. 4.11 that elements 5 and 7 have a high probability of being together in the sample. Moreover, it will not be possible to have the elements 2 and 6 together in the sample.

Depending on the sample size and the values of the auxiliary variable, a problem may occur with "big elements." Big elements are defined as those elements k for which the value X_k is larger than the step length F. Such elements are always selected in the sample whatever be the value of the starting point b. The length of the interval $(T_{k-1}, T_k]$ is so large that it is impossible to jump over it. Their inclusion probability is equal to $\pi_k = 1$. It is not proportional to X_k. If F is much smaller than X_k, element k can even be selected more than once in the same sample. This situation is illustrated in Fig. 4.12.

Suppose four districts are selected with probabilities proportional to the population size. The step length is equal to $T_N/n = 1000/4 = 250$. The district of Mudwater has 255 inhabitants. This is more than the step length. So, Mudwater is a "big element." If the starting value turns out to be $b = 248$, the t-values are equal to $t_1 = 248$, $t_2 = 498$, $t_3 = 748$, and $t_4 = 998$. Both the values t_3 and t_4 are between T_6 and T_7, so element 7 (Mudwater) is even selected twice.

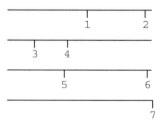

Figure 4.12 A systematic unequal probability sample with "big elements."

Note that the fourth subsegment in Fig. 4.12 completely belongs to element 7. Whatever vertical line is drawn, it will always transect the interval of element 7. So, element 7 will always be selected. Also, the last (small) part of the third segment belongs to element 7. Every vertical line transecting this part will result in a sample containing element 7 twice.

The problem of the "big elements" is solved in practice by first removing these elements from the population and including them in the sample. A somewhat smaller population will remain, and the rest of the sample is selected from this population.

Returning to the example of a sample of size $n=4$ from a population of size $N=7$, it turned out that element 7 (Mudwater) was a big element. Since the number of inhabitants of all other districts is smaller than the step length $F=250$, Mudwater is the only big element. Therefore, this district is removed from the population and included in the sample. The remaining population consists of six districts, with a total population size of 745. A sample of size 3 must be selected from this reduced population. This means a step length of $F=745/3=248.3$. Again, the situation must be checked for big elements. The largest remaining district is Smokeley with 245 inhabitants. This value is smaller than the step length, and therefore there are no big elements any more.

4.4.2 Estimation

If the inclusion probabilities π_k must be proportional to the values X_k of the auxiliary variable X, and at the same time the condition must be satisfied that the inclusion probabilities must add up to n, it follows that

$$\pi_k = n\frac{X_k}{X_T} = n\frac{X_k}{N\bar{X}}, \qquad (4.52)$$

for $k=1,2,\ldots,N$. Substituting (4.52) in expression (2.38) for the Horvitz–Thompson estimator leads to the estimator

$$\bar{y}_{\text{SUP}} = \frac{\bar{X}}{n}\sum_{k=1}^{N} a_k \frac{Y_k}{X_k}. \qquad (4.53)$$

Estimator (4.53) can also be written in a different way. Suppose, a new variable $Z=Y/X$ is defined. The values of Z in the population are denoted by $Z_1, Z_2, \ldots, Z_N$, where $Z_k = Y_k/X_k$, for $k=1,2,\ldots,N$. Note that it is assumed that $X_k > 0$ for all k, and hence the value of Z_k is always defined.

The sample provided values $y_1, y_2, \ldots, y_n$ of Y and values $x_1, x_2, \ldots, x_n$ of X. These values can be used to compute the sample values $z_1, z_2, \ldots, z_n$ of Z, where $z_i = y_i/x_i$, for $i=1,2,\ldots,n$. Estimator (4.42) can now be written as

$$\bar{y}_{\text{SUP}} = \bar{X}\bar{z}, \qquad (4.54)$$

with

$$\bar{z} = \frac{1}{n} \sum_{i=1}^{n} z_i. \tag{4.55}$$

This estimator is equal to the product of two means: population mean of the auxiliary variable X and the sample mean of the Z. Note that the formula for the estimator is exactly equal to that of the estimator for unequal probability sampling (see Section 4.3).

Suppose X is a variable assuming the value 1 for each element in the population. So, X is a constant. Then, expression (4.53) reduces to the simple sample mean. This is correct because it comes down to systematic sampling with equal probabilities.

There is no simple expression for the variance of estimator (4.53). This is caused by the complex nature of the second-order inclusion probabilities. They depend on the both order of the elements in the sampling frame and magnitude of the values of the auxiliary variable. For example, Fig. 4.11 shows that on the one hand elements 2 and 4 will never end up together in the sample, so that their second-order inclusion probability is 0. On the other hand, elements 5 and 7 have a high probability of being selected together in the sample.

In principle, it is possible to process all elements in the sampling frame and to compute all second-order inclusion probabilities, but this can mean a lot of work. Suppose, a survey has to be carried out in a town with a population of 600,000 people. Then even for this relatively small population, the number of second-order inclusion probabilities is equal to $600{,}000 \times (600{,}000 - 1)/2 = 179{,}999{,}700{,}000$.

The magnitude of the variance is determined by several factors. The variance will be small if the inclusion probabilities are approximately proportional to the values of target variable. This property is inherited from unequal probability sampling. Furthermore, the variance will be small if the homogeneity of each sample is similar to that of the population. This property is inherited from systematic sampling.

It is difficult to estimate the variance of the estimator properly. This is caused by problems similar to that of systematic sampling. If there is a specific cyclic structure in the order of the elements in the sampling frame, and the step length is equal to the length of this cycle, the variance in the sample will not be indicative of the true variance of the estimator. This may cause variance estimates to be too small, creating a wrong impression of precise estimators.

In the case of systematic sampling with equal probabilities, the variance expressions of simple random sampling can be used as an approximation if the order of the elements in the sampling frame is completely random. A similar approach can also be used for systematic sampling with unequal probabilities. However, one should be careful because the expression for unequal probability sampling in Section 4.3 applies to with replacement sampling, whereas systematic sampling with unequal probabilities is a form of without replacement sampling. Nevertheless, the approximations may work if the population size is much larger than the sample size.

If it is important to have a good estimate of the variance, an approach similar to that described in Section 4.2 can be applied. Instead of selecting one sample, a number of

Table 4.6 Precision of Estimators for Various Sampling Designs Resulting in a Sample of 10 from the Working Men in Agria

Sampling Design	Variance
Simple random	1178
Systematic	541
Systematic (ordered by income)	201
Probabilities proportional to age	544
Systematic, probabilities proportional to age	292
Systematic, probabilities proportional to age (ordered by income)	155

independent small subsamples are selected. An estimate is computed for each subsample. Application of expressions (4.29) and (4.30) produces a combined estimate and an estimate of the variance of this combined estimate.

It will be assumed in many practical situations that the properties of a systematic sample with unequal probabilities will not differ much from those of an unequal probability sample as described in Section 4.3 and that therefore expression (4.46) can be used. However, one should always be careful.

4.4.3 An Example

The properties of systematic sampling with unequal probabilities are illustrated using a small example. The target population consists of all working men in the Samplonian province of Agria. This population consists of only 58 people. The small population size simplifies the work of computing the second-order inclusion probabilities.

A sample of 10 persons is selected from this population. The objective is to estimate to mean income. Age can be used as an auxiliary variable.

Table 4.6 contains the variance of the estimator of the mean for various sampling designs. For the nonsystematic samples, the formula for the Horvitz–Thompson estimator was used. For the systematic samples, second-order inclusion probabilities were computed for the original order of the elements and also for a sampling frame in which the elements were ordered by increasing income.

The table shows that simple random sampling is not the best way to get a precise estimate of the mean income. Apparently, the structure of the sampling frame is such that systematic sampling reduces the variance by approximately a factor 2. The relationship between income and age is not very strong, but still unequal probability sampling also reduces the variance by a factor 2. The result is even better (reduction by a factor 4) for systematic sampling with probabilities proportional to age. The best results for systematic sampling are achieved by ordering the sampling frame by increasing income. Of course, this is not feasible in practice.

The variance of the estimator based on systematic sampling with unequal probabilities is equal to 292. This means that in terms of a 95% confidence interval in 95 out of 100 cases the estimate will not differ more than $1.96 \times \sqrt{292} = 33$ from the true value. So even for a small sample of size 10, the mean income can be estimated with a reasonably high precision.

EXERCISES

4.1 Use the table with random numbers given below to draw a sample of size $n = 20$ from a population of size $N = 80$. Work row-wise and use the first two digits of each group of five digits. Write down the sequence number of the selected elements.

06966	75356	46464	15180	23367	31416	36083	38160	44008	26146
62536	89638	84821	38178	50736	43399	83761	76306	73190	70916
65271	44898	09655	67118	28879	96698	82099	03184	76955	40133
07572	02571	94154	81909	58844	64524	32589	87196	02715	56356
30320	70670	75538	94204	57243	26340	15414	52496	01390	78802
94830	56343	45319	85736	71418	47124	11027	15995	68274	45056
17838	77075	43361	69690	40430	74734	66769	26999	58469	75469
82789	17393	52499	87798	09954	02758	41015	87161	52600	94263
64429	42371	14248	93327	86923	12453	46224	85187	66357	14125
76370	72909	63535	42073	26337	96565	38496	28701	52074	21346

4.2 A target population consists of 1000 companies. A researcher wants to estimate the percentage of companies exporting their products to other countries. Therefore, he selects a simple random sample of 50 companies without replacement.

a. Compute the variance of the sample percentage of exporting companies, assuming that 360 out of 1000 companies indeed do so.

b. Estimate the variance of the sample percentage, assuming the population percentage is unknown and there are 18 exporting companies in the sample.

c. Estimate the variance of the sample percentage, assuming the population percentage is unknown and there are 14 exporting companies in the sample.

d. Estimate the variance of the sample percentage, assuming the population percentage is unknown and there are 22 exporting companies in the sample.

4.3 A company wants to carry out a customer satisfaction survey among its 10,000 customers. The questionnaire contains a number of questions that are used to compute a satisfaction index. This is value in the interval from 0 to 100, where 0 means extremely unsatisfied and 100 means extremely satisfied. Before really carrying out the survey, a test is conducted. A simple random sample of 20 customers is selected without replacement. The satisfaction scores turn out to be

| 100 | 88 | 72 | 81 | 80 | 69 | 84 | 83 | 65 | 69 | 90 | 65 | 70 | 80 | 90 | 74 | 70 | 96 | 62 | 67 |

a. Compute the sample mean. Compute also an estimate of the standard error of the sample mean.

b. Compute the 95% confidence interval for the mean satisfaction index of all 10,000 customers. The company has a quality management policy aiming at a mean satisfaction index of 80. What conclusion can be drawn from the sample with respect to this target?

c. The company decides to really carry out the customer satisfaction survey. The requirement is that the margin of error of the 95% confidence interval must not exceed a value of 2. Compute the corresponding sample size, using the information obtained in the test survey.

4.4 A camera shop in a town considers including an advertisement for digital cameras in the newsletter of the local tennis club. This club has 1000 members. Before deciding to do so, the shop owner wants to have an estimate of the percentage of tennis club members having a digital camera. To this end, he carries out a survey. He assumes that the percentage of club members with a digital camera does not exceed 30%. Compute the required sample size if the width of the 95% confidence interval may not exceed 3 percent points.

4.5 If a sample has been selected with unequal probabilities, the estimates must be corrected by weighting the observed values of the target variable using

  a. the values of the auxiliary variable;
  b. the selection probabilities;
  c. the square roots of the values of the auxiliary variable;
  d. a combination of the quantities mention in a, b, and c.

4.6 A forestry company experiments with the rate of growth of different types of trees. It has planted 48 trees in one long line alongside a road. The trees are alternately of types A and B. After 2 years, the company wants to know the average height of all trees. The length of all 48 trees can be found in the table below:

---

629 353 664 351 633 314 660 381 640 366 696 348 681 307 633 337 663 331 609 338 675 361 696 304 647 366 669 384 669 389 693 324 698 309 602 341 671 352 663 344 671 342 627 323 612 376 629 363

---

  a. Compute the mean length and the standard deviation of all 48 trees.
  b. Draw a simple random sample without replacement of size $n = 8$. Use the table with random numbers below. Work row-wise and use only the first two digits of each group of five digits. Compute the sample mean and the sample standard deviation.

94830	56343	45319	85736	71418	47124	11027	15995	68274	45056
17838	77075	43361	69690	40430	74734	66769	26999	58469	75469
82789	17393	52499	87798	09954	02758	41015	87161	52600	94263
64429	42371	14248	93327	86923	12453	46224	85187	66357	14125
76370	72909	63535	42073	26337	96565	38496	28701	52074	21346

---

  c. Draw a systematic sample of size 8. Use the value $b = 3$ as starting point. Compute the sample mean and the sample standard deviation.
  d. Compare the results of exercises (b) and (c) with those of exercise. Explain observed differences and/or similarities.

4.7 The table below contains an imaginary population of 20 transport companies. It contains for every company the number of trucks it owns and the amount of goods that have been transported in a specific week. A sample of eight companies is selected. The objective is to estimate the mean of transported goods per company.

Company	Number of trucks (X)	Transported goods (Y)
1	3	35
2	4	37
3	5	48
4	6	64
5	7	75
6	6	62
7	4	39
8	5	46
9	3	29
10	9	93
11	12	124
12	20	195
13	4	42
14	3	28
15	5	46
16	8	83
17	7	71
18	3	25
19	4	41
20	3	27

a. Suppose the sample is selected with equal probabilities and without replacement. Compute the variance and the standard error of the sample mean. Also, compute the margin of error of the 95% confidence interval.

b. Select a sample of eight companies with unequal probabilities. Use the number of trucks per company as auxiliary variable. Select the sample with the cumulative scheme. Use the following values of the randomizer for values in the interval $[0, 1)$: 0.314, 0.658, 0.296, 0.761, 0.553, 0.058, 0.128, and 0.163.

c. Compute the value of the estimator for this sampling design. Make clear how this value is obtained.

d. Estimate the variance and the standard error of the estimator. Also, determine the 95% confidence interval. Compare the margin of error of this estimated interval with the margin of the confidence interval as computed in (a) and draw conclusions with respect to the precision of estimator. Determine whether the estimated confidence interval indeed contains the value of the population parameter.

EXERCISES

4.8 A population consists of $N = 9$ elements. There are three elements for which the value of the target variable Y is equal to 1. For another three elements, the value is equal to 2. The value of Y is equal to 3 for the three remaining elements. Suppose a systematic sample with equal probabilities of size $n = 3$ is selected from this population.

    a. The elements are ordered in the sampling frame such that their values are equal to 1,2,3,1,2,3,1,2,3. Compute the variance of the sample mean for this situation.

    b. Determine the sequence of elements that results in the smallest variance of the sample mean. What is the value of this variance?

    c. Suppose a naive researcher uses the variance formula for a simple random sample without replacement. In which of the situations described under (a) and (b) would this result in the largest mistake? Explain why this is so.

CHAPTER 5

Composite Sampling Designs

5.1 STRATIFIED SAMPLING

Stratified sampling is based on the idea of dividing the population into a number of subpopulations (strata) and to draw a sample from each stratum separately. This idea is as old as sampling theory itself. Sampling theory emerged at the end of the nineteenth century. At that time, researchers were still reluctant to draw samples. It was argued that there was no need to use samples if every element in the population could be observed. Moreover, it was considered improper to replace observations by mathematical calculations. The general idea at that time was that it was impossible to draw reliable conclusions about a population using data that were collected for just a small part of the population.

5.1.1 Representative Samples

The first ideas about sampling were discussed at the meeting of the International Statistical Institute (ISI) in Bern in 1895. It was Anders Kiaer, director of the Norwegian statistical institute, who proposed using sampling instead of complete enumeration. He argued that good results could be obtained with his *Representative Method*. This was a type of investigation where data on a large selection from the population were collected. This selection should reflect all aspects of the population as much as possible. One way to realize such a sample was the "balanced sample." The population was divided into subpopulations by using variables such as gender, age, and region. These subpopulations were called *strata*. The sizes of the strata were supposed to be known. A percentage of persons was taken from each stratum. This percentage was the same for each stratum. Selection of samples took place in some haphazard way (probability sampling had not yet been invented). As a result, the sample distribution of variables such as gender, age, and region was similar

---

Applied Survey Methods: A Statistical Perspective, Jelke Bethlehem
Copyright © 2009 John Wiley & Sons, Inc.

STRATIFIED SAMPLING 101

Figure 5.1 Stratified sampling. Reprinted by permission of Imre Kortbeek.

to the distribution in the population. The sample was representative with respect to these variables.

Probability sampling was introduced in later years, but stratification remained a useful technique to improve the "representativity" of the sample. Stratification also turned out to have the potential to improve the precision of estimates. This particularly works well if strata are chosen such that they are homogeneous with respect to the target variables of the survey (Fig. 5.1).

5.1.2 Sample Selection Schemes

To select a stratified sample, the population is first divided into strata (subpopulations). Next, a sample is selected in each stratum. Researchers are free to choose the sampling design for each stratum, as long as it provides an unbiased estimate of the value of the population parameter in each stratum. Finally, the estimates for all strata are combined into an estimate for whole population.

Stratified sampling has some flexibility. For example, it offers the possibility to say something about each stratum separately. Hence, a national survey could provide information about each province of the country and a business survey could produce statistics for each separate branch. By choosing the proper sample size in each stratum, sufficient precise estimates can be computed.

Stratification can be carried out only if it is known in advance to which stratum each population element belongs. Since a separate sample is selected from each stratum, there must be a sampling frame for each stratum. For example, if a stratified sample must be selected from a large town, where stratification is by neighborhood, a separate sample must be drawn from each neighborhood, requiring a sampling frame for each neighborhood. Lack of availability of sampling frames per strata may prevent stratified sampling.

First, the situation is discussed where samples are selected from all strata using arbitrary sampling designs. Next, the theory will be worked out for the case in which a simple random sample (with equal probabilities and without replacement) is selected from each stratum. Stratification offers not much news from a theoretical point of view. Instead of drawing just one sample, a number of samples are drawn. The only difference is the manner in which stratum estimates are combined into an estimate for the whole population.

5.1.3 Estimation

To be able to write down the formulas for estimators and variances, notations are slightly adapted. Several quantities get an extra index denoting the stratum to which they apply. Suppose the population U is divided into L strata. These strata are denoted by

$$U_1, U_2, \ldots, U_L. \tag{5.1}$$

The strata do not overlap and together cover the complete population U. The number of elements in stratum h is indicated by N_h (for $h = 1, 2, \ldots, L$). Consequently,

$$\sum_{h=1}^{L} N_h = N_1 + N_2 + \cdots + N_L = N. \tag{5.2}$$

The N_h values of the target variable Y in stratum h are denoted by

$$Y_1^{(h)}, Y_2^{(h)}, \ldots, Y_{N_h}^{(h)}. \tag{5.3}$$

The mean of the target variable in stratum h is denoted by

$$\bar{Y}^{(h)} = \frac{1}{N_h} \sum_{k=1}^{N_h} Y_k^{(h)}. \tag{5.4}$$

The mean in the whole population can be written as

$$\bar{Y} = \frac{1}{N} \sum_{h=1}^{L} N_h \bar{Y}^{(h)}. \tag{5.5}$$

So the population mean is equal to the weighted average of the stratum means. The (adjusted) variance in stratum h is equal to

$$S_h^2 = \frac{1}{N_h - 1} \sum_{k=1}^{N_h} \left(Y_k^{(h)} - \bar{Y}^{(h)} \right)^2. \tag{5.6}$$

A sample of size n is drawn from this stratified population. This sample is realized by selecting L subsamples, with respective sample sizes $n_1, n_2, \ldots, n_L$, where $n_1 + n_2 + \cdots + n_L = n$. Whatever sampling design is used for stratum h, with the theory of Horvitz and Thompson (1952), it is always possible to construct an estimator

$$\bar{y}_{\text{HT}}^{(h)} \tag{5.7}$$

STRATIFIED SAMPLING

allowing unbiased estimation of the stratum mean (5.4). The variance of this estimator is denoted by

$$V\left(\bar{y}_{\text{HT}}^{(h)}\right). \tag{5.8}$$

Since estimator (5.7) is an unbiased estimator for the mean of stratum h (for $h = 1, 2, \ldots, L$), expression

$$\bar{y}_S = \frac{1}{N} \sum_{h=1}^{L} N_h \bar{y}_{\text{HT}}^{(h)} \tag{5.9}$$

is an unbiased estimator of the mean of target variable Y in the whole population. Because the subsamples are selected independently, the variance of estimator (5.9) is equal to

$$V(\bar{y}_S) = \frac{1}{N^2} \sum_{h=1}^{L} N_h^2 V\left(\bar{y}_{\text{HT}}^{(h)}\right). \tag{5.10}$$

This expression shows that the variance of the estimator will be small if the variances of the estimators within the strata are small. This offers interesting possibilities for constructing precise estimators.

The theory will now be applied to the case of a stratified sample where a simple random sample without replacement is selected in each stratum. Let the sample size in stratum h be equal to n_h. The n_h observations that become available in stratum h are denoted by

$$y_1^{(h)}, y_2^{(h)}, \ldots, y_{n_h}^{(h)}. \tag{5.11}$$

The sample mean in stratum h is equal to

$$\bar{y}^{(h)} = \frac{1}{n_h} \sum_{i=1}^{n_h} y_i^{(h)}. \tag{5.12}$$

The sample mean in stratum h is an unbiased estimator of the population mean in stratum h. The variance of the estimator is equal to

$$V\left(\bar{y}^{(h)}\right) = \frac{1-f_h}{n_h} S_h^2 \tag{5.13}$$

where $f_h = n_h/N_h$. This variance can be estimated in an unbiased manner by

$$v\left(\bar{y}^{(h)}\right) = \frac{1-f_h}{n_h} s_h^2 \tag{5.14}$$

where

$$s_h^2 = \frac{1}{n_h - 1} \sum_{i=1}^{n_h} \left(y_i^{(h)} - \bar{y}^{(h)}\right)^2. \tag{5.15}$$

Now the estimators for the stratum means can be combined into an estimator for the population mean. This estimator is equal to

$$\bar{y}_S = \frac{1}{N} \sum_{h=1}^{L} N_h \bar{y}^{(h)}. \tag{5.16}$$

This is an unbiased estimator for the population mean. So this estimator is equal to the weighted average of the sample means in the strata. The variance of estimator (5.16) is equal to

$$V(\bar{y}_S) = \frac{1}{N^2} \sum_{h=1}^{L} N_h^2 \frac{1-f_h}{n_h} S_h^2. \tag{5.17}$$

This variance can be estimated in an unbiased manner by

$$v(\bar{y}_S) = \frac{1}{N^2} \sum_{h=1}^{L} N_h^2 \frac{1-f_h}{n_h} s_h^2. \tag{5.18}$$

A closer look at expression (5.17) shows that the variance of the stratification estimator is small if the stratum variances S_h^2 are small. This will be the case if the strata are homogeneous with respect to the target variable, which means there is not much variation in the values of the target variable within strata. The variation in this case is mainly due to the differences in stratum means. So, there is a lot of variation between strata, and not within strata. The conclusion can be drawn that the stratified estimator will be precise if it is possible to construct a stratification with homogeneous strata.

5.1.4 Stratification Variables

The sampling design for a stratified sample is flexible. There are many different ways to construct strata. The only condition is that the stratum sizes must be known and that it must be possible to select a sample in each stratum separately.

Analysis of the variance expression (5.17) has shown that strata should be constructed such that within strata variances are small. This will result in precise estimators. Therefore, the search for proper *stratification variables* (i.e., variables used for the construction of strata) should be aimed at finding variables that have a strong relationship with the target variable. Such a relationship implies that the value of the target variable can be predicted from the value of stratification variables. This comes down to a situation where there is little variation in the values of the target variables, given the values of the stratification variables.

Sometimes, a candidate stratification variable is quantitative. It assumes many different values. An example is the variable age. To be useful as a stratification variable, it has to be transformed into a qualitative variable. For variables such as age, this would mean categorizing age into different groups, for example, 20–29 years, 30–39 years, and so on. The question that remains is "Which grouping is most effective in the sense that the resulting stratified estimator has the smallest variance?"

STRATIFIED SAMPLING

Table 5.1 Construction of Age Strata Using the Cumulative-Square-Root-f Rule

Age	Frequency f	Width w	$\sqrt{fw}$	$\sqrt{fw}$ Cumulated	Strata
0–4	90	5	21	21	0–14
5–9	86	5	21	42	
10–14	79	5	20	62	
15–19	88	5	21	83	15–29
20–29	149	10	39	122	
30–39	110	10	33	155	30–49
40–49	119	10	34	189	
50–59	101	10	32	221	50–69
60–69	74	10	27	248	
70–79	59	10	24	272	70–99
80–99	45	20	30	302	

Dalenius and Hodges (1959) have proposed a rule for this. It is called the *cumulative-square-root-f rule*. First, the frequency distribution of the original variable is determined. The square root of the product of the frequency (f) and the interval width (w) is computed for each value of the variable. Next, values are grouped in such a way that the sum of the computed quantities is approximately the same in each group.

Table 5.1 contains an example of this procedure. It starts from an age distribution in 11 small groups. These 11 groups have to be combined into a smaller number of groups. Five strata have been formed, each containing about one-fifth of the total of all values $\sqrt{f \times w}$.

5.1.5 Sample Allocation

An important aspect of a stratified sample is the *allocation* of the sample. This is the distribution of the total sample size n over the L strata. Conditions on the precision of the estimators in each stratum would determine the sample size in each stratum and therefore the total sample size. However, usually the total sample size n is determined beforehand. This leads to the question how to divide this sample size over the strata.

If the objective is to estimate the population as precise as possible, the best estimator is obtained by means of the so-called *optimal allocation*. This allocation is sometimes also called the *Neyman allocation*. According to this allocation, the sample size n_h in stratum h must be taken equal to

$$n_h = n \frac{N_h S_h}{\sum_{j=1}^{L} N_j S_j}. \tag{5.19}$$

If n_h turns out not to be an integer number, it should be rounded to the nearest integer. The sample size in a stratum will be larger if the variance in a stratum is larger. Not surprisingly, more elements must be observed in less homogeneous strata.

It may happen that the computed value n_h is larger than the total size N_h of the stratum h. In this case, all elements in strata should simply be observed. The remaining sample size can be divided over all other strata by using the optimal allocation rule.

If n_h elements are selected from stratum h of size N_h, the inclusion probabilities of these elements are all equal to n_h/N_h. As a result, the inclusion probabilities are proportional to the stratum standard deviations S_h. Not every element has the same inclusion probability. This is no problem as estimator (5.16) corrects for this. This formula is obtained if the inclusion probabilities n_h/N_h are substituted in the expression for the Horvitz–Thompson estimator.

It is only possible to compute the optimum allocation if the values of the stratum standard deviations S_h are known. Often this is not the case. If estimates from a previous survey are available, they can be used. Sometimes, stratification is applied not only for increasing precision but also for administrative purposes (there is no sampling frame for whole population, but there are sampling frames for each stratum separately) or for obtaining estimates within strata. In this case, it may not be unreasonable to assume that the stratum standard deviations do not differ too much. If all standard deviations are equal, the allocation expression (5.19) reduces to

$$n_h = n \frac{N_h}{N}. \tag{5.20}$$

Allocation according to this formula is called *proportional allocation*. Proportional allocation means that every element in the population has the same inclusion probability n/N. This is why this type of sample is sometimes called a *self-weighting* sample. This is what early survey researchers had in mind when they talked about representative samples.

It has already been said that the choice of the sampling design and the sample size is often a compromise between precision and costs. The costs of data collection can also play a role in determining the allocation of the sample to the strata. There can be situations where interviewing in one stratum is more costly than in another stratum. As a result, the optimal allocation may not be the cheapest allocation.

Suppose that the total costs of the fieldwork may not exceed a specified amount C. Let c_h denote the cost of interviewing one element in stratum h. Then the allocation must be such that

$$C = \sum_{h=1}^{L} c_h n_h. \tag{5.21}$$

This condition replaces the condition that the total sample size must be equal to n. Note that the condition $n_1 + n_2 + \cdots + n_L = n$ is obtained as a special case of condition (5.21) if the costs of interviewing are the same in all strata.

It can be shown that under condition (5.21), the most precise estimator is obtained if the sample size in stratum h is equal to

$$n_h = K \frac{N_h S_h}{\sqrt{c_h}}, \tag{5.22}$$

STRATIFIED SAMPLING

where the constant K is equal to

$$K = \frac{C}{\sum_{h=1}^{L} N_h S_h \sqrt{c_h}}. \tag{5.23}$$

The obvious conclusion can be drawn from this expression that fewer elements are selected in more expensive strata.

5.1.6 An Example of Allocation

The effect of allocation on the precision of the estimator is illustrated using an example. The target population is the working population of Samplonia. The objective is to estimate the mean income. A stratified sample of size 20 is selected. There are two strata: the provinces of Agria and Induston. Table 5.2 summarizes the data.

Note that both stratum variances are smaller than the population variance. Apparently, the strata are more homogeneous than the population as a whole. Therefore, it is worthwhile to consider a stratified sample. Size and variance in the stratum of Agria are small. This will result in only a small sample in this stratum. This can be observed in both the optimal (only two sample elements) and the proportional allocation (seven sample elements). The result for the optimal allocation is different from that of the proportional allocation. This is caused by the fact that the stratum variances are not the same.

Table 5.3 contains the variance of the estimator of the mean income of the working population of Samplonia for each possible allocation of a sample of size 20.

The smallest variance is obtained if 2 persons are selected in Agria and 18 in Induston. This results in a variance equal to 18,595. This is the optimal allocation. The variance is somewhat larger for proportional allocation (7 from Agria and 13 from Induston). Note that a simple random sample of size 20 would result in a variance of 43,757. Many allocations in the table produce a smaller variance. So, even a nonoptimal allocation may help improving precision. However, one should always be careful, as a bad allocation can lead to much less precise estimators.

Table 5.2 Stratification by Province in Samplonia

Stratum	Size	Variance in Incomes	Allocation	
			Optimal	Proportional
Agria	121	47,110	2	7
Induston	220	738,676	18	13
Samplonia	341	929,676	20	20

Table 5.3 The Variance of the Estimator for Each Possible Allocation

| Allocation | | | Allocation | | |
Agria	Induston	Variance	Agria	Induston	Variance
1	19	20,662	11	9	33,244
2	18	18,595	12	8	37,468
3	17	18,611	13	7	42,918
4	16	19,247	14	6	50,204
5	15	20,231	15	5	60,421
6	14	21,497	16	4	75,764
7	13	23,044	17	3	101,356
8	12	24,909	18	2	152,563
9	11	27,155	19	1	306,224
10	10	29,883			

5.2 CLUSTER SAMPLING

The sampling designs discussed until now always assumed a sampling frame for the whole population to be available. This is, unfortunately, not always the case. A way out could be to construct a sampling frame specifically for the survey, but this is very costly and time-consuming. A typical example of such a situation is a survey of individuals where there is no sampling frame containing all individuals in the population.

Sometimes, a sampling frame is available at an aggregate level. The population elements can be grouped into clusters and there is a sampling frame containing all clusters. The idea behind *cluster sampling* is to draw a number of clusters and to include all elements in the selected clusters in the sample.

One example of cluster sampling is to select a number of addresses from an address list and to include all people at the selected addresses (as far as they belong to the target population) in the sample.

Another reason to apply cluster sampling can be to reduce fieldwork costs. Interviewers in a face-to-face survey have to travel less if the people to be interviewed are clustered in regional areas (addresses, neighborhoods).

5.2.1 Selecting Clusters

Cluster sampling assumes that the population can be divided into a number of nonoverlapping subpopulations. These subpopulations are called *clusters*. A number of clusters are selected by using some sampling design. All elements in each selected cluster are included in the sample (Fig. 5.2).

Note that cluster sampling is not the same as stratified sampling. Drawing a stratified sample means that in *every* stratum a *sample* of elements is selected, whereas drawing a cluster sample means that in a *sample* of strata *all* elements are selected. A cluster sample can be seen as a simple random sample at an aggregate level.

CLUSTER SAMPLING

Figure 5.2 Cluster sampling. Reprinted by permission of Imre Kortbeek.

The selected elements are clusters and the values associated with these elements are the totals of the variables for all individual elements in the clusters.

It should be stressed that the choice for cluster sampling is often based on practical arguments. Estimates based on this type of sample design need not necessarily be more precise than estimates based on simple random sampling. On the contrary, including all members of a cluster in the sample may mean that a number of more or less similar elements are observed. This implies that less information is available than if elements would have been selected completely independent of each other. As a result, variances of cluster sample estimators will usually be larger. This phenomenon is called the *cluster effect*.

Clusters can be selected by using all kinds of sampling designs, including the sampling designs that have already been discussed. This chapter describes two such designs. One is to select clusters by means of a simple random sample, with equal probabilities and without replacement. The other is to select clusters with probabilities equal to their size (with replacement).

5.2.2 Selecting Clusters with Equal Probabilities

The notations used for cluster sampling are similar to those used for stratified sampling. It is assumed that the population U can be divided into M clusters

$$U_1, U_2, \ldots, U_M. \tag{5.24}$$

The clusters do not overlap and together cover the complete population U. Let N_h be the size of cluster U_h ($h = 1, 2, \ldots, M$). Consequently,

$$\sum_{h=1}^{M} N_h = N. \tag{5.25}$$

The N_h values of the target variable Y cluster h are denoted by

$$Y_1^{(h)}, Y_2^{(h)}, \ldots, Y_{N_h}^{(h)}. \tag{5.26}$$

If the mean of the target variable in cluster h is denoted by

$$\bar{Y}^{(h)} = \frac{1}{N_h} \sum_{k=1}^{N_h} Y_k^{(h)} \qquad (5.27)$$

and the total of this target variable in cluster h by

$$Y_T^{(h)} = \sum_{k=1}^{N_h} Y_k^{(h)} = N_h \bar{Y}^{(h)}, \qquad (5.28)$$

then the population mean of Y can be written as

$$\bar{Y} = \frac{1}{N} \sum_{h=1}^{M} N_h \bar{Y}^{(h)}. \qquad (5.29)$$

Another population parameter that will be used is the mean cluster total

$$\bar{Y}_T = \frac{1}{M} \sum_{h=1}^{M} Y_T^{(h)} = \frac{1}{M} \sum_{h=1}^{M} N_h \bar{Y}^{(h)}. \qquad (5.30)$$

A simple random sample of m clusters is selected without replacement and with equal probabilities from this population. All elements in the selected clusters are included in the survey. The totals of the target variables in the selected clusters are indicated by

$$y_T^{(1)}, y_T^{(2)}, \ldots, y_T^{(m)}. \qquad (5.31)$$

Assuming the clusters are the elements to be surveyed and the cluster totals are the values of these elements, the Horvitz–Thompson can be used at this aggregated level. By applying the theory provided in Section 4.1, the sample mean

$$\bar{y}_T = \frac{1}{m} \sum_{i=1}^{m} y_T^{(i)} \qquad (5.32)$$

of the cluster totals is an unbiased estimator of the mean cluster total (5.30). Consequently,

$$\bar{y}_{CL} = \frac{M}{N} \bar{y}_T \qquad (5.33)$$

is an unbiased estimator of the population mean of the elements in the target population. The variance of this estimator is, similar to expression (4.5), equal to

$$V(\bar{y}_{CL}) = \left(\frac{M}{N}\right)^2 \frac{1-f}{m} S_C^2 \qquad (5.34)$$

CLUSTER SAMPLING

where $f = m/M$ and

$$S_C^2 = \frac{1}{M-1} \sum_{h=1}^{M} \left(Y_T^{(h)} - \bar{Y}_T \right)^2. \qquad (5.35)$$

This variance can be estimated in an unbiased manner by

$$s_C^2 = \frac{1}{m-1} \sum_{j=1}^{m} \left(y_T^{(h)} - \bar{y}_T \right)^2. \qquad (5.36)$$

Note that the variance of the values of the target variable within the clusters does not play a role in this formula. The variance is determined by the variation between clusters and not within clusters.

5.2.3 Selecting Cluster Probabilities Proportional to Size

Expressions (5.34) and (5.35) show that the variance of the estimator is determined by the variation in the cluster totals of the target variable. The more they differ, the larger the variance will be. In populations where the values of the target variable show little variation, the cluster totals are largely determined by the numbers of elements in the clusters. The total of a large cluster will be large and that of a small cluster will be small. Consequently, the estimator will have a large variance in such situation.

The effect of the cluster sizes on the variance of the estimator can be reduced by drawing an unequal probability sample. Instead of drawing the clusters with equal probabilities, they can be drawn with probabilities proportional to their size. To be able to apply the theory provided in Section 4.3, the clusters have to be selected with replacement. Again, the clusters are seen as the elements to be surveyed, and the cluster totals are the values of the elements. The selection probability q_h of cluster h is taken equal to

$$q_h = \frac{N_h}{N}, \qquad (5.37)$$

for $h = 1, 2, \ldots, M$. By substituting $X_h = N_h$ in expression (4.43) the Horvitz–Thompson estimator for the mean of the cluster totals becomes

$$\frac{N}{M} \bar{\bar{y}}, \qquad (5.38)$$

where

$$\bar{\bar{y}} = \frac{1}{m} \sum_{j=1}^{m} \bar{y}^{(j)} \qquad (5.39)$$

is the mean of the observed cluster means. Hence,

$$\bar{y}_{CL} = \frac{M}{N} \frac{N}{M} \bar{\bar{y}} = \bar{\bar{y}} \qquad (5.40)$$

is an unbiased estimator for the population mean of the target variable Y. By applying formula (4.45), the variance of estimator (5.40) can be written as

$$V(\bar{y}_{\text{CL}}) = \frac{1}{Nm} \sum_{h=1}^{M} N_h \left(\bar{Y}^{(h)} - \bar{Y} \right)^2. \qquad (5.41)$$

This variance can be estimated in an unbiased manner by

$$v(\bar{y}_{\text{CL}}) = \frac{1}{m(m-1)} \sum_{j=1}^{m} \left(\bar{y}^{(j)} - \bar{\bar{y}} \right)^2. \qquad (5.42)$$

Expression (5.41) shows that the variance of the estimator is determined by the variation in the cluster means of the target variable and not by the cluster totals. The variance will be small if there is little variation in the cluster means. The more the cluster means differ, the larger the variance will be.

5.2.4 An Example

The use of cluster sampling is shown using data from Samplonia. A sample of persons is to be selected. The seven districts are used as clusters. A sample is drawn by selecting two districts and by including all their inhabitants in the sample. The relevant data are listed in Table 5.4. The sample size depends on the clusters selected. For example, if the two smallest clusters (Newbay and Oakdale) are selected, the sample size is equal to 49, whereas the sample size would be 145 if the two largest clusters (Smokeley and Mudwater) were selected.

If two clusters are selected with equal probabilities and without replacement, the standard error of the estimator for the mean income is equal to 565. If two clusters are selected with probabilities equal to their size (and with replacement), the standard error of the estimator turns out to be equal to 602. So, this variance is even larger. Apparently, the cluster size is not (approximately) proportional to the cluster total.

The standard error of the equal probability cluster sample is so large because there is a substantial amount of variation in the cluster totals. This can be seen in Table 5.4.

Table 5.4 Income by District in Samplonia

District	Size	Total Income	Mean Income
Wheaton	60	21,371	356
Greenham	38	12,326	324
Newbay	23	7,910	344
Oakdale	26	91,872	3534
Smokeley	73	117,310	1607
Crowdon	49	66,425	1356
Mudwater	72	103,698	1440
Samplonia	341	420,913	1234

The standard error of the unequal probability cluster sample is so large because there is a substantial amount of variation in the cluster means. It makes a lot of difference whether two clusters are selected from the province of Agria or Induston. Incomes in Agria are on average much lower than in Induston.

Note that a standard error in the order of magnitude of 600 would also have been obtained with a simple random sample of three elements. Although many more elements are observed in the cluster sample (it varies between 49 and 145), the precision of the estimator is equal to only that of a simple random sample of size 3. This is a disappointing result. Apparently, it does not matter so much whether two elements are observed or two clusters of elements are observed. This is caused by the phenomenon that elements within clusters are very similar. Observing more elements in a cluster does not provide much more information. This phenomenon is called the *cluster effect*.

The effectiveness of a sampling design is indicated sometimes also by means of the *effective sample size*. This is the sample size of a simple random sample of elements that would produce an estimator with the same precision. For both cluster sampling designs, the effective sample size is 3.

5.3 TWO-STAGE SAMPLING

It has been shown in the previous section that the cluster samples may not perform very well from the point of view of precision. Due to the cluster effect, the variances of estimators may be much larger than those based on simple random samples of the same size. One could say that more elements are observed than are really necessary for such a precision. The performance can be improved by not including all elements in the selected clusters in the sample, but just a sample of elements. This is the principle underlying the *two-stage sample design*.

5.3.1 Selection in Stages

To select a two-stage sample, first a sample of clusters is drawn. Next, a sample of elements is drawn from each selected cluster. The clusters are called *primary units* in the terminology of two-stage sampling and the elements in the clusters are called *secondary units*.

Sampling need not be restricted to two stages. It is very well possible to draw a three-stage sample. For example, the first stage may consist of drawing municipalities (the primary units), followed in the second stage by drawing addresses (the secondary units) in the selected municipalities, and finally persons (the tertiary units) are drawn at the selected addresses. Two-stage samples occur much more than three-stage samples and samples with even more stages. Only two-stage samples are described here (Fig. 5.3).

A number of choices have to be made to define a two-stage sampling design. First, a sampling design must be chosen for selecting primary units. Second, a sampling design must be defined to draw secondary units from the selected primary units.

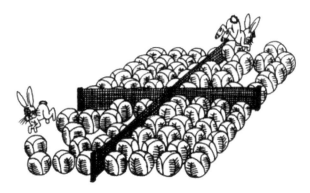

Figure 5.3 Two-stage sampling. Reprinted by permission of Imre Kortbeek.

Finally, a sample allocation decision must be made. On the one hand, small samples of elements could be drawn from a large sample of clusters, and on the other hand, large samples of elements could be drawn from a small sample of clusters. Generally, the former will increase the precision of estimators, but it will also increase the cost of the fieldwork.

5.3.2 Notations

Notations for two-stage sampling are similar to those of cluster sampling. It is assumed the population U can be divided into M clusters (primary units)

$$U_1, U_2, \ldots, U_M. \tag{5.43}$$

The clusters do not overlap and together cover the complete population U. Let N_h be the size of cluster U_h ($h = 1, 2, \ldots, M$). Consequently,

$$\sum_{h=1}^{M} N_h = N. \tag{5.44}$$

The N_h values of the target variable Y in cluster h are denoted by

$$Y_1^{(h)}, Y_2^{(h)}, \ldots, Y_{N_h}^{(h)}. \tag{5.45}$$

If the mean of the target variable in cluster h is denoted by

$$\bar{Y}^{(h)} = \frac{1}{N_h} \sum_{k=1}^{N_h} Y_k^{(h)} \tag{5.46}$$

and the total of this target variable in cluster h by

$$Y_T^{(h)} = \sum_{k=1}^{N_h} Y_k^{(h)} = N_h \bar{Y}^{(h)}, \tag{5.47}$$

TWO-STAGE SAMPLING

then the population mean of Y can be written as

$$\bar{Y} = \frac{1}{N} \sum_{h=1}^{M} N_h \bar{Y}^{(h)}. \tag{5.48}$$

Another population parameter that will be used is the mean cluster total

$$\bar{Y}_T = \frac{1}{M} \sum_{h=1}^{M} Y_T^{(h)} = \frac{1}{M} \sum_{h=1}^{M} N_h \bar{Y}^{(h)}. \tag{5.49}$$

A sampling design and a sample size have been determined in advance for each possible primary unit, in case it is selected. The sample sizes in the M primary units are denoted by

$$n_1, n_2, \ldots, n_M. \tag{5.50}$$

The values of the n_h sampled elements in primary unit h are denoted by

$$y_1^{(h)}, y_2^{(h)}, \ldots, y_{n_h}^{(h)}. \tag{5.51}$$

Not only the sampling design and the sample size are determined in advance for each primary unit but also the estimator to be used for estimating the total of the target variable Y. These estimators are denoted by

$$y_T^{(1)}, y_T^{(2)}, \ldots, y_T^{(M)}. \tag{5.52}$$

A sample of m primary units is selected from this population. The sample is denoted by the vector

$$b = (b_1, b_2, \ldots, b_M) \tag{5.53}$$

of indicators. If this sample is drawn without replacement, the indicators can only assume the value 1 (selected) or 0 (not selected). If the sample is selected with replacement, the value of the indicator b_h is equal to the frequency of element h in the sample (for $h = 1, 2, \ldots, M$).

5.3.3 Selection of Primary Sampling Units Without Replacement

First, the case is described in which primary units are selected without replacement. The first-order inclusion probability of primary unit h is denoted by τ_h, for $h = 1, 2, \ldots, M$. The second-order inclusion probability of primary units g and h is denoted by τ_{gh}, for $g, h = 1, 2, \ldots, M$.

The estimator defined by

$$\bar{y}_{TS} = \frac{1}{N} \sum_{h=1}^{M} b_h \frac{y_T^{(h)}}{\tau_h} \tag{5.54}$$

is an unbiased estimator for the population mean of the target variable. The subscript TS denotes two-stage sampling. The variance of this estimator is equal to

$$V(\bar{y}_{TS}) = \frac{1}{2N^2} \sum_{g=1}^{M} \sum_{h=1}^{M} (\tau_g \tau_h - \tau_{gh}) \left(\frac{Y_T^{(g)}}{\tau_g} - \frac{Y_T^{(h)}}{\tau_h} \right)^2 + \frac{1}{N^2} \sum_{h=1}^{M} \frac{V\left(y_T^{(h)}\right)}{\tau_h}. \quad (5.55)$$

The variance consists of two components. The first component covers the variation in clusters totals. It measures the variation between clusters. The second component covers the variation of the values of the elements in the clusters. So, it measures variation within clusters.

The cluster effect occurs if elements within a cluster are similar to one another. In that case, the variance (5.55) reduces to that of the estimator for the cluster sample. Apparently, it does not matter very much whether all elements in a cluster are observed or just a sample of elements is observed. The precision of the estimator is mainly determined by the number of the primary units in the sample.

The simplest two-stage sampling design is that of simple random sampling in both stages: primary and secondary units are selected with equal probabilities and without replacements. If n_h secondary units are drawn from primary unit h ($h = 1, 2, \ldots, M$), the first-order inclusion probability of all secondary units is equal to n_h/N_h. Therefore, the Horvitz–Thompson estimator for the total of primary unit h is equal to

$$y_T^{(h)} = N_h \bar{y}^{(h)} = \frac{N_h}{n_h} \sum_{i=1}^{n_h} y_i^{(h)}. \quad (5.56)$$

This is the sample mean multiplied by the size of the primary unit.

If m primary units are selected with equal probabilities and without replacement, the first-order inclusion probability of primary unit h is equal to $\tau_h = m/M$. Substitution in expression (5.54) results in the unbiased estimator for the population mean

$$\bar{y}_{TS} = \frac{M}{N} \frac{1}{m} \sum_{h=1}^{M} b_h N_h \bar{y}^{(h)}. \quad (5.57)$$

So, the estimator is equal to the mean of the estimators for the population totals in the selected primary units.

Substitution of the first- and second-order inclusion probabilities in expression (5.55) produces the variance of estimator (5.57). This variance is equal to

$$V(\bar{y}_{TS}) = \left(\frac{M}{N} \right)^2 \left(1 - \frac{m}{M} \right) \frac{S_1^2}{m} + \frac{M}{mN^2} \sum_{h=1}^{M} N_h^2 \left(1 - \frac{n_h}{N_h} \right) \frac{S_{2,h}^2}{n_h}, \quad (5.58)$$

where

$$S_1^2 = \frac{1}{M-1} \sum_{h=1}^{M} \left(Y_T^{(h)} - \bar{Y}_T \right)^2 \quad (5.59)$$

TWO-STAGE SAMPLING

is the variance of the totals of the primary units and

$$S_{2,h}^2 = \frac{1}{N_h-1}\sum_{k=1}^{N_h}\left(Y_k^{(h)} - \bar{Y}^{(h)}\right)^2 \qquad (5.60)$$

is the variance within primary unit h, for $h = 1, 2, \ldots, M$. The variance (5.58) can be estimated in an unbiased manner by

$$v(\bar{y}_{TS}) = \left(\frac{M}{N}\right)^2\left(1 - \frac{m}{M}\right)\frac{s_1^2}{m} + \frac{M}{mN^2}\sum_{h=1}^{M}b_h N_h^2\left(1 - \frac{n_h}{N_h}\right)\frac{s_{2,h}^2}{n_h}, \qquad (5.61)$$

where

$$s_1^2 = \frac{1}{m-1}\sum_{h=1}^{M}b_h\left(N_h\bar{y}^{(h)} - N\bar{y}_{TS}\right)^2 \qquad (5.62)$$

is the sample variance of the estimated totals of the primary units and

$$s_{2,h}^2 = \frac{1}{n_h-1}\sum_{k=1}^{N_h}\left(y_k^{(h)} - \bar{y}^{(h)}\right)^2 \qquad (5.63)$$

is the sample variance within primary unit h.

The two-stage sample design can be chosen such that the sample becomes *self-weighting*. This means that all elements in the population have the same probability of being selected in the sample. The inclusion probability of an element in a two-stage sample (with simple random sampling in both stages) is equal to the product of the inclusion probability of the cluster it is part of and the inclusion probability of the element within the cluster. So the inclusion probability of element k in cluster h is equal to

$$\frac{m}{M}\frac{n_h}{N_h}. \qquad (5.64)$$

To obtain a self-weighting sample, the same proportion of elements must be drawn in all clusters. This means that n_h/N_h must be constant over all clusters. Assuming the total sample size to be equal to n, the sample size n_h in cluster h must be equal to

$$n_h = \frac{n}{N}\frac{M}{m}N_h. \qquad (5.65)$$

Substitution of expression (5.65) in estimator (5.57) leads to

$$\bar{y}_{TS} = \frac{1}{n}\sum_{h=1}^{M}b_h\sum_{i=1}^{n_h}y_i^{(h)}. \qquad (5.66)$$

It is clear from expression (5.66) that all selected elements are assigned the same weight. The estimator is simply computed by adding all sample values and dividing this sum by the total sample size n.

5.3.4 Selection of Primary Sampling Units with Replacement

Expressions (5.59) and (5.60) show that the variance of the estimator is determined to a large extent by the differences in the totals of the target variable of the primary units. Particularly if the means of the primary units are more or less the same, differences in sizes of the primary units may lead to a large variance. This effect was already described for cluster sampling. Here also, its impact can be reduced by drawing primary units with probabilities proportional to their size. To implement this, the sample has to be selected with replacement. Let

$$q_1, q_2, \ldots, q_M \tag{5.67}$$

be the selection probabilities of the primary units. If a sample of m primary units is drawn, the inclusion expectations are $\tau_h = E(b_h) = mq_h$. This notation is similar to that for selecting primary units without replacement.

In a two-stage sample, where primary units are drawn with replacement, the estimator defined by

$$\bar{y}_{TS} = \frac{1}{mN} \sum_{h=1}^{M} b_h \frac{y_T^{(h)}}{q_h} \tag{5.68}$$

is an unbiased estimator for the population mean of the target variable. According to a theorem by Raj (1968), the variance of this estimator is equal to

$$V(\bar{y}_{TS}) = \frac{1}{mN^2} \sum_{h=1}^{M} q_h \left(\frac{Y_T^{(h)}}{q_h} - Y_T \right)^2 + \frac{1}{mN^2} \sum_{h=1}^{M} \frac{V\left(y_T^{(h)}\right)}{q_h}. \tag{5.69}$$

This variance can be estimated in an unbiased manner by

$$v(\bar{y}_{TS}) = \frac{1}{m(m-1)N^2} \sum_{h=1}^{M} b_h (z_h - \bar{z})^2, \tag{5.70}$$

where

$$z_h = \frac{y_T^{(h)}}{q_h} \tag{5.71}$$

and

$$\bar{z} = \frac{1}{m} \sum_{h=1}^{M} b_h z_h. \tag{5.72}$$

TWO-STAGE SAMPLING

This theorem is applied to the situation in which primary units are drawn with probabilities proportional to their size. Furthermore, it is assumed that secondary units are drawn with equal probabilities and without replacement. Consequently,

$$q_h = \frac{N_h}{N}, \qquad (5.73)$$

for $h = 1, 2, \ldots, M$. The Horvitz–Thompson estimator for the populations is now equal to

$$\bar{y}_{TS} = \frac{1}{m} \sum_{h=1}^{M} b_h \bar{y}^{(h)}. \qquad (5.74)$$

So, the estimator is simply equal to the mean of the sample means in the clusters. The variance of this estimator is equal to

$$V(\bar{y}_{TS}) = \frac{1}{mN} \sum_{h=1}^{M} N_h \left(\bar{Y}^{(h)} - \bar{Y} \right)^2 + \frac{1}{mN} \sum_{h=1}^{M} N_h \left(1 - \frac{n_h}{N_h} \right) S_{2,h}^2. \qquad (5.75)$$

The first component in this variance is the variance of the Horvitz–Thompson for a cluster sample where clusters are drawn proportional to their size (see expression (5.41)). The second component is contributed by the within-cluster sampling variation. Variance (5.75) can be estimated in an unbiased manner by

$$v(\bar{y}_{TS}) = \frac{1}{m(m-1)} \sum_{h=1}^{M} b_h \left(\bar{y}^{(h)} - \bar{\bar{y}} \right)^2, \qquad (5.76)$$

where

$$\bar{\bar{y}} = \frac{1}{m} \sum_{h=1}^{M} b_h \bar{y}^{(h)} \qquad (5.77)$$

is the mean of the sample means of the observed elements in the clusters.

The two-stage sampling design can be tuned such that samples are *self-weighting*. The inclusion probabilities of all elements can be made equal if the inclusion probability of secondary element k in primary unit h is taken equal to

$$\pi_k = m \frac{N_h}{N} \frac{n_h}{N_h} = m \frac{n_h}{N}. \qquad (5.78)$$

To obtain a self-weighting sample, the same number of elements must be drawn in each selected cluster. If n_0 is the sample size in each cluster, the estimator (5.72) turns into

$$\bar{y}_{TT} = \frac{1}{mn_0} \sum_{h=1}^{M} b_h \sum_{i=1}^{n_0} y_i^{(h)}. \qquad (5.79)$$

Expressions for the variance and for the estimator of the variance can be obtained by substituting $n_h = n_0$ in expressions (5.75) and (5.74).

Selecting the primary units with probabilities proportional to their size has some advantages over selecting primary units with equal probabilities. It has already been shown that this type of sampling may lead to estimators with a higher precision. This is the case when there is not much variation in the means of the primary units. An additional advantage is that the workload is more evenly spread, as the same number of elements are observed in each primary unit. The last advantage is that the sample size in each cluster is the same. Sample sizes do not depend anymore on the primary units selected. So, the total sample size in known in advance.

5.3.5 An Example

The properties of a two-stage sample are illustrated in an example where the mean income of the working population of Samplonia is estimated. The seven districts are primary units ($M = 7$). A sample of districts is drawn with probabilities proportional to size. The individual persons are the secondary units. Samples of persons are selected with equal probabilities and without replacement. The same number of persons is drawn in each selected sample, thereby making the sample self-weighting. The variance of the estimator is equal to expression (5.75).

Table 5.5 contains the variance of the estimator for all kinds of combinations of sample sizes for primary and secondary units, all resulting in a total sample size of approximately 20. Note that districts have been selected with replacement. Therefore, it may happen that a district is selected more than once in the sample. In this case, a new sample of secondary units is drawn for each occurrence in the sample.

It should be kept in mind that the variance of the estimator in a simple random sample of size 20 is equal to 43,757. All two-stage samples provided in Table 5.5 have a larger variance. Only if 20 districts are selected, and one person per district, the variance is of the same order of magnitude. This will not come as a surprise as the two-stage sample resembles the simple random sample very much in this situation.

Table 5.5 The Variance of the Estimator of the Mean Income in a Self-Weighting Two-Stage Sample

Number of Selected District	Number of Selected Persons per District	Sample Size	Variance in the Estimator
1	20	20	731,697
2	10	20	370,987
3	7	21	250,261
4	5	20	190,632
5	4	20	154,561
7	3	21	112,847
10	2	20	82,418
20	1	20	46,347

5.3.6 Systematic Selection of Primary Units

There is one type of two-stage sampling that is sometimes used by national statistical institutes and that is a design in which both primary and secondary units are selected by means of systematic sampling. It is described how this was done at Statistics Netherlands. This sampling design was developed at a time when there was no sampling frame available for the total population of The Netherlands. However, The Netherlands was divided into municipalities and each municipality separately had its own population register. Therefore, it was decided to select a two-stage sample where the municipalities were the primary units. The first stage of the sample selection process consisted of selecting a systematic sample of municipalities with probabilities proportional to the population size of the municipalities. The second stage of the process consisted of selecting a systematic sample of persons (with equal probabilities) from each selected municipality.

To reduce travel costs, there was the additional condition that in each selected municipality a minimum number of persons must be selected. This minimum number is indicated by n_0. If n is the total sample size, the number of municipalities to be selected must be equal to $m = n/n_0$. Recipe 4.5 can be used to draw a systematic sample of size m with unequal probabilities. The step length is equal to $F = N/m$, where N is the total population of The Netherlands.

Municipalities h for which $N_h \geq F$ are "big" elements. They are always selected in the sample. Not n_0 persons but $n_h = nN_h/N$ persons are to be selected from "big" elements. This change is required to keep inclusion probabilities the same for all persons in the population. The inclusion probability π_k of a person k in a "big" municipality h is now equal to

$$\pi_k = 1 \times \frac{n_h}{N_h} = \frac{nN_h/N}{N_h} = \frac{n}{N}. \tag{5.80}$$

The total size of all "big" municipalities together is denoted by N_B. If the number of persons to be selected from these municipalities is denoted by n_B, then

$$n_B = n\frac{N_B}{N}. \tag{5.81}$$

The remaining municipalities together have a total size of $N - N_B$. A sample of size $n - n_B$ persons must be selected from these municipalities. Since n_0 persons have to be observed in each selected municipality, the number of still-to-be-selected municipalities must be equal to $(n - n_B)/n_0$. The inclusion probability of such a municipality h is equal to

$$\frac{n - n_B}{n_0} \frac{N_h}{N - N_B} \tag{5.82}$$

and the inclusion probability π_k of a person in such a municipality is equal to

$$\pi_k = \frac{n - n_B}{n_0} \frac{N_h}{N - N_B} \frac{n_0}{N_h} = \frac{n}{N}. \tag{5.83}$$

It turns out that all persons, in whatever municipality they live, have the same inclusion probability. So, this produces a self-weighting sample.

Note that it is assumed that several computations above produce integer results. This does not always have to be the case in practical situations. Some rounding will be required, and this may result in slight deviations.

Since this sampling design produces self-weighting samples, the estimator for the population mean of the target variable is simply obtained by computing the mean of all observed elements in the sample.

Sample selection is systematic in both stages. This makes it difficult to properly estimate the variance of the estimator. At least, two strata have to be distinguished, a stratum of "big" municipalities that are selected with certainty and a stratum containing all other municipalities. Assuming the order of elements in the sampling frames of the municipalities is unrelated to the target variables of the survey, sampling within municipalities can be assumed to be simple random. If not too many municipalities are selected, and the order of the municipalities in the sampling frame is arbitrary, sampling of municipalities can be seen as with unequal probabilities and with replacement. Expressions (5.75) and (5.76) can then be used for approximate variance computations.

5.4 TWO-DIMENSIONAL SAMPLING

Thus far, the elements in the population can be identified by a single unique sequence number. Consequently, elements could be represented as points on a straight line. Such a population is called one-dimensional population. There are, however, situations in which it is meaningful to see a target population as two-dimensional. Suppose, the objective of a research project is to investigate how many plants of a specific type grow in the field. Counting all plants in the field is a time-consuming job. Therefore, the map of the area is divided into squares of 1 m × 1 m by means of a rectangular grid. A sample of squares is selected and the plants are counted in the selected grid. The squares are the units of measurement. Each square is now identified by two sequence numbers: a row number and a column number. Therefore, the population of squares could be called a two-dimensional population.

5.4.1 Two-Dimensional Populations

An element of a two-dimensional population is identified by two sequence numbers, one for each dimension. The sequence numbers in the first dimension run from 1 to N and the sequence numbers in the second dimension run from 1 to M. A two-dimensional population can be represented by Table 5.6.

Element (k, h) denotes the element with row number k and column number h in the table. The value of the target variable for this element is denoted by Y_{kh} ($k = 1, 2, \ldots, N$ and $h = 1, 2, \ldots, M$). Furthermore, the notation

$$Y_{k+} = \sum_{h=1}^{M} Y_{kh} \qquad (5.84)$$

Table 5.6 A Two-Dimensional Population

	Dimension 2			
Dimension 1	1	2	...	M
1	(1, 1)	(1, 2)	...	(1, M)
2	(2, 1)	(2, 2)	...	(2, M)
⋮	⋮	⋮		⋮
N	(N, 1)	(N, 2)	...	(N, M)

is introduced for the kth row total and

$$Y_{+h} = \sum_{k=1}^{N} Y_{kh} \tag{5.85}$$

for the hth column total. Therefore, the population total Y_T is equal to

$$Y_T = \sum_{k=1}^{N} Y_{k+} = \sum_{h=1}^{M} Y_{+h}. \tag{5.86}$$

All information with respect to the target variable is summarized in Table 5.7.

5.4.2 Sampling in Space and Time

Two-dimensional sampling may be an option if a phenomenon to be investigated has a geographical spread, such as the occurrence of plants or animals in an area. Maybe more often, two-dimensional sampling is used in survey where time is one of the dimensions. An example is a budget survey. The objective of such a survey is to estimate yearly expenditures of households. At first sight, the way to do this may be selecting a sample of households and asking them to keep track of all their expenditures for a year. This requires a very large (if not impossible) effort for households. Therefore, a two-dimensional population is constructed where the first dimension

Table 5.7 The Target Variable in a Two-Dimensional Population

	Dimension 2				Total
Dimension 1	1	2	...	M	
1	Y_{11}	Y_{12}	...	Y_{1M}	Y_{1+}
2	Y_{21}	Y_{22}	...	Y_{2M}	Y_{2+}
⋮	⋮	⋮		⋮	
N	Y_{N1}	Y_{N2}	...	Y_{NM}	Y_{N+}
Total	Y_{+1}	Y_{+2}	...	Y_{+M}	Y_T

Table 5.8 Two-Dimensional Sampling for a Budget Survey

Household	Month			
	1	2	...	12
1	(1, 1)	(1, 2)	...	(1, 12)
2	(2, 1)	(2, 2)	...	(2, 12)
⋮	⋮	⋮		⋮
N	(N, 1)	(N, 2)	...	(N, 12)

consists of households and the second dimension divides the year into a number of time periods, for example, months. The elements to be investigated are households in specific months (see Table 5.8). Consequently, a selected household needs to keep track of its expenditures in 1 month only.

Sampling from two-dimensional populations where time is one dimension is usually called *sampling in space and time* (Fig. 5.4).

There are many ways to select a sample from a two-dimensional population. Assuming the sample size to be equal to r, here is a list of some possible sampling designs.

- *Method A*. Transform the two-dimensional population into a one-dimensional population by ordering all elements in some way. This could, for example, be done row-wise or column-wise. The two-dimensional nature of the population is ignored, and one of the sampling designs of Chapter 4 can be applied.
- *Method B1*. First, draw a simple random sample of n rows from the N rows. Next, draw a simple random sample of r/n elements (assuming r is divisible by n) from each selected row. This comes down to selecting a two-stage sample where the rows are the primary units and the elements in the rows are the secondary elements.
- *Method B2*. First, draw a simple random sample of m columns from the M columns. Next, draw a simple random sample of r/m elements (assuming r is divisible by m) from each selected column. This comes down to selecting a two-stage sample where the columns are the primary units and the elements in the columns are the secondary elements.
- *Method C*. First, draw a simple random sample of n rows from the N rows. Next, draw a simple random sample of m columns from the M columns. Finally, draw a simple random sample (ignoring the two-dimensional character) of size r from the resulting $n \times m$ elements. This guarantees that not too many elements are selected from one row or one column.
- *Method D*. First, draw a simple random sample of n rows from the N rows. Next, draw a simple random sample of m columns from the M columns. Finally, apply a fixed filter that selects r elements from a matrix of $n \times m$ elements. This filter can be designed such that the sample is spread over a fixed number of rows and columns.

Figure 5.4 Sampling in space and time. Reprinted by permission of Imre Kortbeek.

Method A offers no guarantees for a balanced spread of the sample observations over rows and columns. It may very well happen that one element has to report about more periods than another element. Also, the amount of fieldwork in one period may be much more than in another period.

Methods B1 and B2 are both two-stage samples. This makes it possible to control the distribution of the sample in one dimension. Method B1 allows controlling the number of elements in the sample. Since periods are selected at random for elements, it may happen that in certain periods of the year much more data are collected than in another period of the year. Method B2 allows controlling the number of periods in which data collection takes place, but it may happen that one element has to report on more periods than another element.

Methods C and D give more control over the distribution of the sample over both dimensions. The first step is to select n rows and m columns. This results in a subtable consisting of n rows and m columns. The elements in this subtable together form the *donor table*.

The next step for method C is to select a simple random sample of size r from the donor table. The next step for method D is to apply the so-called *filter table*. See Table 5.9 for an example. A filter table consists, like the donor table, of n rows and m columns. The value r_{ij} in cell (i, j) can either by 0 or 1. The corresponding element in the donor table is selected in the sample if $r_{ij} = 1$ and it is not selected if $r_{ij} = 0$.

The filter table must be composed such that its total is equal to r. Furthermore, each row total must be equal to s and each column total must be equal to t.

Table 5.9 A Filter Table

Dimension 1	Dimension 2				Total
	1	2	...	m	
1	r_{11}	r_{12}	...	r_{1m}	r_{1+}
2	r_{21}	r_{22}	...	r_{2m}	r_{2+}
⋮	⋮	⋮		⋮	⋮
n	r_{n1}	r_{n2}	...	r_{nm}	r_{n+}
Total	r_{+1}	r_{+2}	...	r_{+m}	r

Table 5.10 Transported Passengers on Bus Lines in Induston

Bus Line	Time Period					
	7–9	9–11	11–13	13–15	15–17	17–19
Oakdale–Smokeley	3	8	2	4	4	5
Oakdale–Crowdon	4	5	4	3	3	4
Oakdale–Mudwater	2	9	1	5	3	6
Smokeley–Crowdon	22	6	11	5	3	23
Smokeley–Mudwater	12	8	2	8	4	14
Crowdon–Mudwater	19	4	7	2	1	21

Method D is illustrated by means of an example. Suppose a survey is carried out to estimate the number of passengers on a specific day in the buses of the public transport system of the province of Induston in Samplonia. Starting point is a two-dimensional population where the six rows represent the six bus lines and the six columns represent 2-h time periods. The population data can be found in Table 5.10.

A sample of size $r = 5$ must be selected from this population. First, a donor table consisting of three rows and three columns is selected. So, the donor table contains nine elements. To select a sample of five elements from this donor table, the following filter table could be applied:

1	1	0
0	1	1
0	0	1

Suppose, rows 4, 1, and 2 and columns 1, 5, and 3 are selected. Then, the donor table will contain the following elements:

(4,1)	(4,5)	(4,3)
(1,1)	(1,5)	(1,3)
(2,1)	(2,5)	(2,3)

Application of the filter table produces a sample consisting of the elements (4, 1), (4, 5), (1, 5), (1, 3), and (2, 3).

5.4.3 Estimation of the Population Mean

The first-order inclusion probability of an element does not depend on the composition of the filter table. It can be shown that the first-order inclusion probability of every element (k, h) is equal to

$$\pi_0 = \pi_{(k,h)} = \frac{r}{NM}. \qquad (5.87)$$

The second-order inclusion probabilities do depend on the composition of the filter table. This will not come as a surprise as a filter may exclude specific combinations of elements. For example, if the filter table only has 1s at the diagonal, elements in the same row or in the same column can never be selected together.

The second-order inclusion probability of two elements (k, h) and (k', h') with $k \neq k'$ and $h = h'$ is denoted by π_1. This inclusion probability is equal to

$$\pi_1 = \pi_{(k,h)(k',h)} = \frac{1}{N(N-1)M} \left(\sum_{j=1}^{m} r_{+j}^2 - r \right). \qquad (5.88)$$

The second-order inclusion probability of two elements (k, h) and (k', h') with $k = k'$ and $h \neq h'$ is denoted by π_2. This inclusion probability is equal to

$$\pi_2 = \pi_{(k,h)(k,h')} = \frac{1}{M(M-1)N} \left(\sum_{i=1}^{n} r_{i+}^2 - r \right). \qquad (5.89)$$

Finally, the second-order inclusion probability of two elements (k, h) and (k', h') with $k \neq k'$ and $h \neq h'$ is denoted by π_3. This inclusion probability is equal to

$$\pi_3 = \pi_{(k,h)(k',h')} = \frac{1}{M(M-1)N(N-1)} \left(r^2 + r - \sum_{i=1}^{n} r_{i+}^2 - \sum_{j=1}^{m} r_{+j}^2 \right). \qquad (5.90)$$

The values of the target variable Y in the donor table are denoted by z_{ij} (for $i = 1, 2, \ldots, n$ and $j = 1, 2, \ldots, m$). Furthermore, the notation

$$y_{ij} = r_{ij} z_{ij} \qquad (5.91)$$

is introduced. So, the value y_{ij} is equal to the value of the target variable if the corresponding element in the donor table is selected in the sample. And $y_{ij} = 0$, if the element is not selected in the sample. The Horvitz–Thompson estimator for the population mean of Y can now be written as

$$\bar{y}_{\mathrm{TD}} = \frac{1}{\pi_0 MN} \sum_{i=1}^{n} \sum_{j=1}^{m} y_{ij}. \qquad (5.92)$$

This is an unbiased estimator. To be able to write down the variance of this estimator, three quantities D_1, D_2, and D_3 are introduced:

$$D_1 = N \sum_{k=1}^{N} \sum_{h=1}^{M} Y_{kh}^2 - \sum_{h=1}^{M} Y_{+h}^2, \qquad (5.93)$$

$$D_2 = M \sum_{k=1}^{N}\sum_{h=1}^{M} Y_{kh}^2 - \sum_{k=1}^{N} Y_{k+}^2, \qquad (5.94)$$

and

$$D_3 = NM \sum_{k=1}^{N}\sum_{h=1}^{M} Y_{kh}^2 - Y^2 - D_1 - D_2. \qquad (5.95)$$

The variance of estimator (5.92) is now equal to

$$V(\bar{y}_{\text{TD}}) = \frac{\pi_0^2 - \pi_1}{\pi_0^2} D_1 + \frac{\pi_0^2 - \pi_2}{\pi_0^2} D_2 + \frac{\pi_0^2 - \pi_3}{\pi_0^2} D_3. \qquad (5.96)$$

To be able to write down the variance of this estimator by using the sample data, the sample analogues of D_1, D_2, and D_3 are introduced:

$$d_1 = \sum_{j=1}^{m} r_{+j} \sum_{i=1}^{n} y_{ij}^2 - \sum_{j=1}^{m} y_{+j}^2, \qquad (5.97)$$

$$d_2 = \sum_{i=1}^{n} r_{i+} \sum_{j=1}^{m} y_{ij}^2 - \sum_{i=1}^{n} y_{i+}^2, \qquad (5.98)$$

and

$$d_3 = r \sum_{i=1}^{n}\sum_{j=1}^{m} y_{ij}^2 - y_{++}^2 - d_1 - d_2, \qquad (5.99)$$

where

$$y_{i+} = \sum_{j=1}^{m} y_{ij}, \qquad (5.100)$$

$$y_{+j} = \sum_{i=1}^{n} y_{ij}, \qquad (5.101)$$

and

$$y_{++} = \sum_{i=1}^{n}\sum_{j=1}^{m} y_{ij}. \qquad (5.102)$$

TWO-DIMENSIONAL SAMPLING 129

Table 5.11   Possible Samples of Size 3 from the Two-Dimensional Population of Bus Lines in Induston

Filter Table	Variance of the Estimator
$\begin{array}{\|c\|c\|c\|}\hline 1 & 0 & 0 \\ \hline 0 & 1 & 0 \\ \hline 0 & 0 & 1 \\ \hline\end{array}$	12,153
$\begin{array}{\|c\|c\|}\hline 1 & 0 \\ \hline 1 & 0 \\ \hline 0 & 1 \\ \hline\end{array}$	14,922
$\begin{array}{\|c\|}\hline 1 \\ \hline 1 \\ \hline 1 \\ \hline\end{array}$	20,459
$\begin{array}{\|c\|c\|c\|}\hline 1 & 1 & 0 \\ \hline 0 & 0 & 1 \\ \hline\end{array}$	14,035
$\begin{array}{\|c\|c\|}\hline 1 & 1 \\ \hline 1 & 0 \\ \hline\end{array}$	16,804
$\begin{array}{\|c\|c\|c\|}\hline 1 & 1 & 1 \\ \hline\end{array}$	17,800

The estimator of variance (5.96) is now equal to

$$v(\bar{y}_{TD}) = \frac{\pi_0^2 - \pi_1}{\pi_0^2} \frac{d_1}{\pi_1} + \frac{\pi_0^2 - \pi_2}{\pi_0^2} \frac{d_2}{\pi_2} + \frac{\pi_0^2 - \pi_3}{\pi_0^2} \frac{d_3}{\pi_3}. \quad (5.103)$$

5.4.4 An Example

The effect of the composition of the filter table on the variance of the estimator is illustrated using the example of the bus lines in Induston. All population data are provided in Table 5.10. The objective is to estimate the number of passengers on a specific day in the buses of the province of Induston. Suppose, sample of size $r = 3$ must be selected. Table 5.11 contains a number of different filter tables that result in such a sample. For each filter table, the variance of the estimator has been computed.

Note that all other possible filter tables can be obtained from the filter tables in Table 5.11 by permuting either the rows or the columns. The value of the variance does not change under such permutations.

The smallest value of the variance is obtained if the sample is distributed over as many rows and columns as possible. This can be explained by the lack of a cluster effect. If several elements are selected within a row or within a column, there will be a cluster effect resulting in larger variances.

Drawing a simple random sample of size 3 from the filter table (method C) would result in an estimator with a variance equal to 14,146. Comparison with the variances in Table 5.11 leads to the conclusion that the precision can only be improved if the sample is forced over as many rows and columns as possible.

Most standard works about sampling theory do not discuss two-dimensional sample. A source of more information on this type of sampling is De Ree (1978).

EXERCISES

5.1 Anders Kiaer, the director of the Norwegian national statistical office, proposed a sampling technique in 1895 that
   a. was similar to a simple random sample;
   b. resulted in a sample that resembled the population as much as possible;
   c. was similar to a two-stage sampling design;
   d. was similar to a stratified sample with simple random sampling within strata.

5.2 If it is assumed that the costs of interviewing are the same for every person in the population, the optimal allocation in a stratified sample is determined
   a. with the cumulative-square-root-f rule;
   b. by taking the sample sizes in the strata proportional to the standard deviations of the target variable in the strata;

EXERCISES

c. by taking the sample sizes in the strata proportional to the product of the size and the standard deviation of the target variable in the strata;

d. by taking the sample sizes in the strata proportional to the product of the size and the variances of the target variable in the strata.

5.3 A stratified sample is obtained by

a. randomly drawing strata and randomly drawing elements from the selected strata;

b. randomly drawing strata and selecting all elements in the selected strata;

c. randomly drawing elements from all strata;

d. randomly selecting elements from the population and afterward establishing from which strata the selected elements came.

5.4 A sampling design must be defined for an income survey in the town of Rhinewood. The town consists of two neighborhoods Blockmore and Glenbrook. The table below contains some available information that can be used.

Town	Variance of the Variable Income	Number of Inhabitants
Blockmore	40,000	15,000
Glenbrook	640,000	10,000
Rhinewood	960,000	25,000

The variance estimates have been obtained in an earlier survey. They can be used as indicators of the actual (adjusted) population variances.

a. Suppose a simple random sample without replacement of size 400 is selected. Compute the variance and the standard error of the sample mean of the income variable. Also, compute the margin of the 95% confidence interval.

b. The researcher decides to draw a stratified sample (with simple random sampling without replacement within strata). The sample sizes are allocated by means of proportional allocation. Compute the variance and the standard error of the estimator of the mean income for this sampling design. Also compute the margin of the 95% confidence interval.

c. Since indications of stratum variances are available, it is possible to apply optimal allocation. Compute the variance and the standard error of the estimator of the mean income for this sampling design. Also, compute the margin of the 95% confidence interval.

d. Compare the results of exercises (a), (b), and (c). Explain the observed differences and/or similarities.

5.5 The town of Ballycastle has been struck by a flood. The town is divided into three neighborhoods Balinabay, Oldbridge, and Roswall with 10,000, 5000, and 20,000 houses, respectively. To establish the value of the damage, a stratified sample of 140 houses is selected. One estimates the ratios of the standard deviations of damage in the three neighborhoods as 10: 7:3.

a. Compute the proportional allocation.

b. Compute the optimal allocation assuming equal costs.

5.6 A population consisting of 30 elements can be divided into six subpopulations labeled A, B, ..., F. The table below contains the values of the target variable for each subpopulation, the sum of these values, the mean of these values, and the adjusted population variance of these values. The population mean is equal to 4.5 and the adjusted population variance is equal to 6.258621.

Subpopulation	Values	Sum	Mean	Population Variance
A	1 2 3	6	2	1.0
B	1 3 5	9	3	4.0
C	1 1 1 3 3 3 5 5 5	27	3	3.0
D	7 7 8 9 9	48	8	0.8
E	4 4 5 5 6 6	30	5	0.8
F	2 5 8	15	5	9.0

a. Compute the variance of the estimator of the population mean if a simple random sample of size 6 is selected without replacement.

b. Compute the variance of the estimator if a stratified sample is selected (with simple random sampling within strata) where one element is drawn from each subgroup.

c. Explain why it is impossible to compute an estimate for the variance of the estimator for the sampling design in exercise (b).

d. Suppose a cluster sample is selected for two subpopulations. Selection of cluster is with equal probabilities and without replacement. Compute the variance of the estimator for this sampling design.

5.7 Peaches are grown for commercial purposes in one part of Samplonia. The area consists of 10 small villages. There are a number of peach growers in each village. There are in total 60 peach growers. A sample survey is carried out to obtain insight into the yearly production of peaches. The table below contains all population data. The mean production per grower is measured in bushels (approximately 35 L).

Village	Number of Growers	Mean Production	Total Production	Variance S^2
1	3	158	474	25
2	4	149	596	17
3	5	137	685	35
4	6	130	780	24
5	10	112	1120	18
6	3	162	486	33
7	4	151	604	25

8	5	143	715	26
9	8	119	952	34
10	12	101	1208	26
Total area	60	127	7620	424

    a. Suppose a simple random sample of 12 growers is selected without replacement. Compute the variance of the population mean (mean production per grower).

    b. A cluster sample could be selected to reduce travel costs. Since the average cluster size is equal to $60/10 = 6$, a sample of 12 growers requires two clusters to be drawn with equal probabilities and without replacement. Compute the variance of the estimator for the mean production per grower for this sampling design.

    c. To reduce the effect of the cluster sizes on the variance, it is also possible to draw the two clusters with replacement and with probabilities equal to their size. Compute the variance of the estimator for the mean production per grower for this sampling design.

5.8 A two-stage sample is selected from a population. A number of m primary units are selected with equal probabilities and without replacement. A number of n_h secondary units are selected from each selected primary unit h with equal probabilities and without replacement.

    a. Suppose all values of the target variable are the same within each primary unit. Write down the variance of the estimator of the population for this sampling design in this situation.

    b. What can be learnt from this formula with respect to the sample size and the distribution of the sample size over primary and secondary units?

5.9 A two-stage sample is selected from a population. A number of m primary units are selected with replacement and with probabilities equal to their size. A number of n_h secondary units are selected from each selected primary unit h with equal probabilities and without replacement.

    a. Suppose the mean of the target variable is the same in each primary unit. Write down the variance of the estimator of the population for this sampling design in this situation.

    b. What can be learnt from this formula with respect to the sample size and the distribution of the sample size over primary and secondary units?

CHAPTER 6

Estimators

6.1 USE OF AUXILIARY INFORMATION

Some sampling designs were described in Chapters 4 and 5 that improved the precision of the Horvitz–Thompson estimator by using an auxiliary variable. For example, if a quantitative auxiliary variable has a strong correlation with the target variable, sampling with probabilities proportional to the values of the auxiliary variable will lead to a precise estimator. This will also be the case, if there is a qualitative auxiliary variable that is highly correlated with the target variable. Such an auxiliary variable can be used in a stratified sampling design.

Auxiliary information can also be used in a different way. Instead of taking advantage of auxiliary information in the sampling design, it is possible to improve the estimation procedure. This will be the topic of this chapter. To keep things simple, it is assumed that the sample is selected by means of simple random sampling (with equal probabilities and without replacement).

The theory is also restricted to estimators that incorporate information of one auxiliary variable only (Fig. 6.1). It is possible to use more auxiliary variables. An example can be found in Chapter 10, where the *generalized regression estimator* is described.

Two estimators will be discussed that use a quantitative auxiliary variable: the *ratio estimator* and the *regression estimator*. One estimator will be described that uses a qualitative auxiliary variable: the *poststratification estimator*. Similar to sampling designs, estimators will perform better as the relationship between target variable and auxiliary variable is stronger.

6.2 A DESCRIPTIVE MODEL

All estimators described in this section are special cases of a model describing the relationship between the target variable and the auxiliary variable.

Applied Survey Methods: A Statistical Perspective, Jelke Bethlehem
Copyright © 2009 John Wiley & Sons, Inc.

A DESCRIPTIVE MODEL

Figure 6.1 Estimation using auxiliary information. Reprinted by permission of Imre Kortbeek.

Definition 6.1 A *descriptive model* F assumes that the value Y_k of the target variable for element k can be written as

$$Y_k = F(X_k; \theta) + R_k, \qquad (6.1)$$

where F is a function that depends only on the value X_k of the auxiliary variable and the values of a limited number of *model parameters* denoted by the vector θ. The function F must have been chosen such that the mean of the *residuals* $R_1, R_2, \ldots, R_N$ is equal to 0:

$$\bar{R} = \frac{1}{N} \sum_{k=1}^{N} R_k = 0. \qquad (6.2)$$

The objective is to estimate the population mean of the target variable Y. By applying expressions (6.1) and (6.2) this mean can be written as

$$\bar{Y} = \frac{1}{N} \sum_{k=1}^{N} (F(X_k; \theta) + R_k) = \frac{1}{N} \sum_{k=1}^{N} F(X_k; \theta). \qquad (6.3)$$

Hence, it is possible to compute the population mean if the exact form of the function F and the values $X_1, X_2, \ldots, X_N$ of the auxiliary variable are known. The form of the function F is often known but not the values of the model parameters. An example of a descriptive model is the linear model

$$Y_k = A + BX_k + R_k, \qquad (6.4)$$

in which A and B are the model parameters, so $\theta = (A, B)$.

A perfect descriptive model would be able to predict the values of Y without error. All residuals R_k would be equal to 0. Unfortunately, such models are seldom encountered in practice. Most models are only partly able to explain the behavior of the target variable. Nevertheless, they can be useful. To measure the predictive power of a descriptive model, the *residual sum of squares* SS_R is used. This quantity is defined as

$$SS_R = \sum_{k=1}^{N} (Y_k - F(X_k; \theta))^2 = \sum_{k=1}^{N} R_k^2. \qquad (6.5)$$

The predictive power of a model is better as the residual sum of squares is smaller. The model parameters of some models are already fixed by condition (6.2). If this is not the case, the additional condition is imposed that the values of the model parameters must minimize the residual sum of squares SS_R. However, these values can only be computed if all values of X_k and Y_k in the population are known. This is not the case. The solution is to estimate the model parameters by using the sample data. First, the condition is imposed that the mean of the sample residuals must be equal to 0:

$$\bar{r} = \frac{1}{n}\sum_{i=1}^{n} r_i = \frac{1}{n}\sum_{i=1}^{n}(y_i - F(x_i; \theta)) = 0. \quad (6.6)$$

If this is not sufficient to obtain estimates of the model parameters, the residual sum of squares

$$SS_R = \sum_{i=1}^{n} r_i^2 = \sum_{i=1}^{n}(y_i - F(x_i; \theta))^2 \quad (6.7)$$

in the sample is minimized. Let t be the vector of estimators of the model parameters θ that has been obtained in this way. If the value X_k of the auxiliary variable is known for every element in the population, the population mean of Y can be estimated by substituting the function values $F(X_k; \theta)$ with the estimated function values $F(X_k; t)$ in expression (6.3). This leads to the estimator

$$\bar{y}_F = \frac{1}{N}\sum_{k=1}^{N} F(X_k; t). \quad (6.8)$$

The subscript F indicates that the estimator is based on a descriptive model with function F. It turns out that estimator (6.8) is unbiased, or approximately unbiased for large samples, for all specific models described in the next sections:

$$E(\bar{y}_F) \approx \bar{Y}. \quad (6.9)$$

The variance of the model-based estimator is equal to, or approximately equal to,

$$V(\bar{y}_F) \approx \left(\frac{1}{n} - \frac{1}{N}\right)\frac{SS_R}{N-1}. \quad (6.10)$$

This variance can be estimated (approximately) unbiased by

$$v(\bar{y}_F) = \left(\frac{1}{n} - \frac{1}{N}\right)\frac{1}{n-1}\sum_{i=1}^{n}(y_i - F(x_i; t))^2. \quad (6.11)$$

It will be clear from expression (6.10) that there is a close relationship between the variance of the estimator the predictive power of the descriptive model. A model that is able to predict the values of the target variable without much error will result in a precise estimator.

Use of a model-based estimator will only be effective if it results in a smaller variance than that of the Horvitz–Thompson estimator. Since simple random sampling

THE DIRECT ESTIMATOR

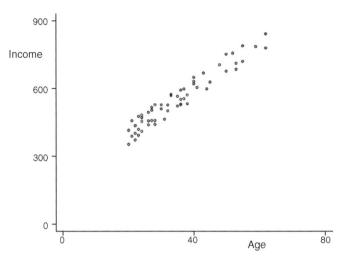

Figure 6.2 The relationship between income and age for the working males in Agria.

is assumed in this chapter, the Horvitz–Thompson estimator is equal to the sample mean. To measure of the improvement in precision of the model-based estimator, the *efficiency of the estimator*, is introduced. It is defined by

$$\text{Eff}(\bar{y}_F) = \frac{V(\bar{y})}{V(\bar{y}_F)}. \tag{6.12}$$

A value of the efficiency larger than 1 indicates that model-based estimator is better than the sample mean. A value smaller than 1 means the simple sample mean is superior to the model-based estimator.

A number of specific estimators will be described in the next sections. They are all based on some descriptive model. In fact, they are all based on the assumption of some kind of linear relationship between the target variable and the auxiliary variable.

All these estimators will be illustrated using a small example. The objective is to estimate the mean income of working males in the province of Agria. The population consists of 58 persons only. The variable age will be used as auxiliary variable. Figure 6.2 shows a scatter plot of the relation between income and age in this population. It will be clear that it is not unreasonable to assume some kind of linear relationship.

6.3 THE DIRECT ESTIMATOR

The simplest estimator that can be based on a descriptive model is the one that uses no auxiliary variable at all. This comes down to a model that always predicts the same value for the target variable. This model can be written as

$$F(X_k; A) = A, \tag{6.13}$$

in which A is the only model parameter. Imposing the condition that the average of all residuals R_k must be equal to 0 results in

$$A = \bar{Y}. \tag{6.14}$$

This is exactly the population mean to be estimated. The value of A can be estimated by the sample mean

$$a = \bar{y}. \tag{6.15}$$

The sample-based estimator of the descriptive model can now be written as

$$F(X_k; a) = \bar{y}. \tag{6.16}$$

Substitution of (6.16) in expression (6.8) results in an estimator that could be called the *direct estimator*:

$$\bar{y}_D = \bar{y}. \tag{6.17}$$

It can be concluded that application of a descriptive model without an auxiliary variable produces nothing new. It is the Horvitz–Thompson estimator for the case of simple random sampling without replacement. This is an unbiased estimator for the population mean of Y. The variance of this estimator is equal to

$$V(\bar{y}_D) = \frac{1-f}{n} S_Y^2, \tag{6.18}$$

where

$$S_Y^2 = S^2 = \frac{1}{N-1} \sum_{k=1}^{N} (Y_k - \bar{Y})^2 \tag{6.19}$$

is the (adjusted) population variance as defined in Chapter 2. Since the residual sum of squares is equal to

$$SS_R = \sum_{k=1}^{N} (Y_k - F(X_k; A))^2 = \sum_{k=1}^{N} (Y_k - \bar{Y})^2, \tag{6.20}$$

it is clear that for the direct estimator the expression

$$V(\bar{y}_D) = \frac{1-f}{n} \frac{SS_R}{N-1} \tag{6.21}$$

holds exactly. The sample-based estimator for the variance is

$$v(\bar{y}_D) = \frac{1-f}{n} s_Y^2, \tag{6.22}$$

where

$$s_Y^2 = s^2 = \frac{1}{n-1} \sum_{i=1}^{n} (y_i - \bar{y})^2 \tag{6.23}$$

is the sample variance.

THE RATIO ESTIMATOR

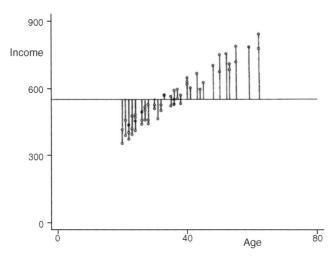

Figure 6.3 The direct estimator.

It is possible to show the variance of the direct estimator graphically. The data about the incomes of working males in Agria are used for this. The value of the population mean of the target variable Y is equal to 551. The descriptive model for the direct estimator is, in fact, the line $Y = F(X; A) = A$. In this case, it is the line $Y = F(X) = 551$. It is the horizontal line in the scatter plot in Fig. 6.3.

The residuals can be interpreted as the distances of the points (representing the values of the elements) to the line. These distances have been drawn in Fig. 6.3 as vertical line segments. The variance of the estimator can be seen (apart from a constant) as the average of the squared distances. So the estimator is more precise as the points are closer to the line. The variance of the estimator in the example of Fig. 6.3 is equal to 1178. It is clear from the plot that direct estimator is not the best estimator here. One can think of lines that are much closer to the points.

6.4 THE RATIO ESTIMATOR

Traditionally, the ratio estimator is probably the most used model-based estimator. Its popularity may be due to the simplicity of computation, while at the same time it takes advantage of auxiliary information. The ratio estimator is effective in situations where the ratios Y_k/X_k of the values of the target variable and the auxiliary variable vary less than the values Y_k of the target variable themselves. If this situation occurs, it is better to first estimate the population ratio

$$R = \frac{\bar{Y}}{\bar{X}} \qquad (6.24)$$

after which multiplication with the (known) population mean of the auxiliary variable results in an estimator for the population mean of the target variable. The assumption that the ratios Y_k/X_k show little variation can be translated into a descriptive model. This assumption implies that the points with coordinates (X_k, Y_k) lie approximately in a straight line that goes through the origin of the scatter plot. The corresponding descriptive model can be written as

$$F(X_k; B) = BX_k. \qquad (6.25)$$

This model contains only one parameter B. Therefore, the condition that the population mean of residuals must be equal to 0 fixes the value of the model parameter:

$$B = \frac{\bar{Y}}{\bar{X}}. \qquad (6.26)$$

In line with the analogy principle, an estimator b for B is obtained by replacing the population quantities by the corresponding sample quantities:

$$b = \frac{\bar{y}}{\bar{x}}. \qquad (6.27)$$

The estimator of the descriptive model becomes

$$F(X_k; b) = bX_k = \frac{\bar{y}}{\bar{x}} X_k. \qquad (6.28)$$

Substitution of (6.28) in (6.8) produces the *ratio estimator*:

$$\bar{y}_R = \bar{y} \frac{\bar{X}}{\bar{x}}. \qquad (6.29)$$

So, the ratio estimator is equal to the sample mean multiplied by a correction factor. This correction factor adjusts the estimator for a difference between the sample mean and the population mean of the auxiliary variable. For example, if the sample values of X are relatively small, the sample values of Y are probably also relatively small. The correction factor will be larger than 1. In this case, the sample mean will be corrected in the proper direction; that is, its value will be increased.

The ratio estimator is not an unbiased estimator, but it can be shown (see, for example, Cochran, 1977) that it is approximately unbiased. There is a small bias caused by the fact that the expected value of the ratio of two random variables is not equal to the ratio of the expected values of the random variables. This implies that b is not an unbiased estimator of B. However, the ratio estimator is asymptotically design unbiased (ADU), which means that for large sample sizes the bias is so small that it can be ignored.

There is no exact expression for the variance of the estimator. By using expression (6.10), an approximation is obtained that works well for large sample sizes. Working out this expression for model (6.25) leads to

$$V(\bar{y}_R) \approx \frac{1-f}{n} \frac{SS_R}{N-1} = \frac{1-f}{n} \frac{1}{N-1} \sum_{k=1}^{N} \left(Y_k - \frac{\bar{Y}}{\bar{X}} X_k \right)^2. \qquad (6.30)$$

THE RATIO ESTIMATOR

This variance expression can be rewritten as

$$V(\bar{y}_R) \approx \frac{1-f}{n}\left(S_Y^2 - 2R_{XY}S_X S_Y \frac{\bar{Y}}{\bar{X}} + S_X^2 \left(\frac{\bar{Y}}{\bar{X}}\right)^2\right), \tag{6.31}$$

where

$$S_X^2 = \frac{1}{N-1}\sum_{k=1}^{N}(X_k - \bar{X})^2 \tag{6.32}$$

is the adjusted population variance of auxiliary variable X, where

$$R_{XY} = \frac{S_{XY}}{S_X S_Y} \tag{6.33}$$

is the population correlation between the two variables X and Y, and

$$S_{XY} = \frac{1}{N-1}\sum_{k=1}^{N}(X_k - \bar{X})(Y_k - \bar{Y}) \tag{6.34}$$

is the adjusted population covariance.

The variance (6.30) is smaller if the values Y_k and X_k are better proportional. The variance can be estimated (approximately unbiased) by using the sample data with the expression

$$v(\bar{y}_R) \approx \frac{1-f}{n}\frac{1}{n-1}\sum_{i=1}^{n}\left(y_i - \frac{\bar{y}}{\bar{x}}x_i\right)^2 \tag{6.35}$$

In addition, this estimator is ADU. So, the bias vanishes for large sample sizes.

Suppose variance (6.31) is a good approximation of the true variance of the ratio estimator. By comparing expressions (6.18) and (6.31), a condition can be determined under which the ratio estimator is more precise than the direct estimator. This is the case if

$$R_{XY} > \frac{S_X/\bar{X}}{2S_Y/\bar{Y}}. \tag{6.36}$$

So, the ratio estimator has a smaller variance than the direct estimator if the correlation between the target variable Y and the auxiliary variable X is sufficiently large. The quantity

$$\frac{S_X}{\bar{X}} \tag{6.37}$$

is called the *coefficient of variation*. It is an indicator of the relative precision of the variable. Suppose the auxiliary variable is taken to be the target variable, but measured in a previous survey. Then it is not unlikely to assume the coefficients of variation of X and Y are approximately equal. Consequently, the ratio estimator is better than the direct estimator if the value of the correlation coefficient is larger than 0.5. The ratio estimator only performs worse than the direct estimator if the coefficient of variation of the auxiliary variable is at least twice as large as that of the target variable.

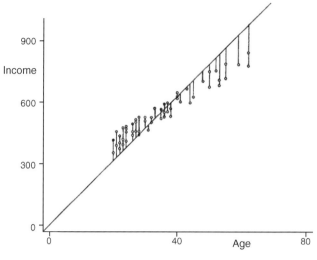

Figure 6.4 The ratio estimator.

The properties of the ratio estimator are shown graphically using the sample of working males in the Samplonian province of Agria (see Fig. 6.4). The descriptive model corresponds to a straight line in the scatter plot, where the incomes of the persons are plotted against their ages. The line goes through the origin and the center of gravity (the point with the mean of the auxiliary variable X and the mean of the target variable Y as coordinates). The slope of the line is equal to B. Again, the variance of the estimator is determined by the sum of squares of the distances of the point from the line. By comparing Fig. 6.4 with Fig. 6.3, it will become clear that the ratio estimator is a better estimator than the direct estimator. The distances to the line are much smaller. Indeed, the variance of this estimator turns out to be 482. Note that the variance of the direct estimator is equal to 1178.

The effect of using the ratio estimator is also shown in a simulation experiment. The target population consists of 200 dairy farms in the rural part of Samplonia. The objective of the survey is to estimate the average daily milk production per farm. Two estimators are compared: the direct estimator and the ratio estimator. The ratio estimator uses the number of cows per farm as the auxiliary variable. This seems not unreasonable as one may expect milk production per farm to be more or less proportional to the number of cows per farm.

Selection of a sample of size 40 and computation of the estimator has been repeated 500 times for both estimators. This gives 500 values of each estimator. The distribution of these values has been plotted in a histogram in Fig. 6.5. The histogram on the left shows the distribution of the direct estimator. The distribution of the ratio estimator is shown on the right.

The ratio estimator performs much better than the direct estimator. The distribution of its values concentrates much more around the true value. The standard error of the direct estimator here is equal to 35.6, whereas it is 12.5 for the ratio estimator.

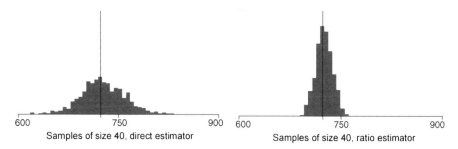

Figure 6.5 Simulating the sample distribution of the direct and the ratio estimator.

So, a much more precise estimator can be obtained with the same sample size if proper auxiliary information is available.

6.5 THE REGRESSION ESTIMATOR

The regression estimator is based on a linear descriptive model in its most general form. It assumes the points in the scatter plot of the target variable against the auxiliary variable to approximately lie on a straight line. This corresponds to the model

$$F(X_k; A, B) = A + BX_k. \qquad (6.38)$$

It is assumed that the values of both model parameters A and B are unknown. The condition that the population mean of the residuals must be 0 does not yet fix the values of the model parameters. So, the second condition comes into play, and that is the sum of squares of residuals must be minimized. Application of least squares theory results in

$$B = R_{XY} \frac{S_Y}{S_X} = \frac{S_{XY}}{S_X^2} = \frac{\sum_{k=1}^{N}(X_k - \bar{X})(Y_k - \bar{Y})}{\sum_{k=1}^{N}(X_k - \bar{X})^2} \qquad (6.39)$$

and

$$A = \bar{Y} - B\bar{X}. \qquad (6.40)$$

The quantity R_{XY} is the population correlation coefficient between X and Y (as defined in (6.33)), and S_{XY} is the population covariance between X and Y (as defined in (6.34)).

Of course, these optimal values of A and B cannot be computed in practice. So, they have to be estimated using the sample data. The model parameter B is estimated by

$$b = \frac{\sum_{i=1}^{n}(x_i - \bar{x})(y_i - \bar{y})}{\sum_{i=1}^{n}(x_i - \bar{x})^2} \qquad (6.41)$$

and the model parameter A is estimated by

$$a = \bar{y} - b\bar{x}. \tag{6.42}$$

The estimator for the descriptive model becomes

$$F(X_k; a, b) = \bar{y} - b(\bar{x} - X_k). \tag{6.43}$$

Substitution of (6.43) in (6.8) produces the *regression estimator*

$$\bar{y}_{LR} = \bar{y} - b(\bar{x} - \bar{X}). \tag{6.44}$$

Like the ratio estimator, the regression estimator can be seen as a correction of the simple sample mean (the direct estimator). The regression estimator corrects the difference between sample mean and population mean of the auxiliary variable.

The regression estimator is not an unbiased estimator. The reason is that b is not an unbiased estimator of B and therefore $\bar{x}b$ is not an unbiased estimator of $\bar{X}B$. However, all these estimators are ADU. So, the bias vanishes for a large sample size.

The variance of the regression estimator can be determined by using expression (6.10). This results in

$$V(\bar{y}_{LR}) \approx \frac{1-f}{n} \frac{SS_R}{N-1} = \frac{1-f}{n} \frac{1}{N-1} \sum_{k=1}^{N} (Y_k - \bar{Y} - B(X_k - \bar{X}))^2. \tag{6.45}$$

The variance (6.45) can be rewritten as

$$V(\bar{y}_{LR}) \approx \frac{1-f}{n} S_Y^2 (1 - R_{XY}^2). \tag{6.46}$$

This expression makes clear that a high correlation between target variable and auxiliary results in a small variance. The stronger the relationship between X and Y, the closer the correlation will be to $+1$ or -1, and the smaller the factor $1 - R_{XY}^2$ in (6.46) will be.

As the variance cannot be computed in practice, it is estimated using the sample data by

$$v(\bar{y}_{LR}) = \frac{1-f}{n} \frac{1}{n-1} \sum_{i=1}^{n} (y_i - \bar{y} - b(x_i - \bar{x}))^2. \tag{6.47}$$

Cochran (1977) suggests replacing the denominator $n-1$ in (6.47) by $n-2$. This suggestion comes from the theory of linear regression estimation. Assuming the sample is selected from an infinite population, the quantity

$$\frac{1}{n-2} \sum_{i=1}^{n} (y_i - \bar{y} - b(x_i - \bar{x}))^2 \tag{6.48}$$

is an unbiased estimator of

$$S_Y^2 (1 - R_{XY}^2). \tag{6.49}$$

THE REGRESSION ESTIMATOR

Both expression (6.47) and the adjustment suggested by Cochran lead to asymptotically unbiased estimator of the variance. Similar to (6.46), variance estimator (6.47) can be rewritten as

$$v(\bar{y}_{LR}) \approx \frac{1-f}{n} s_Y^2 (1 - r_{XY}^2), \qquad (6.50)$$

where

$$r_{XY} = \frac{S_{XY}}{S_X S_Y} = \frac{\frac{1}{n-1}\sum_{i=1}^{n}(x_i - \bar{x})(y_i - \bar{y})}{\sqrt{\frac{1}{n-1}\sum_{i=1}^{n}(x_i - \bar{x})^2 \frac{1}{n-1}\sum_{i=1}^{n}(y_i - \bar{y})^2}} \qquad (6.51)$$

is the correlation coefficient in the sample.

Assuming that (6.46) is a good approximation of the true variance, it can be concluded that the variance of the regression estimator is never larger than the variance of the direct estimator. This becomes clear by rewriting the variance of the regression estimator as

$$V(\bar{y}_{LR}) \approx V(\bar{y}_D)(1 - R_{XY}^2). \qquad (6.52)$$

Therefore, the efficiency of the regression estimator is equal to

$$\text{Eff}(\bar{y}_{LR}) = \frac{V(\bar{y}_D)}{V(\bar{y}_{LR})} = \frac{1}{1 - R_{XY}^2}. \qquad (6.53)$$

As soon as there is some linear relationship between X and Y, the correlation coefficient will differ from 0, resulting in a regression estimator that is more efficient than the direct estimator. The regression estimator will only be as precise as the direct estimator if there is no linear relationship at all; that is, the correlation coefficient is 0.

If the variance of the regression estimator is compared with that of the ratio estimator, it turns out that the regression estimator is more precise if

$$\left(R_{XY} \frac{S_Y}{S_X} - \frac{\bar{Y}}{\bar{X}}\right)^2 > 0. \qquad (6.54)$$

This condition is not satisfied only if

$$R_{XY} \frac{S_Y}{S_X} = \frac{\bar{Y}}{\bar{X}}. \qquad (6.55)$$

This is the case if the linear descriptive model for the estimator coincides with the model for the ratio estimator. To say it differently, the regression estimator and the ratio estimator have the same precision of the regression line that goes through the origin.

The result of comparing the various estimators is that the use of the regression estimator should always be preferred, as the variance of the regression estimator is never larger than the variance of the other estimators. However, the ratio estimator is still often used. The reason is that the computations for the regression estimator are

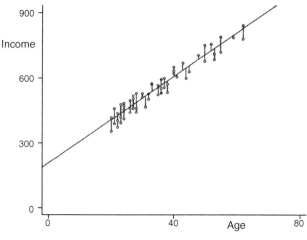

Figure 6.6 The regression estimator.

much more cumbersome. A ratio estimator can be computed by hand or with a simple hand calculator.

The properties of the regression estimator can be shown graphically using the sample of working males in the Samplonian province of Agria (see Fig. 6.6). The descriptive model corresponds to the regression line in the scatter plot, where the incomes of the persons are plotted against their ages. The slope of the line is equal to B and its intercept is equal to A.

Again, the distances from the point to the line represent the residuals. By comparing Figs 6.4 and 6.6, it will become clear that the regression line is the "best" line. The residuals are very small and there do not seem to be other lines that come "closer" to the points. The (approximate) variance of the regression estimator is equal to 85. This is much smaller than the variance of ratio estimator (482) or of the direct estimator (1178).

Section 6.4 contains a second example. It is a description of a simulation experiment in which the performance of the estimator is explored for estimating the mean milk production of 200 dairy farms. This experiment can be repeated for the regression estimator. The results would turn out to be comparable to that of the ratio estimator. The variance of the ratio estimator is 12.5 and the variance of the regression estimator is 12.4. The reason is that the regression line almost goes through the origin. Hence, the descriptive models of the ratio estimator and the regression estimator are almost the same.

6.6 THE POSTSTRATIFICATION ESTIMATOR

The ratio estimator and the regression estimator both use a quantitative auxiliary variable. Quantitative variables measure a phenomenon at a numerical scale. Examples are age and income. It is meaningful to carry out computations with its values, such as calculating totals and means. There are also qualitative variables. They just label elements so that they can be divided into groups. Examples of such variables are gender, marital status, and province of residence. Computations are not meaningful. Therefore, they cannot be used in the regression or ratio estimator.

The *poststratification* estimator is an estimator making use of a qualitative auxiliary variable. A special trick is used to be able to incorporate such a variable in a descriptive model. The quantitative variable is replaced by a set of dummy variables. A *dummy variable* is a variable that can only assume the values 0 and 1. Suppose, this qualitative variable has L categories; that is, it divides the population into L groups (here called *strata*). The qualitative variable is now replaced by L dummy variables. The values of the L dummy variables for element k are denoted by

$$X_k^{(1)}, X_k^{(2)}, \ldots, X_k^{(L)}, \tag{6.56}$$

for $k = 1, 2, \ldots, N$. The value of the hth dummy variable is equal to

$$X_k^{(h)} = \begin{cases} 1, & \text{if element } k \text{ is in stratum } h, \\ 0, & \text{if element } k \text{ is not in stratum } h. \end{cases} \tag{6.57}$$

So, always one dummy variable has the value 1 for an element, while all other dummy variables are 0. The total number of elements in a stratum can now be written as

$$N_h = \sum_{k=1}^{N} X_k^{(h)}. \tag{6.58}$$

It is assumed that stratum sizes $N_1, N_2, \ldots, N_L$ of all L strata are known. To predict the values of the target variables using the L dummy variables, the following descriptive model is used:

$$F(X_k; \theta) = F(X_k^{(1)}, X_k^{(2)}, \ldots, X_k^{(L)}; B_1, B_2, \ldots, B_L) = \sum_{h=1}^{L} B_k X_k^{(h)} \tag{6.59}$$

in which $B_1, B_2, \ldots, B_L$ are the model parameters, the values of which have to be determined. If an element k is a member of stratum h, the predicted value of the target variable is equal to B_h (for $h = 1, 2, \ldots, L$). In fact, this model assumes that the target variable shows no or little variation within strata; that is, the strata are homogeneous with respect to the target variable. To say it otherwise, elements within a stratum are similar.

Minimizing the residual sum of squares results in

$$B_h = \bar{Y}^{(h)} = \frac{1}{N_h} \sum_{k=1}^{N} X_k^{(h)} Y_k. \tag{6.60}$$

for $h = 1, 2, \ldots, L$. So, the optimal value of model parameter B_h is equal to the mean of the target variable in corresponding stratum h.

Application of the optimal model requires knowledge of the stratum means of the target variable. This will not be the case in practice, as this would enable computation of the population mean using the relation

$$\bar{Y} = \frac{1}{N} \sum_{h=1}^{L} N_h \bar{Y}^{(h)} \qquad (6.61)$$

As usual, the solution is the estimate of the parameters $B_1, B_2, \ldots, B_L$ using the sample data. Minimizing the residual sum of squares for the sample results in an estimator

$$b_h = \bar{y}^{(h)}, \qquad (6.62)$$

for B_h. This is the sample mean of the target variable in the corresponding stratum. Note that the sampling design does not fix the number of sample elements in each stratum. So, the sample mean is based on a random number of observations. It is theoretically possible that no observations at all become available in a stratum. In practical situations, the probability of empty strata is usually so small that it can be ignored.

By using expression (6.62), the descriptive model can be estimated by

$$F\left(X_k^{(1)}, X_k^{(2)}, \ldots, X_k^{(L)}; b_1, b_2, \ldots, b_L\right) = \sum_{h=1}^{L} \bar{y}^{(h)} X_k^{(h)}. \qquad (6.63)$$

Substitution of (6.63) in (6.8) results in the *poststratification estimator* for the population mean

$$\bar{y}_{PS} = \frac{1}{N} \sum_{h=1}^{L} N_h \bar{y}^{(h)}. \qquad (6.64)$$

So, the estimator is equal to the weighted mean of the estimators for the stratum means. This estimator is unbiased provided there is at least one observation in each stratum. There is no simple, exact analytical expression for the variance of poststratification estimator. However, there is a large sample approximation:

$$V(\bar{y}_{PS}) = \frac{1-f}{n} \sum_{h=1}^{L} W_h S_h^2 + \frac{1}{n^2} \sum_{h=1}^{L} (1 - W_h) S_h^2, \qquad (6.65)$$

where $W_h = N_h/N$ is the relative size of stratum h and S_h^2 is the (adjusted) population variance of the target variable in stratum h. Variance (6.65) can be estimated by replacing the population variances in (6.65) with their sample estimates. This results in

$$v(\bar{y}_{PS}) = \frac{1-f}{n} \sum_{h=1}^{L} W_h s_h^2 + \frac{1}{n^2} \sum_{h=1}^{L} (1 - W_h) s_h^2. \qquad (6.66)$$

The poststratification estimator is precise if the strata are homogeneous with respect to the target variable. This implies that variation in the values of the target variable is typically caused by differences in means between strata and not by variation within strata (Fig. 6.7).

The use of the poststratification estimator is illustrated by using an example based on Samplonian data. The objective is to estimate the mean income of the working

Figure 6.7 The poststratification estimator. Reprinted by permission of Imre Kortbeek.

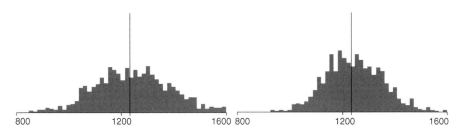

Figure 6.8 Simulating the sample distribution of the direct estimator and the poststratification estimator.

population of Samplonia ($N = 341$). A large number of samples of size 40 are simulated. Figure 6.8 contains the results. The histogram on the left contains the (simulated) distribution of the direct estimator. The histogram on the right contains the distribution for the poststratification estimator. Province of residence is used as auxiliary variable. Since Samplonia consists of the two provinces of Agria and Induston, there are two strata.

The poststratification estimator seems to perform better, although the differences with the direct estimator are not very spectacular. This is confirmed by comparing the standard errors, which are 143 for the direct estimator and 105 for the poststratification estimator. Apparently, the strata are not so homogeneous with respect to income.

EXERCISES

6.1 To be useful as an auxiliary variable in a model-based estimator, at least the following information must be available for a variable X:
   a. The distribution of the variable X in the target population.
   b. The regression coefficients of the regression model predicting the target variable Y from the variable X.

c. The value of correlation between target variable Y and the variable X.

d. Both the regression coefficients and the correlation coefficient.

6.2 The efficiency of an estimator $\bar{y}_F$ based on a descriptive function F is defined as

a. $\mathrm{Eff}(\bar{y}_F) = \dfrac{V(\bar{y})}{V(\bar{y}_F)}$.

b. $\mathrm{Eff}(\bar{y}_F) = \dfrac{V(\bar{y}_F)}{V(\bar{y})}$.

c. $\mathrm{Eff}(\bar{y}_F) = \dfrac{S(\bar{y})}{S(\bar{y}_F)}$.

d. $\mathrm{Eff}(\bar{y}_F) = \dfrac{S(\bar{y}_F)}{S(\bar{y})}$.

6.3 A retail organization wants insight into the amount of shoplifting in the 5000 shops of its members. Target variable is the total value of stolen goods in a shop. Auxiliary variable is the floor space of the shop. A simple random sample of 100 shops is selected without replacement. The sample results are summarized in the table below.

Mean value of shoplifting (euro)	500
Standard deviation of shoplifting (euro)	300
Mean floor space (square meters)	4,900
Standard deviation of floor space (square meters)	3,200
Covariance between shoplifting and floor area	770,000

Furthermore, the information is available that the average floor size of all 5000 shops is equal to 5000 m².

a. Estimate the standard error of the sample mean of the shoplifting values.

b. Compute the value of the regression estimator, assuming floor size is used as auxiliary variable.

c. Estimate the standard error of the regression estimator. To do this, first compute the correlation coefficient.

d. Compare the standard errors computed under (a) and (c), and explain the differences.

6.4 An income survey has been carried out in the town of Woodham. A simple random sample of 100 households has been selected from the total population of 3410 households. Woodham consists of two neighborhoods: Old North and New South. The number of households in each neighborhood is known. All available information is summarized in the table below.

	Old North	New South	Woodham
Sample households	25	75	100
Sample mean of income	348	1,692	
Sample variance of income	48,218	724,649	895,455

EXERCISES

    a. Compute the sample mean of all 100 households.
    b. Estimate the standard error of the sample mean.
    c. Estimate the 95% confidence interval for the mean income in the population. Interpret this interval in general term.
    d. Suppose the additional information is available that there are 1210 households in the Old North and 2200 households in the New South. Use the poststratification estimator to compute an estimate of the mean income in the population, using neighborhood as auxiliary variable.
    e. Estimate the standard error of the poststratification estimator.
    f. Compare the values of the estimators computed under (a) and (d). Explain the differences.

6.5 There is a relationship between the income of a household and the total floor space of the home of a household in a certain region of Samplonia. A simple random sample of size 4 has been selected without replacement. The table below contains the sample data.

Household	1	2	3	4
Floor space	116	81	73	99
Income	1200	950	650	1050

The mean floor space of all houses in the population is 103.7 m^2.

    a. Compute the ratio estimator for the mean income using floor space as auxiliary variable.
    b. Compute the regression estimator for the mean income using floor space as auxiliary variable.

6.6 Local elections will be held in the town of Springfield. There is new political party, Forward Springfield, that takes part in elections for the first time. To get some indication of the popularity of this party, an opinion poll is carried out. The total population of voters consists of 40,000 people. The town consists of two neighborhoods: Northwood (with 30,000 voters) and Southfield (with 10,000 voters). A simple random sample of 2000 voters is drawn. Each selected person is asked for which party he or she will vote. The sample results are summarized in the table below.

Votes for Forward Springfield	Northwood	Southfield
Yes	1338	58
No	182	422

    a. Estimate the percentage of voters that will vote for Forward Springfield. Estimate the variance of this estimator and compute the 95% confidence interval.

There are substantial differences between the two neighborhoods of Northwood and Southfield. Typically, poorer people live in Northwood and richer people in Southfield. It is not unlikely that there is a relationship between voting behavior and socioeconomic status. So it might be a good idea to use the poststratification estimator.

b. Estimate the percentage of voters that will vote for Forward Springfield poststratifying the sample by neighborhood. Also, estimate the variance of this estimator.

It is very likely that this opinion poll will be repeated in the future. Then, a stratified sample will be selected. The variable neighborhood will be used as stratification variable. Costs of interviewing are different in the two neighborhoods. The costs per interview are €16 in Northwood and € 25 in Southfield. Suppose the variance of the target variable (votes for Forward Springfield) is 900 in Northwood and 1600 in Southfield.

c. Compute the optimal allocation for this sampling design under the condition that the total interviewing costs may not exceed €20,000.

CHAPTER 7

Data Collection

7.1 TRADITIONAL DATA COLLECTION

The first step in the survey process concentrates on design issues. The target population is defined, the population parameters to be estimated are determined, and a questionnaire is designed. In addition, a sampling design is specified and the sample is selected accordingly from the sampling frame. The next step in the survey process is collecting the survey data. The questions in the questionnaire have to be answered by the selected elements. This phase of the survey is sometimes called the *fieldwork*. This term refers to the interviewers who go into the "field" to visit the persons selected in the sample. However, there are more means to collect the data. This section describes three traditional *modes of data collection*: face-to-face interviewing, telephone interviewing, and mail interviewing. Section 7.2 is devoted to more modern ways of data collection.

Mail interviewing is the least expensive of the three data collection modes. Paper questionnaires are sent by mail to the elements (e.g., persons, households, or companies) selected in the sample. They are invited to answer the questions and to return the completed questionnaire to the survey agency. A mail survey does not involve interviewers. Therefore, it is a cheap mode of data collection. Data collection costs include only mailing costs (letters, postage, and envelopes). Another advantage is that the absence of interviewers can be considered less threatening by potential respondents. As a consequence, respondents are more inclined to answer sensitive questions.

The absence of interviewer also has a number of disadvantages. They cannot provide additional explanation or assist the respondents in answering the questions. This may cause respondents to misinterpret questions, which has a negative impact on the quality of the data collected. Also, it is not possible to use show cards. *Show cards* are typically used for answering closed questions. Such a card contains the list of all possible answers to a question. It allows respondents to read through the list at their

---

Applied Survey Methods: A Statistical Perspective, Jelke Bethlehem
Copyright © 2009 John Wiley & Sons, Inc.

own pace and select the answer that reflects their situation or opinion. Mail surveys put high demands on the design of the paper questionnaire. It should be clear to all respondents how to navigate through the questionnaire and how to answer questions.

Since the persuasive power of the interviewers is absent, response rates of mail surveys tend to be low. Of course, reminder letters can be sent, but this is often not very successful. More often survey documents end up in the pile of old newspapers.

In summary, the costs of a mail survey are relatively low, but often a price has to be paid in terms of data quality: response rates tend to be low, and the quality of the collected data is also often not very good. However, Dillman (2007) believes that good results can be obtained by applying his *Tailored Design Method*. This is a set of guidelines for designing and formatting mail survey questionnaires. They pay attention to all aspects of the survey process that may affect response rates or data quality.

Face-to-face interviewing is the most expensive of the three data collection modes. Interviewers visit the homes of the persons selected in the sample. Well-trained interviewers will be successful in persuading many reluctant persons to participate in the survey. Therefore, response rates of face-to-face surveys are usually higher than those of a mail survey. The interviewers can also assist respondents in giving the right answers to the questions. This often results in better quality data. However, the presence of interviewers can also be a drawback. Research suggests that respondents are more inclined to answer sensitive questions if there are no interviewers in the room.

The survey organization may consider sending a letter announcing the visit of the interviewer. Such a letter can also give additional information about the survey, explain why it is important to participate, and assure that the collected information is treated confidentially. As a result, the respondents are not taken by surprise by the interviewers. Such an announcement letter may also contain the telephone number of the interviewer. This makes it possible for the respondent to make an appointment for a more appropriate day and/or time. Of course, the respondent can also use this telephone number to cancel the interview.

The response rate of a face-to-face survey is higher than that of a mail survey, and so is the quality of the collected data. But a price has to be paid literally: face-to-face interviewing is much more expensive. A team of interviewers has to be trained and paid. Also, they have to travel a lot, and this costs time and money.

A third mode of data collection is *telephone interviewing*. Interviewers are also needed for this mode, but not as many as needed for face-to-face interviewing. They do not lose time traveling from one respondent to the next. They can remain in the call center of the survey agency and conduct more interviews in the same amount of time. Therefore, the interviews cost less. An advantage of telephone interviewing over face-to-face interviewing is that often respondents are more inclined to answer sensitive questions because the interviewer is not present in the room.

Telephone interviewing also has some drawbacks. Interviews cannot last too long and questions may not be too complicated. Another complication may be the lack of a proper sampling frame. Telephone directories may suffer from severe undercoverage because more and more people do not want their phone number to be listed in the directory. Another development is that increasingly people replace their landline phone by a mobile phone. Mobile phone numbers are not listed in directories in many countries. For example, according to Cobben and Bethlehem (2005) only between 60% and 70% of the Dutch population can be reached through a telephone directory.

Telephone directories may also suffer from overcoverage. For example, if the target population of the survey consists of households, then only telephone numbers of private addresses are required. Telephone numbers of companies must be ignored. It is not always clear whether a listed number refers to a private address or a company address (or both).

A way to avoid the undercoverage problems of telephone directories is to apply *random digital dialing* (RDD) to generate random phone numbers. A computer algorithm computes valid random telephone numbers. Such an algorithm is able to generate both listed and unlisted numbers. So, there is complete coverage. Random digital dialing also has drawbacks. In some countries, it is not clear what an unanswered number means. It can mean that the number is not in use. This is a case of overcoverage. No follow-up is needed. It can also mean that someone simply does not answer the phone, a case of nonresponse, which has to be followed up. Another drawback of RDD is that there is no information at all about nonrespondents. This makes correction for nonresponse very difficult (see also Chapter 9 about nonresponse and Chapter 10 about weighting adjustment).

The fast rise of the use of mobile phones has not made the task of the telephone interviewer easier. More and more landline phones are replaced by mobile phones. A landline phone is a means to contact a household whereas a mobile phone makes contact with an individual person. Therefore, the chances of contacting any member of the household are higher in case of landline phones. And if persons can only be contacted through their mobile phones, it is often in a situation not fit for conducting an interview. In addition, it was already mentioned that sampling frames in many countries do not contain mobile phone numbers. And a final complication is that in countries such as The Netherlands, people often switch from one phone company to another. As a result, they get a different phone number. For more information about the use of mobile phones for interviewing, see, for example, Kuusela et al. (2006).

The choice of the mode of data collection is not an easy one. It is usually a compromise between quality and costs. In a large country such as the United States, it is almost impossible to collect survey data by means of face-to-face interviewing. It requires so many interviewers to do so much traveling that the costs would be very high. Therefore, it is not surprising that telephone interviewing emerged here as a major data collection mode. In a very small and densely populated country such as The Netherlands, face-to-face interviewing is much more attractive. Coverage problems of telephone directories and low response rates also play a role in the choice for face-to-face interviewing. More about data collection issues can be found in Couper et al. (1998).

7.2 COMPUTER-ASSISTED INTERVIEWING

Collecting survey data can be a complex, costly, and time-consuming process, particularly if high-quality data are required. One of the problems of traditional data collection is that the completed paper questionnaire forms usually contain many errors. Therefore, substantial resources must be devoted to make these forms error free. Extensive data editing is required to obtain data of acceptable quality. Rapid developments in information technology since the 1970s have made it possible to use microcomputers for data collection. Thus, *computer-assisted interviewing* (CAI) was born. The paper questionnaire was replaced by a computer program containing the questions to be asked. The computer took control of the interviewing process, and it also checked answers to questions on the spot.

Like traditional interviewing, computer-assisted interviewing has different modes of data collection. The first mode of data collection was *computer-assisted telephone interviewing* (CATI). Couper and Nicholls (1998) describe how it was developed in the United States in the early 1970s. The first nationwide telephone facility for surveys was established in 1966. The driving force was to simplify sample management. These systems evolved in subsequent years into full-featured CATI systems. Particularly in the United States, there was a rapid growth in the use of these systems. However, CATI systems were used little in Europe until the early 1980s.

Interviewers in a CATI survey operate a computer running interview software. When instructed so by the software, they attempt to contact a selected person by phone. If this is successful and the person is willing to participate in the survey, the interviewer starts the interviewing program. The first question appears on the screen. If this is answered correctly, the software proceeds to the next question on the route through the questionnaire.

Many CATI systems have a tool for call management. Its main function is to offer the right phone number at the right moment to the right interviewer. This is particularly important when the interviewer has made an appointment with a respondent for a specific time and date. Such a call management system also has facilities to deal with special situations such as a busy number (try again after a short while) or no answer (try again later). This all helps to increase the response rate as much as possible.

More about the use of CATI in the United States can be found in Nicholls and Groves (1986). De Bie et al. (1989) give an overview of the available software in the early stages of development.

The emergence of small portable computers in the 1980s made *computer-assisted personal interviewing* (CAPI) possible. It is a form of face-to-face interviewing in which interviewers take their laptop computer to the home of the respondents. There they start the interview program and attempt to get answers to the questions.

Statistics Netherlands started experiments with this mode of data collection in 1984. Computers were first tried in a price survey. In this survey, interviewers visit shops and record prices of products. It turned out that interviewers were able to handle the hardware and software. Moreover, respondents (shopkeepers) did not object to this kind of data collection.

The outcome of this experiment provided insight into the conditions laptop computers had to satisfy to be useful for this kind of work. First, they should not be too heavy. A weight of 3 kg was considered the maximum (often women) interviewers could handle. Second, the readability of the screen should always be sufficient, even in bad conditions, such as a sunny room. Third, battery capacity should be sufficient to allow a day of interviewing without recharging. And if this was not possible, interviewers should have easy-to-replace spare batteries. The situation should be avoided in which the interviewer has to plug a power cable into wall socket in the home of the respondent. Finally, the interviewers preferred a full-size keyboard. They considered small keys too cumbersome and error-prone.

After the success of the first experiment, a second experiment was carried out. This time, the laptops were tested in a real interview situation in the homes of the respondents. The aim of this experiment was to test whether respondents accepted this type of data collection. Respondents were randomly assigned to a group that was interviewed in the traditional way or a group that was interviewed with laptops. It turned out there was no effect on response rates. Respondents simply accepted it as a form of progress in survey taking. At that time, there was some concern about "big brother" effects. This form of electronic data collection might cause anxiety among respondents that they might have become part of a large government operation to collect large amounts of data about people and that therefore their privacy was at stake. However, no such "big brother" effects could be observed. Another conclusion was that interviewers very rapidly became accustomed to using the new technology for their work.

The success of these experiments convinced Statistics Netherlands that is was possible to use CAPI in its regular surveys. In 1987, the Dutch Labor Force Survey (LFS) became a CAPI survey. Approximately, 400 interviewers were equipped with a laptop computer. It was an EPSON PX-4, running under the operating system CP/M (see Fig. 7.1). Each month, the interviewers visited 12,000 addresses and conducted around 30,000 interviews. After the day of work, they returned home and connected their computers to the power supply to recharge the batteries. They also connected their laptop to a telephone and modem. At night, when the interviewers

Figure 7.1 The Epson PX-4 laptop computer that was used in the 1987 Labor Force Survey of Statistics Netherlands.

were asleep, their computers automatically called Statistics Netherlands and uploaded the collected data. New address data were downloaded in the same session. In the morning, the computer was ready for a new day of interviewing.

It is interesting to compare the old Labor Force Survey with the new one. The old LFS was carried out each year from 1973 to 1985. During the course of the fieldwork spanning a number of weeks, approximately 150,000 respondents were visited. There were no professional interviewers. Interviews were carried by civil servants of the municipality. They used paper questionnaire forms. The fieldwork for the new LFS was spread over 12 months and was carried out by professional interviewers equipped with laptops. So, the old and new LFS differed in several ways. Therefore, it is not easy to determine to what extent computer-assisted interviewing led to improvements. Still, some conclusions could be drawn. First, CAPI has considerably reduced the total data processing time. The period between the completion of the fieldwork and the publication of the first results could be many months for the old LFS. For the new LFS, the first tables were published only a few weeks after the completion of the fieldwork. Of course, these timely statistics were much more valuable. Second, the quality of the collected data improved. This was to be expected due to the checks that were incorporated in the interview program. Third, respondents completely accepted computer-assisted interviewing as a mode of survey data collection. There was no increase in nonresponse rates. Fourth, interviewers had no problems using laptops for interviewing. They needed only a moderate amount of training and supervision. Finally, the conclusion was that, with the exception of the financial investments required, CAPI had advantages.

The CAPI system of Statistics Netherlands is called Blaise. It was developed by Statistics Netherlands. It evolved in the course of time in a system running under MS-DOS and later under Windows. All surveys of this institute, and of many other national statistical institutes, are now carried out with Blaise. More about the early years of CAPI at Statistics Netherlands can be found in CBS (1987) and Bethlehem and Hofman (2006). More information about CAPI in general can be found in Couper et al. (1998).

The computer-assisted mode of mail interviewing also emerged. It is called *computer-assisted self-interviewing* (CASI), or sometimes also *computer-assisted self-administered questionnaires* (CASAQ). The electronic questionnaire is sent to the respondents. They answer the questions and send it back to the survey agency. Early CASI applications used diskettes or a telephone and modem to send the questionnaire, but nowadays it is common practice to download it from the Internet. The answers are returned electronically in the same fashion.

A CASI survey is only feasible if all respondents have a computer on which they can run the interview program. Since the use of computers was more widespread among companies than households in the early days of CASI, the first CASI applications were business surveys. An example is the production of fire statistics in The Netherlands. These statistics were collected in the 1980s by means of CASI. Diskettes were sent to the fire brigades. They ran the questionnaire on their (MS-DOS) computers. The answers were stored on diskettes. After completing the questionnaire, the diskettes were returned to Statistics Netherlands.

Another early application of CASI was data collection for the foreign trade statistics of Statistics Netherlands. Traditionally, data for these statistics were collected through customs at the borders of the country. However, borders have vanished within the European Union. So, data collection at the borders with the neighboring countries of The Netherlands came to an end. Now, data are collected by a survey among the companies exporting and importing goods. To do this as efficiently as possible, a CASI survey was conducted. The interviewing program was sent (once) to the companies. On a regular basis, they ran the program and sent back the data to Statistics Netherlands.

An early application of CASI in social surveys was the *Telepanel* (see Saris, 1998). The Telepanel was founded in 1986. It was a panel of 2000 households that agreed to regularly fill in questionnaires with the computer equipment provided to them by the survey organization. A home computer was installed in each household. It was connected to the telephone with a modem. It was also connected to the television in the household so that it could be used as a monitor. After a diskette was inserted into the home computer, it automatically established a connection with the survey agency to exchange information (downloading a new questionnaire or uploading answers of the current questionnaire). Panel members had agreed to fill in a questionnaire each weekend.

The rapid development of the Internet in 1990s led to a new mode of data collection. Some call it *computer-assisted web interviewing* (CAWI). The questionnaire is offered to respondents through the Internet. Therefore, such a survey is sometimes also called a *web survey* or *online survey*. In fact, such an online survey is a special type of a CASI survey. At first sight, online surveys have a number of attractive properties. Now that so many people are connected to the Internet, it is an easy way to get access to a large group of potential respondents. Furthermore, questionnaires can be distributed at very low costs. No interviewers are needed, and there are no mailing and printing costs involved. Finally, surveys can be launched very quickly. Little time is lost between the moment the questionnaire is ready and the start of the fieldwork. As a result, it is a cheap and fast means to get access to a large group of people.

However, online surveys also have some serious drawbacks. These drawbacks are mainly caused by undercoverage (not everyone has access to Internet) and the lack of proper sampling designs (often self-selection is applied). Because of the increasing popularity of online surveys and the associated methodological problems, a special chapter is devoted to this mode of data collection (Chapter 11).

Application of computer-assisted interviewing for data collection has three major advantages:

- It simplifies the work of interviewers. They do not have to pay attention any more to choosing the correct route through the questionnaire. Therefore, they can concentrate on asking questions and assisting respondents in getting the answers.
- It improves the quality of the data collected because answers can be checked and corrected during the interview. This is more effective than having to do it afterward in the survey agency.

- Data are entered in the computer during the interview resulting in a clean record, so no more subsequent data entry and data editing are necessary. This considerably reduces time needed to process the survey data and thus improves the timeliness of the survey results.

7.3 MIXED-MODE DATA COLLECTION

It is clear that surveys can be conducted by using various data collection modes. This raises the question which mode to use in a specific survey. Biemer and Lyberg (2003) discuss optimal designs for data collection using one mode. This is also called *single-mode data collection*. Modes differ in various aspects, particularly data quality, costs, and timeliness. Face-to-face interviewing is expensive. Every household does not have a telephone or Internet connection and therefore cannot be approached by a telephone or online survey. Mail surveys have a low response rate and take a lot of time to process. Thus, each individual data collection mode has its advantages and disadvantages. Mixing data collection modes provides an opportunity to compensate for the weakness of each individual mode. This can reduce costs and at the same time increase response rates and data quality. Sampled elements can be allocated to a specific mode on the basis of known background characteristics. If there are persons that do not cooperate in one mode and are willing to participate in another mode, this can reduce the selectivity of the response.

For example, Dutch statistics show that 90% of the children between 12 and 14 years had access to Internet in 2005. For men over 65 years, this percentage was much lower, 34%, and for women over 65 years, it was only 21% (*source: Statline*, Statistics Netherlands). Consequently, the elderly would be severely underrepresented in an online survey. However, elderly persons are known to be cooperative when interviewed face-to-face. So, one might consider approaching the elderly face-to-face and using online interviewing for young people. This is a form of what is called *mixed-mode data collection*.

Mixed-mode data collection consists of a combination of two or more data collection modes. De Leeuw (2005) describes two mixed-mode approaches. The first is a *concurrent* approach. The sample is divided in groups that are approached by different modes, at the same time (see Fig. 7.2). The concurrent approach aims at maximizing response rates by selecting the proper mode for each group.

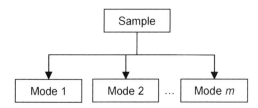

Figure 7.2 A concurrent mixed-mode approach.

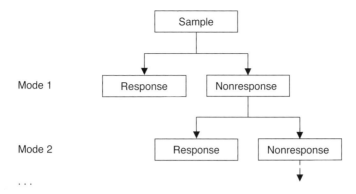

Figure 7.3 A sequential mixed-mode approach.

A second mixed-mode approach is the *sequential* approach. All sample elements are approached by one mode. The nonrespondents are then followed up by a different mode than the one used in the first approach. This process can be repeated for several modes of data collection (see Fig. 7.3).

Another form of a mixed-mode data collection is to let respondents select their preferred data collection mode. They are given a choice, which is a very flexible and respondent-friendly alternative. There is a potential drawback though, as this option provides persons with an extra opportunity to refuse participation. For example, if the choice is presented in the form of a postcard that they have to return to indicate their preference. This requires an effort from the respondent, and not returning the postcard can be regarded as refusal.

Longitudinal or panel surveys often use some kind of mixed approach. Persons fill in a questionnaire a number of times. Often, the first questionnaire is a large one, in which a lot of information about the respondents is selected. The follow-up questionnaires are short. They only record possible changes in the situation of the respondents. It is common practice to use a more expensive mode in the first wave to maximize response. A less costly mode is used in the follow-up interviews. For example, most Labor Force Surveys in Europe are panel surveys where the first wave is conducted face-to-face and the subsequent waves are conducted by telephone.

The Labor Force Survey approach is different from the sequential approach as displayed in Fig. 7.3. The entire sample is approached in the same mode. In subsequent waves, a different mode is used, but this is still the same mode for all the sample persons. In the situation displayed in Fig. 7.3, the nonrespondents of the initial sample are followed up in a different mode from the one used for the entire sample.

There are some studies into the effects of nonresponse that use the sequential mixed-mode design. At Statistics Netherlands, a large-scale follow-up of nonrespondents from the Dutch Labor Force Survey has been performed (see Schouten, 2007). The Social and Cultural Planning Office in The Netherlands also used a mixed-mode design to follow up nonrespondents in the Dutch Amenities and Services Utilization Survey (see Stoop, 2005). But these are merely experiments, no regular fieldwork practices. They serve a methodological purpose: validating the methods that are used and the assumptions that are made when adjusting for nonresponse.

It is not easy to decide which mixture of modes to use. Several factors need to be considered before choosing the optimal design. Some of them are discussed here.

A major concern in mixed-mode data collection is that data quality may get affected by the occurrence of *mode effects*, a phenomenon in which asking a person the same question in different data collection modes would lead to different answers. An example is asking a closed question with a substantial number of answer options. The respondent in a face-to-face survey would be presented a show card with all possible answers. In case of a telephone survey, the interviewer would read all possibilities to the respondents. Research indicates that this results in a preference for the last options in the list. Respondents in a web survey have to read through the list themselves. This seems to lead to a preference for answers early in the list.

Sequential mixed-mode data collection may help increase response rates. However, nonresponse is not the only source of errors in surveys. Chapter 8 presents an overview of possible errors. The effects of these errors may differ for each mode. For example, a mail survey is affected more by processing errors than a computer-assisted telephone survey. Generally, data collection modes with the most serious errors also tend to be the cheapest modes. So, it comes down to a trade-off between data quality and costs.

The topic of the survey may limit mixed-mode possibilities. Some topics may be less suited for a survey that is interviewer assisted. Typically, answers to sensitive questions may be closer to the truth when there are no interviewers involved. A mail survey or web survey may therefore be more appropriate for this type of questions.

Time is also an important aspect. The fieldwork of a sequential mixed-mode approach will take longer because modes follow each other in time. So, much time may not be available. This survey design also requires decisions when to move on to the next data collection mode. Should such a decision be time dependent only? Or should it be based on the response rate of the current mode? The latter strategy will make it uncertain how long the fieldwork period will be.

Data collection costs also depend on the mode chosen. A telephone survey is much cheaper than a face-to-face survey. A mail survey is even cheaper than a telephone survey. The cheapest mode is probably a web survey. If there is only a limited budget available, face-to-face interviewing may be out of question, and a choice has to be made for one or more less costly modes.

Last but certainly not the least, attention has to be paid to case management. Sample elements have to be assigned to the proper mode. In the course of the fieldwork, they may have to be reassigned to another mode. This requires an effective and reliable *case management system*. It has to see to it that the cases are assigned to the proper mode, cases are not assigned to multiple modes, or cases are not assigned to any mode at all. Unfortunately, there are no general-purpose case management systems for mixed-mode surveys. This means that tailor-made systems have to be developed.

7.4 ELECTRONIC QUESTIONNAIRES

The elements of paper questionnaires were rather straightforward, containing questions for respondents and instructions for interviewers to jump to other questions or to the end of the questionnaire. Application of some form of computer-assisted interviewing requires the questionnaires to be defined. Such questionnaires can have more elements than paper questionnaires. Here is a nonexhaustive list:

- *Questions*. Each question may have an identification (number or name), a question text, a specification of the type of answer that is expected (text, number, selection from a list, etc.), and a field in which the answer is stored.
- *Checks*. This is a logical expression describing a condition that must be fulfilled, and an error message (which is displayed when the condition is not met).
- *Computations*. They may involve answers to previous questions and other information. Computations can be used to compute the answer to another question, as a component in a check, or to compute the route to the following question.
- *Route Instructions*. These instructions describe the order in which questions are processed and also under which conditions they are processed.

Route instructions can take several forms. This is illustrated using a simple example of a fragment of a questionnaire. Figure 7.4 shows how this fragment could look like in paper form.

```
1. Are you male or female?
     Male . . . . . . . . . . . . . . . . . . . 1    Skip to question 3
     Female . . . . . . . . . . . . . . . . . 2

2. Have you ever given birth?
     Yes  . . . . . . . . . . . . . . . . . . 1
     No . . . . . . . . . . . . . . . . . . . 2

3. How old are you?                    _ _ years

Interviewer: If younger than 17 then goto END

4. What is your marital status?
     Never been married . . . . . . . . . . . 1    Skip to question 6
     Married  . . . . . . . . . . . . . . . . 2
     Separated  . . . . . . . . . . . . . . . 3
     Divorced . . . . . . . . . . . . . . . . 4    Skip to question 6
     Widowed  . . . . . . . . . . . . . . . . 5    Skip to question 6

5. What is your spouse's age?          _ _ years

6. Are you working for pay or profit?
     Yes  . . . . . . . . . . . . . . . . . . 1
     No . . . . . . . . . . . . . . . . . . . 2

END OF QUESTIONNAIRE
```

Figure 7.4 A paper questionnaire.

The questionnaire contains two types of routing instructions. First, there are skip instructions attached to answer codes of closed questions. This is the case for questions 1 and 4. The condition deciding the next question asked depends only on the answer to the current question. Second, there are instructions for the interviewer that are included in the questionnaire between questions. These instructions are typically used when the condition deciding the next question depends on the answers to several questions, or on the answer to a question that is not the current question. Figure 7.4 contains an example of such an instruction between questions 3 and 4.

Specification languages of CAI systems (so-called *authoring languages*) usually do not contain interviewer instructions. Skip instructions appear in different formats. Figure 7.5 contains a specification of the sample questionnaire of Fig. 7.4 in the authoring language of the *CASES* system. This system was developed by the University of California in Berkeley.

Route instructions are goto oriented in *CASES*. There are two types:

- Skips attached to answer codes are called *unconditional gotos*. An example is the jump to question "Age" if the answer to the question "Sex" is "Male".
- Interviewer instructions are translated into *conditional gotos*. An example is the instruction at the end of the question "Age." There is a jump to the end of the questionnaire if answer to the question "Age" is less than 16.

An example of a CAI system with a different authoring language is the *Blaise system* developed by Statistics Netherlands. The authoring language of this system uses IF-THEN-ELSE structures to specify routing instructions. Figure 7.6 contains the Blaise code for the sample questionnaire.

There has been an intensive debate on the use of goto instructions in programming languages. A short paper by Edsger Dijkstra in 1968 ("Go To Statement Considered Harmful") was the start of the structured programming movement. It has become clear that this also applies to questionnaires. Use of goto instructions in questionnaires makes these instruments very hard to test and to document.

The way in which the routing structure is specified is not the only difference between Figs 7.5 and 7.6. The developers of Blaise have considered a clear view on the routing structure so important that routing is specified in a separate section of the specification (the rules section).

Note that in the example shown in Fig. 7.6 only questions have been used. It contains no checks or computations.

Several CAI software systems offer a modular way of specifying electronic questionnaires. This means the questionnaire is split into a number of subquestionnaires, each with its own question definitions and routing structure. Subquestionnaires can be developed and tested separately. It is possible to incorporate such modules as a standard module in several surveys, thereby reducing development time and promoting consistency between surveys.

There can also be routing instructions at the level of subquestionnaires. Answers to questions in one subquestionnaire may determine whether or not another subquestionnaire is executed. Furthermore, subquestionnaires can be used to implement

ELECTRONIC QUESTIONNAIRES

```
>Sex<
    Are you male or female?
    <1> Male                    [goto Age]
    <2> Female
    @

>Birth<
    Have you ever given birth?
    <1> Yes
    <2> No
    @

>Age<
    How old are you ?
    <12-20>
    @
[@] [if Age lt <16> goto End]

>MarStat<
    What is your marital status?
    <1> Never been married      [goto Work]
    <2> Married
    <3> Separated
    <4> Divorced                [goto Work]
    <5> Widowed                 [goto Work]
    @

>Spouse<
    What is your spouse's age?
    <16-20>
    @

>Work<
    Are you working for pay or profit?
    <1> Yes
    <2> No
    @
```

Figure 7.5 The sample questionnaire in CASES.

hierarchical questionnaires. Such questionnaires allow a subquestionnaire to be executed a number of times. A good example of a hierarchical questionnaire is a household questionnaire. There are questions at the household level, and then there is a set of questions (subquestionnaire) that must be repeated for each eligible member of the household.

On the one hand, a subquestionnaire can be seen as one of the objects in a questionnaire. It is part of the routing structure of the questionnaire, and it can be executed just like a question or a check. On the other hand, a subquestionnaire contains a questionnaire of its own. By zooming into a subquestionnaire, its internal part becomes visible, and that is a questionnaire with its objects and routing conditions.

Interviewing software can have different modes of behavior. The first aspect is routing behavior. This determines how the software leads interviewers or respondents

```
DATAMODEL Example
FIELDS
  Sex        "Are you male or female?": (Male, Female)
  Birth      "Have you ever given birth?": (Yes, No)
  Age        "How old are you?: 0..120
  MarStat    "What is your marital status?":
             (Never Mar "Never been married",
             Married   "Married",
             Separate  "Separated",
             Divorced  "Divorced",
             Widowed   "Widowed")
  Spouse     "What is your spouse's age?": 0..120
  Work       "Are you working for pay or profit?":(Yes,No)

RULES
  Sex
  IF Sex = Female THEN
     Birth
  ENDIF
  Age
  IF Age >= 17 THEN
     MarStat
     IF MarStat = Married) OR (MarStat = Separate) THEN
        Spouse
     ENDIF
     Work
  ENDIF

ENDMODEL
```

Figure 7.6 The sample questionnaire in Blaise.

through the questionnaire. There are two types of routing: *dynamic routing* and *static routing*.

- *Dynamic routing* means that one is forced to follow the route through the questionnaire as defined by the implemented branching and skipping instructions. One is always on the route as it was programmed by the developer. It is not possible to go to questions that are off the route. CAPI and CATI almost always apply dynamic routing.
- *Static routing* means that one has complete freedom to move to any question in the questionnaire, whether it is on or off the route. This is usually inappropriate for interviewing systems, but it is often applied when entering or editing data collected on paper forms (computer-assisted data input, CADI).

CAI software often has the possibility to include checks in interviewing programs. There can be range errors and consistency errors. A *range error* occurs if a given answer is outside the valid set of answers, for example, an age of 348 years. A *consistency error* indicates an inconsistency in the answers to a set of questions. An age of 8 years may be valid, a marital status "married" is not uncommon, but if the same person gives both answers, there is probably something wrong. To detect these types of errors, conditions can be specified that have to be satisfied. Checking behavior determines how these conditions are checked. There are two types of checking: *dynamic checking* and *static checking*.

- *Dynamic checking* means that all relevant conditions are immediately checked after an answer has been entered. This is usually appropriate in an interviewing situation.
- *Static checking* means that conditions are checked only after the program is instructed to do so, for example, by pressing a special function key. Static checking may be appropriate when entering data from paper forms. First, all data are copied from the form to the computer and then checks are activated.

Sometimes, it is also possible to set error reporting behavior of an interviewing program. It determines if and how errors are displayed to the interviewer or respondent:

- *Dynamic error reporting* means that a message is displayed on the screen immediately after an error has been encountered. The interviewer or respondent cannot continue with the interview. First, the problem has to be solved. This type of error reporting is often applied in CAI software.
- *Static error reporting* means that no immediate action is required when errors are detected. Questions involved in errors are marked. One can continue answering questions. Error messages can be viewed at any time by moving to a specific question and asking for error reports involving this question. This form of error reporting can be applied in data entry situations.

7.5 DATA COLLECTION WITH BLAISE

7.5.1 What is Blaise?

There are many software packages for survey data collection. One of these packages has more or less become a de facto standard in the world of data collection for official statistics. The name of this package is Blaise. The first version of the Blaise system was developed in 1986 by Statistics Netherlands. The aim was to tackle the disadvantages of traditional data collection with paper questionnaire forms. See Bethlehem (1997) for more background information.

The Blaise language is the basis of the Blaise system. This language is used to define a questionnaire. The Blaise questionnaire definition contains all possible questions, route instructions, checks to be carried out on the answers, and computations that may be needed in the course of the interview. Therefore, this definition can be seen as *metadata* definition. It describes the data to be collected. It acts as a knowledge base from which the system extracts information it needs for its various modules for data collection or data processing. Therefore, Blaise enforces consistency of data and metadata in all steps of the survey process.

Development of a Blaise survey starts with the questionnaire definition in the Blaise language. Such a definition is called a *data model* in Blaise. Once the data model has been entered in the system, it is checked for correctness. If so, it is translated into a *metadata file*. The most important module of the system is the *data entry program* (DEP). It is used for entering data. The DEP can do this in various ways depending on the data collection mode selected:

- The DEP can enable easy entry and correction of data that have been collected by means of traditional paper questionnaire forms. This data collection and processing mode is called *computer-assisted data input* in Blaise.
- The DEP can enable computer-assisted interviewing. It supports CAPI, CATI, and CAWI.

Blaise is not a data analysis package. However, it helps in preparing data and metadata for existing statistical analysis. The system contains a tool for generating system files for packages such as SPSS and Stata.

The first version of Blaise was released in 1986. It ran under the operating system MS-DOS. The first Windows version came on the market in 1999. In 2003, a version was released allowing online data collection. For more information about Blaise, see Statistics Netherlands (2002).

Blaise derives its name from the famous French theologian and mathematician Blaise Pascal (1623–1662). Pascal is famous not only for his contributions to science but also for the fact that his name was given to the well-known programming language. The Blaise language has its roots, for a large part, in this programming language.

7.5.2 A Simple Blaise Questionnaire

Here it is shown how a very simple questionnaire is created in the Blaise system. This example contains only a few basic features of the Blaise language. Figure 7.7 contains the questionnaire as it could have been designed in the traditional paper form.

The questionnaire contains only seven questions, and they are about work. There are a few things in this questionnaire worth mentioning. There are various types of questions. Questions 1, 2, and 5 are *closed questions*. An answer has to be selected from a list. Question 7 is also a closed question, but here more than one answer is allowed. Such a question is sometimes called a *check-all-that-apply question*. Furthermore, questions 3 and 4 are *numerical questions*. They require a number as answer. Finally, question 6 is an example of an *open question*. On an open question, any text is accepted as an answer.

The questionnaire contains *route instructions*. These instructions are necessary to prevent respondents from answering irrelevant questions. Route instructions appear in two forms. First, some questions are followed by jump instructions. For example, if a respondent is still going to school, no job description is required, so he skips to question 6. Second, there are written instructions for the interviewer, for example, "If male, then skip question 3."

DATA COLLECTION WITH BLAISE   **169**

```
THE NATIONAL COMMUTER SURVEY

1. Are you male or female?
      Male  ..........................................  1
      Female  ........................................  2

2. What is your marital status?
      Never married  ................................  1
      Married  ......................................  2
      Divorced  .....................................  3
      Widowed  ......................................  4

Interviewer: If male, then skip question 3.

3. How many children have you had?           ...... children

4. What is your age?                         ...... years

5. What is your main activity?
      Going to school  ..............................  1     → 7
      Working  ......................................  2     → 6
      Keeping house  ................................  3     → Stop
      Something else  ...............................  4     → Stop

6. Give a short description of your job
      ...........................................................................................

7. How do you travel to your work or school?
   (Check at most 3 answers)
      Public bus, tram or metro  ...................  1
      Train  ........................................  2
      Car or motor cycle  ..........................  3
      Bicycle  ......................................  4
      Walked  .......................................  5
      Other  ........................................  6
```

Figure 7.7 A simple paper questionnaire.

Figure 7.8 contains a possible definition of this questionnaire in the Blaise language. Some words, such as QUESTIONNAIRE and ENDIF, are printed in upper case. These words have a special meaning in the Blaise language. Their use is reserved for special situations and therefore are called reserved words. To emphasize this special meaning, they are printed in boldface and capitals. However, reserved words may also be typed in lowercase and normal face.

The first line of the questionnaire in Fig. 7.8 is the identification of the questionnaire. The end of the specification is indicated by the reserved word ENDQUEST. The questionnaire definition contains two sections: the *fields section* and the *rules section*. The fields section contains the definition of all questions to be asked (together with a description of what type of answer is expected). So, it defines the fields in the database to contain the survey data. The rules section defines what is to be done with the questions (fields). It contains the order of the questions, checks to be carried out on the answers, and computations.

```
DATAMODEL Commut "The National Commuter Survey"
FIELDS
  Gender   "Are you male or female?": (Male, Female)
  MarStat  "What is your marital status?":
    (NevMarr  "Never married",
     Married  "Married",
     Divorced "Divorced",
     Widowed  "Widowed")
  Children "How many children have you had?": 0..25
  Age "What is your age?": 0..120
  Activity "What is your main activity?":
    (School   "Going to school".
     Working  "Working",
     HousKeep "Keeping house",
     Other    "Something else")
  Descrip "Give a short description of your job": STRING[40]
  Travel "How do you travel to your school or work?":
    SET [3] OF
    (NoTravel "Do not travel, work at home",
     PubTrans "Public bus, tram or metro",
     Train    "Train",
     Car      "Car or motor cycle",
     Bicycle  "Bicycle",
     Walk     "Walk",
     Other    "Other means of transport")
RULES
  Gender MarStat
  IF Gender = Female THEN
    Children
  ENDIF
  Age Activity
  IF Activity = Working THEN
    Descrip
  ENDIF
  IF (Activity = Working) OR (Activity = School) THEN
    Travel
  ENDIF

  IF (Age < 15) "If age less than 15" THEN
    MarStat = NevMarr "he/she is too young to be married!"
  ENDIF

ENDMODEL
```

Figure 7.8 A simple Blaise questionnaire.

7.5.3 The Fields Section

All questions are defined in the *fields section*. This section starts with the reserved word FIELDS. Every question definition follows a simple scheme. It starts with *question name*. It identifies the question. The question name is followed by the *question text*. This is the text of the question to be presented to the respondents (on screen or on paper). The question text must be placed between quotes and must be followed by a colon. The last part of the question definitions is the *answer definition*. It describes the valid answers to that question.

The difference with traditional questionnaires is that the *question numbers* in Fig. 7.7 are replaced by question names in Fig. 7.8. So one does not talk about question 2 but about question "Age", and one refers to question "Travel" instead of question 6. It is important to identify questions by names instead of numbers. It improves the readability of the questionnaire, and problems are avoided in case questions have to be added or deleted.

The fields section in Fig. 7.8 introduces eight questions. The Blaise language offers different types of questions. It is possible to define an *open question*. For such a question, any text is accepted as an answer provided the length of the text does not exceed the specified maximum length. An example is the question Descrip. The answer may not be longer than 40 characters:

```
Descrip "Give a short description of your job": STRING [40]
```

A *numerical question* expects a number as an answer. This number must be in the specified range. The question Age is an example. The answer must be in the range from 0 to 120:

```
Age "What is your age?": 0..120
```

For a *closed question*, an answer must be picked from a specified list of answer options. The question Activity in Fig. 7.8 is an example of such a question :

```
Activity "What is your main activity?":
  (School "Going to school".
   Working "Working",
   HousKeep "Keeping house",
   Other "Something else")
```

Each possible answer is defined by a short answer name (e.g., School) and, optionally, a longer answer text (e.g., "Going to school"). The answer name is used internally to identify the answer. The answer text is presented to the respondent. Note that possible answers are identified by names instead of by numbers. Just like in the case of question names, using answer names improves readability and maintainability of questionnaire specifications.

Sometimes, the respondent must be allowed to select more than one answer from a list. For this case, the reserved words SET OF can be added to a closed question. Then the question becomes a check-all-that-apply question. Optionally, the maximum number of answers to be selected may be specified between square brackets.

The question `Travel` is such a closed question. At most, three options can be selected to answer this question:

```
Travel "How do you travel to your school or work?":
  SET [3] OF
  (NoTravel "Do not travel, work at home",
  PubTrans  "Public bus, tram or metro",
  Train     "Train",
  Car       "Car or motor cycle",
  Bicycle   "Bicycle",
  Walk      "Walk",
  Other     "Other means of transport")
```

There are two predefined answers that can always be given in response to a question of any type: DONTKNOW and REFUSAL. These answer possibilities do not have to be defined because they are implicitly available. They can be entered with special function keys.

This simple example does not exhaust all possible question types that can be used in Blaise. More information can be found in Statistics Netherlands (2002).

7.5.4 The Rules Section

The rules section starts with the keyword RULES. This section describes what the system must do with the questions. There are four types of rules:

- *Route instructions*. describe the order of the questions and the conditions under which they will be asked.
- *Checks*. determine whether a specified statement is true for the answers of the questions involved. If it is false, the system will generate an error message. Subsequent action depends on the application at hand. Two kinds of checks are supported. A check can detect a *hard error*. This is a real error that must be fixed before the form can be considered clean. A check can also detect a *soft error*. Such an error may point to a possible problem. Soft errors may be suppressed if it is decided there is nothing wrong.
- *Computations*. on the values of questions and other data can be used to determine the proper route through the questionnaire, to carry out complex checks, or to derive values of questions that are not asked or that are corrected.
- *Layout instructions*. determine the layout of the questions and the entry fields displayed on the screen.

The example shown in Fig. 7.8 contains a number of route instructions. Writing down the name of a question in the rules section means asking the question.

DATA COLLECTION WITH BLAISE   **173**

The rules section of the example starts with the two question names `Gender` and `MarStat`:

```
RULES
   Gender MarStat
```

The two questions will be processed in this order. Questions can also be asked, subject to a condition. The question `Children` will only be asked in the example if the answer `Female` has been given to the question `Gender`:

```
IF Sex = Female THEN
   Children
ENDIF
```

Checks are conditions that have to be satisfied. Checks are stated in terms of what the correct relationship between fields should be. An example from the questionnaire in Fig. 7.8 is as follows:

```
MarStat = NevMarr
```

The specification instructs the system to check whether the field `MarStat` has the value `NevMarr`. If not, an error message will be produced. A label, any text between double quotes, can be attached to a condition. Such a text will be used as an error message if the condition is not satisfied.

```
MarStat = NevMarr "he/she is too you to be married!"
```

Checks can be subject to conditions:

```
IF (Age > 15) THEN
   MarStat = NevMarr
ENDIF
```

The check `MarStat = NevMarr` will only be carried out if the answer to the question `Age` has a value less than 15. The application will reject entries in which people younger than 15 years are married.

The example in Fig. 7.8 does not contain any computation or layout instruction.

7.5.5 Dynamic Question Texts

It is important that respondents always understand the questions in the questionnaire. Sometimes, it helps to adapt the question text to the situation at hand. For paper questionnaires this is, of course, not possible, but computer-assisted interviewing software usually has this feature.

The Blaise system can adapt all question texts in the data model. This is accomplished by including the name of a question in the text of another question. The question name has to be preceded by a ^. When the system displays the text on the screen, the question name will be replaced by its answer.

For example, suppose the field section of a data model contains the questions `Travel` and `Reason` as defined below. Note that the name `Travel` is included both in the text of the question `Reason` and in the text of its answer:

```
FIELDS
  Travel
      "How do you travel to work?":
      (train, bus, metro, car, bicycle)
  Reason
      "Why do you go by ^Travel and not by car?":
      (Jam     "By ^Travel no traffic jams",
      Comfort  "Going by ^Travel is more comfortable",
      Environ  "Going by ^Travel is better for the
                environment",
      Health   "Going by ^Travel is healthier",
      NoCar "Does not have a car")
RULES
  Travel
  IF NOT (Travel = Car) THEN
     Reason
  ENDIF
```

Suppose, a CAPI interview is conducted, and the respondent gives the answer `bicycle` to the question `Travel`. Then the question `Reason` will be displayed as follows:

```
Why do you go by bicycle and not by car?
1:  By bicycle no traffic jams
2:  Going by bicycle is more comfortable
3:  Going by bicycle is better for the environment
4:  Going by bicycle is healthier
5:  Does not have a car
```

7.5.6 Subquestionnaires

Simple questionnaires all follow the scheme described above: first a set of questions is defined in the rules section. Then the order in which they will be asked is defined in the rules section. In many situations, however, questionnaires have a much more complex structure. First of all, some survey situations call for a hierarchical approach. Suppose a labor force survey is carried out. The questionnaire starts by asking a few general questions about the composition of the household. Then, members of the household are asked about their activities. People who work will be asked about their jobs, and those looking for work will be asked how they go about looking for work. In fact, a new questionnaire is filled in for each relevant household member. And it is not clear in advance how many of these subquestionnaires are needed.

There is another problem. If a comprehensive survey is to be conducted, the questionnaire will tend to become large and complex. In such a case, it is wise to take an approach commonly used in software development: the best way to build a large program is to analyze it in subproblems and solve each one of them separately. This approach makes it possible to build modular systems. Blaise is designed to stimulate such a structured design of large questionnaires. It is possible to distribute the questions over several subquestionnaires, keeping together questions that are logically related to each another. A simple questionnaire is designed for every group of questions. Next, the subquestionnaires are combined into one large questionnaire.

7.5.7 Integrated Survey Processing

The Blaise system promotes integrated survey processing. The Blaise language is, in fact, a metadata language. It is used to specify relevant information about the data to be collected and processed. The system is able to exploit this knowledge. It can automatically generate various data collection and data editing applications. Moreover, it can prepare data and metadata for other data processing software, for example, for adjustment weighting, tabulation, and analysis. This avoids having to specify the data more than once, each time in a different "language." It also enforces data consistency in all data processing steps.

The data model is the knowledge base of the Blaise system. It forms the backbone of an integrated survey processing system (see Fig. 7.9). The data model is created in the design phase of the survey. If data are to be collected by means of a paper questionnaire form, the form and the corresponding data entry and data editing program (CADI) can be generated from the data model. If data are to be collected by means of some form of computer-assisted interviewing, there is a choice for CAPI, CATI, CASI, or CAWI. Blaise can also be used for mixed-mode data collection. It can generate the data collection instruments for several modes simultaneously. Since all these instruments are generated from the same metadata source, consistency between modes is guaranteed.

Whatever form of data collection is used, the result will be a "clean" data file. The next step in the process will often be the computation of adjustment weights to correct a possible bias due to nonresponse (see also Chapters 9 and 10). The Blaise tool Bascula can take care of this. It is able to read the Blaise data files directly and extract the information about the variables (the metadata) from the Blaise specification. Running Bascula will cause an extra variable to be added to the data file containing the adjustment weight for each case.

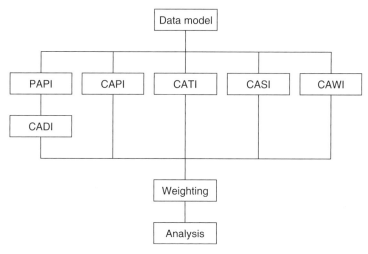

Figure 7.9 Integrated survey processing.

Now the data are ready for analysis. There are many packages available for this. Well-known examples are SAS, SPSS, and Stata. They all require a data file and a description of the data (metadata). Blaise has tools to put the data in the proper format (Manipula). Furthermore, there is a tool (Cameleon) to create setup files with data descriptions for a number of statistical packages.

EXERCISES

7.1 What was the effect on response rates of the introduction of computers for face-to-face interviewing (CAPI)?
   a. The response rates were lower than those of traditional face-to-face interviewing.
   b. The response rates were higher than those of traditional face-to-face interviewing.
   c. No significant changes in response rates were observed.
   d. Because of all other advantages of computer-assisted interviewing, effects on response rates have not been investigated.

7.2 Which of the following effects is not caused by a change from traditional face-to-face interviewing to CAPI?
   a. It is easier to follow the correct route through the questionnaire.
   b. Nonresponse rates decrease.
   c. Less time is required to process the survey data.
   d. It is possible to carry out complex checks on the answers to the questions.

EXERCISES **177**

7.3 To carry out a telephone survey among households, a sample of phone numbers is selected from a telephone directory. This procedure can lead to
   a. overcoverage, but not to undercoverage;
   b. undercoverage, but not to overcoverage;
   c. both to overcoverage and undercoverage;
   d. all kinds of problems (such as nonresponse) but not to overcoverage or undercoverage.

7.4 Design a small questionnaire in the Blaise language. The questionnaire must contain four questions:
   - Ask whether the respondent has a PC at home.
   - If so, ask the respondent whether there is Internet access at home.
   - If so, ask whether the Internet is used for e-mail and/or surfing on the World Wide Web.
   - Ask which browser is used (Internet Explorer, Firefox, etc.).
   - Pay attention to question texts, question types, and routing.

7.5 Which of the following statements about random digit dialing is not true?
   a. Also people with nonlisted phone numbers can be contacted.
   b. There are no coverage problems.
   c. There is no information about people not answering the call.
   d. It may be impossible to distinguish nonexisting numbers from nonresponse due to "not at home."

CHAPTER 8

The Quality of the Results

8.1 ERRORS IN SURVEYS

When conducting a survey, a researcher is confronted with all kinds of phenomena that may have a negative impact on the quality, and therefore the reliability, of the outcomes. Some of these disturbances are almost impossible to prevent. So, efforts will have to be aimed at reducing their impact as much as possible. Notwithstanding all these efforts, final estimates of population parameters may be distorted. All phenomena causing these distortions are called *sources of error*. The impact these together have on the estimates is called the *total error*.

Sources of error will, if present, increase the uncertainty with respect to the correctness of estimates. This uncertainty can manifest itself in two ways in the distribution of an estimator: (1) it can lead to a systematic deviation (bias) from the true population value or (2) it can increase the variation around the true value of the population parameter.

Let Z be a population parameter that has to be estimated and let z be an estimator that is used for this purpose. Chapter 6 discussed the properties of a good estimator. One was that an estimator must be *unbiased*. This means its expected value must be equal to the value of the population parameter to be estimated:

$$E(z) = Z. \tag{8.1}$$

If an estimator is not unbiased, it is said to have a bias. This bias is denoted by

$$B(z) = E(z) - Z. \tag{8.2}$$

Another desirable property of an estimator is that its variance is as small as possible. This means that

$$V(z) = E(z - E(z))^2 \tag{8.3}$$

must be small. An estimator with a small variance is called *precise*.

Applied Survey Methods: A Statistical Perspective, Jelke Bethlehem
Copyright © 2009 John Wiley & Sons, Inc.

A precise estimator may still be biased. Therefore, just the value of the variance itself is not a good indicator of how close estimates are to the true value. A better indicator is the *mean square error*. This quantity is defined by

$$M(z) = E(z - Z)^2. \tag{8.4}$$

It is the expected value of the squared difference of the estimator from the value to be estimated. Writing out this definition leads to a different expression for the mean square error:

$$M(z) = V(z) + B^2(z). \tag{8.5}$$

Now, it is clear that the mean square error contains both sources of uncertainty: a variance component and a bias component. The mean square error of an estimator is equal to its variance if it is unbiased. A small mean square error can be achieved only if both the variance and the bias are small. Figure 8.1 distinguishes four different situations that may be encountered in practice.

The distribution on the upper left side is the ideal situation of an estimator that is precise and unbiased. All possible outcomes are close to the true value, and there is no systematic overestimation or underestimation. The situation in the lower left graph is less attractive. The estimator is still unbiased but has a substantial variance. Hence, confidence intervals will be wider. Reliability is not affected. The confidence level of a 95% confidence interval remains 95%. The situation is completely different for the graph in the upper right corner. The estimator is precise but has a substantial bias. As a result, a confidence interval computed using the survey data would almost certainly not

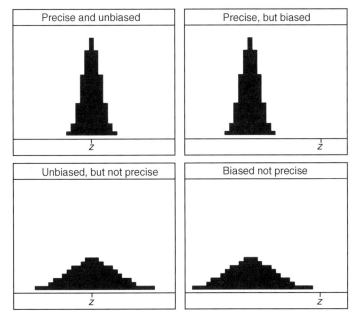

Figure 8.1 The relation between total error, bias, and precision.

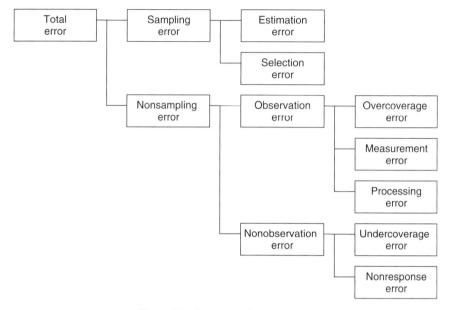

Figure 8.2 Taxonomy of survey errors.

contain the true value. The confidence level is seriously affected. Estimates will be unreliable. Wrong conclusions will be drawn. The graph in the lower right corner offers the highest level of uncertainty. The estimator is biased and moreover it is also not precise. This is the situation in which the mean square error has its largest value.

Survey estimates will never be exactly equal to the population characteristics they intend to estimate. There will always be some error. This error can have many causes. Bethlehem (1999) describes taxonomy of possible causes. It is reproduced in Fig. 8.2. The taxonomy is a more extended version of one given by Kish (1967).

The ultimate result of all these errors is a discrepancy between the survey estimate and the population parameter to be estimated. Two broad categories of phenomena can be distinguished contributing to this total error: sampling errors and nonsampling errors.

Sampling errors are introduced by the sampling design. They are due to the fact that estimates are based on a sample and not on a complete enumeration of the population. Sampling errors vanish if the complete population is observed. Since only a sample is available for computing population characteristics, and not the complete data set, one has to rely on estimates. The sampling error can be split into a selection error and an estimation error.

The *estimation error* denotes the effect caused by using a probability sample. Every new selection of a sample will result in different elements and thus in a different value of the estimator. The estimation error can be controlled through the sampling design. For example, the estimation error can be reduced by increasing the sample size or by taking selection probabilities proportional to the values of some well-chosen auxiliary variable.

A *selection error* occurs when wrong selection probabilities are used in the computation of the estimator. For example, true selection probabilities may differ from anticipated selection probabilities if elements have multiple occurrences in the sampling frame. Selection errors are hard to avoid without thorough investigation of the sampling frame.

Nonsampling errors may even occur if the whole population is investigated. They denote errors made during the process of obtaining answers to questions asked. Nonsampling errors can be divided into observation errors and nonobservation errors.

Observation errors are one form of nonsampling errors. They refer to errors made during the process of obtaining and recording answers. An *overcoverage error* is caused by elements that are included in the survey but do not belong to the target population. A *measurement error* occurs when respondents do not understand a question or do not want to give the true answer, or if the interviewer makes an error in recording the answer. In addition, interview effects, question wording effects, and memory effects belong to this group of errors. A measurement error causes a difference between the true value and the value processed in the survey. A *processing error* denotes an error made during data processing, for example, data entry.

Nonobservation errors are made because the intended measurements cannot be carried out. *Undercoverage* occurs when elements of the target population do not have a corresponding entry in the sampling frame. These elements can and will never be contacted. Another nonobservation error is *nonresponse* error when elements selected in the sample do not provide the required information.

The taxonomy discussed above makes it clear that a lot can go wrong during the process of collecting survey data, and usually it does. Some errors can be avoided by taking preventive measures at the design stage. However, some errors will remain. Therefore, it is important to check collected data for errors, and when possible, to correct these errors. This activity is called *data editing*. Data editing procedures are not able to handle every type of survey error. They are most suitable for detecting and correcting measurement errors, processing errors, and possibly overcoverage. Phenomena such as selection errors, undercoverage, and nonresponse require a different approach. This approach often leads to the use of adjustment weights in estimation procedures, and not to the correction of individual values in records.

8.2 DETECTION AND CORRECTION OF ERRORS

A survey is a fallible instrument, subject to many forms of bias and error. Data editing is one means of controlling and reducing survey errors, especially those arising from the interchange between respondents and interviewers, or between respondents and self-administered forms, during the data collection process.

Data editing is the process of detecting errors in survey data and correcting the detected errors, whether these steps take place in the interview or in the survey office after data collection. Traditionally, statistical organizations, especially those in government, have devoted substantial amounts of time and major resources to data editing in the belief that this was a crucial process in the preparation of accurate statistics. Current data editing tools have become so powerful that questions are now raised as to whether too much data editing occurs. A new objective for some is to minimize the amount of data editing performed while guaranteeing a high level of data quality.

Data editing taking place at the level of individual forms is called *microediting*. Questionnaire forms are checked and corrected one at a time. The values of the variables in a form are checked without using information in other forms. Microediting typically is an activity that can take place during the interview or during data capture.

Data editing taking place at the level of aggregated quantities, obtained by using all available cases, is called *macroediting*. Macroediting requires a file of records. This means it is typically an activity that takes place after data collection, after data entry, and possibly after microediting. According to Pierzchala (1990), data editing can be seen as addressing four principal types of data errors:

- *Completeness Errors*. The first thing to be done when filled-in forms come back to the survey agency is to determine whether they are complete enough to be processed. Forms that are blank or unreadable, or nearly so, are unusable. They can be treated as cases of unit nonresponse (see Chapter 9), scheduled for callback, deleted from the completed sample, imputed in some way (see Section 8.3), depending on the importance of the case.
- *Domain Errors*. Each question has a domain (or range) of valid answers. An answer outside this domain is considered an error. Such an error can easily be detected for numeric questions, since domain errors are defined as any answer falling outside the allowable range. For questions asking for values or quantities, it is sometimes possible to specify improbable as well as impossible values. For example, if the age of respondents is recorded, a value of 199 would certainly be unacceptable. A value of 110 is unlikely but not impossible. For a closed question, the answer has to be chosen from a list (or range) of alternatives. The error may consist of choosing no answer, more answers than allowed, or an answer outside the allowable range. For open questions, the domain imposes no restrictions. Any text is accepted as an answer.
- *Consistency Errors*. Consistency errors occur when the answers to two or more questions contradict each other. Each question may have an answer in its valid domain, but a combination of answers may be impossible or unacceptable. A completed questionnaire may report a person as being an employee or less than 5 years of age, but the combination of these answers for the same person is probably an error. For instance, a firm known to have 10 employees should not report more than 10,000 person-days worked in the past year. Consistency errors usually occur for combinations of closed questions and/or numeric questions.

    When a consistency error is detected, the answer causing the error is not always obvious. A correction may be necessary in one, two, or more questions. Moreover, resolving one inconsistency may produce another. So, it is easier to detect consistency errors than to solve them.

- *Routing Errors (Skip Pattern Errors)*. Many questionnaires contain routing instructions. These instructions specify conditions under which certain questions must be answered. In most cases, closed and numeric questions are used in these instructions. In paper questionnaires, routing instructions usually take the form of skip instructions attached to the answers of questions, or of printed instructions to the interviewer. Routing instructions ensure that all applicable questions are asked, while inapplicable questions are omitted.

    A routing error occurs when an interviewer or respondent fails to follow a route instruction, and a wrong path is taken through the questionnaire. Routing errors are also called *skip pattern errors*. As a result, the wrong questions are answered or applicable questions are left unanswered.

When errors are detected in completed questionnaire forms, they have to be corrected. One obvious way to accomplish this is to recontact the respondents and confront them with the errors. They may then provide the correct answers. Unfortunately, this approach is not feasible in daily practice. Respondents consider completing a questionnaire form already a burden in the first place. Having to reconsider their answers would in most cases lead to a refusal to do so. Moreover, this approach is time consuming and costly. Therefore, survey agencies rely on other techniques to deal with errors in the collected data. They start with a survey data file in which all wrong answers are removed, so that the corresponding questions are considered to be unanswered.

It should be noted that analyzing a survey data set with missing data items is not without risks. First, the "holes" in the data set may not be missing at random (MAR). If data are missing in some systematic way, the remaining data may not properly reflect the situation in the target population. Second, many statistical analysis techniques are not able to properly cope with missing data and therefore may produce misleading results. Some techniques even require all data to be there and interpret codes for missing values as real values.

There are two approaches that ignore missing data in the statistical analysis. The first one is called *list-wise deletion*. This approach simply omits all records from the analysis in which at least one value is missing. So, only complete records are used. Application of list-wise deletion assumes the remaining records to be a random sample from all records. So, there are no systematic differences between the records with missing data and the complete records. Unfortunately, this is often not the case in practice. It is not uncommon that specific groups in the population have problems answering a question. Moreover, a consequence of this approach is that a lot of information is not used. The whole questionnaire form is thrown away if the answer to just one question is missing.

Another, less drastic approach is *pair-wise deletion*. To compute a statistical quantity all records are used for which the relevant variables have a value. This means that different sets of records may be used for different statistics.

A simple example shows the effects of list-wise deletion and pair-wise deletion. Table 8.1 contains a small data set, containing the values of three variables X, Y, and Z for four records.

Table 8.1 A Data Set with Missing Values

Record	X	Y	Z
1	4	4	5
2	5	–	6
3	–	5	4
4	6	6	–

The data set contains a missing value in three different records. Suppose, the objective is to compute the correlation between the two variables X and Y. Application of list-wise deletion would lead to omitting three records, including record 4 that contains values for both X and Y. Since only one record remains, it is not possible to compute the correlation coefficient.

The correlation coefficient R_{XY} is defined as $R_{XY} = S_{XY}/(S_X \times S_Y)$, where S_{XY} is the covariance between X and Y, and S_X and S_Y are the respective standard deviations. Computation of the covariance requires records with the values of both X and Y. Records 2 and 4 can be used for this. This results in $S_{XY} = 2$. Computation of S_X requires records with a value for X. There are three such records (1, 2, and 4), resulting in $S_X = 1$. Likewise, records 1, 3, and 4 can be used for S_Y. This gives $S_Y = 1$. Hence, the correlation coefficient is equal to $R_{XY} = 2/(1 \times 1) = 2$. Since, by definition, the correlation coefficient is constrained to the interval $[-1, +1]$, this value is impossible! The cause of this inconsistency is that computation of the various components is based on different sets of records.

The approach probably most often applied in practical situations is *imputation*. A wrong or missing value is replaced by a *synthetic value*. This synthetic value is the outcome of a technique that attempts to predict the unknown value as accurately as possible using the available information. Imputation techniques are discussed in more detail in Section 8.3.

Checks for domain errors involve only one question at the time. In case an error is detected, it is clear which question is causing this error and therefore which answer must be corrected. The situation is different with respect to checks for consistency and routing errors. Several questions are usually involved in such checks. If an error is detected, it will often not be clear which question caused the error. The answer to one question can be wrong, but it is also possible that there are errors in the answers to more questions. Without more information it is often impossible to detect the source of the error. Fellegi and Holt (1976) have developed a theory to solve this problem. Their theory is based on the principle that the values of the variables in a record should be made to satisfy all checks by changing the fewest possible number of values. The number of synthetic values should be as small as possible. Real data should be preferred over synthetic data. It is assumed that, generally, errors are rare and therefore it must be possible to get rid of errors by changing a few data values. Consequently, a useful rule of thumb is to first locate the variable that is involved in many errors in a record. Changing the value of just that variable may cause many errors to disappear.

8.3 IMPUTATION TECHNIQUES

To avoid missing data problems, often some kind of imputation technique is applied. Imputation means that missing values are replaced by synthetic values. This synthetic value is obtained as the result of some technique that attempts to estimate the missing values. After applying an imputation technique, there are no more "holes" in the data set. So, all analysis techniques can be applied without having to worry about missing values. However, there is a downside to this approach.

There is no guarantee that an imputation technique will reduce a bias that may have been caused by the missing data. It depends on the type of missing data pattern and the specific imputation technique that is applied. Three types of missing data mechanisms can be distinguished. Let X represent a set of auxiliary variables that are completely observed and let Y be a target variable of which some values are missing. The variable Z represents causes of missingness unrelated to X and Y, and the variable R indicates whether or not a value of Y is missing.

In case of *missing completely at random* (MCAR), missingness is caused by a phenomenon Z that is completely unrelated to X and Y. Estimates for parameters involving Y will not be biased. Imputation techniques will not change this.

In case of *missing at random* (MAR), missingness is caused partly by an independent phenomenon Z and partly by the auxiliary variable X. So, there is an indirect relationship between Y and R. This leads to biased estimates for Y. Fortunately, it is possible to correct such a bias by using an imputation technique that takes advantage of the availability of all values of X, both for missing and for nonmissing values of Y.

In case of *not missing at random* (NMAR), there may be a relationship between Z and R and between X and R, but there is also a direct relationship between Y and R that cannot be accounted for X. This situation also leads to biased estimates for Y. Unfortunately, imputation techniques using X are not able to remove the bias.

There are many imputation techniques available. A number of them are described in this chapter. These are all *single-imputation* techniques. This means that a missing value is replaced by one synthetic value. Another approach is *multiple imputation*. This technique replaces a missing value by a set of synthetic values. A summary of technique is given at the end of this section.

8.3.1 Single-Imputation Techniques

Assuming sampling without replacement, the sample is represented by the set of indicators $a_1, a_2, \ldots, a_N$. The value of the indicator a_k is equal to 1 if element is selected in the sample, otherwise it is equal to 0.

Let Y be the target variable for which some values are missing in the sample. Missingness is denoted by the set of indicators $R_1, R_2, \ldots, R_N$. Of course, $R_k = 0$ if $a_k = 0$. A missing value of a sampled element k is indicated by $a_k = 1$ and $R_k = 0$.

Let X be an auxiliary variable for which no missing values occur in the sample. So, the value X_k is always available if $a_k = 1$.

Sometimes, the value of a missing item can be logically deduced with certainty from the nonmissing values of other variables. This is called *deductive imputation*.

If strict rules of logic are followed, this technique has no impact on the properties of the distribution of estimators.

For example, if we know a girl is 5 years old, we can be certain she has had no children. Likewise, if a total is missing but the subtotals are not missing the total can easily be computed.

Although deductive imputation is the ideal form of imputation, it is frequently not possible to apply it.

Imputation of the mean implies that a missing value of a variable is replaced by the mean $\bar{y}_R$ of the available values of this variable. Let k be an element in the sample for which the value Y_k is missing. The imputed value is defined by

$$\hat{Y}_k = \bar{y}_R = \frac{\sum_{k=1}^{N} a_k R_k Y_k}{\sum_{k=1}^{N} a_k R_k}. \qquad (8.6)$$

Since all imputed values will be equal to the same mean, the distribution of this variable in the completed data set will be affected. It will have a peak at the mean of the distribution.

For imputation of the mean within groups, the sample is divided into a number of nonoverlapping groups. Qualitative auxiliary variables are used for this. Within a group, a missing value is replaced by the mean of the available observations in that group.

Imputation of the mean within groups will perform better than imputation of the mean if the groups are homogeneous with respect to the variable being imputed. Since all values are close to each other, the imputed group mean will be a good approximation of the true, but unknown, value.

Random imputation means that a missing value is replaced by a value that is randomly chosen from the available values for the variable. The set of available values is equal to

$$\{Y_k | a_k = 1 \wedge R_k = 1\}. \qquad (8.7)$$

This imputation is sometimes also called *hot-deck imputation*. It is a form of *donor imputation*: a value is taken from an existing record where the value is not missing.

The distribution of the values of the variable for the complete data set will look rather natural. However, this distribution does not necessarily resemble the true distribution of the variable. Both distributions may differ if the missing values are not randomly missing.

Random imputation within groups divides the sample into a number of nonoverlapping groups. Qualitative auxiliary variables are used to create these groups. Within a group, a missing value is replaced by a randomly chosen value from the set of available values in that group.

Random imputation within groups will perform better than random imputation if the groups are homogeneous with respect to the variable being imputed. Since all values are close to each other, the randomly selected value will be a good approximation of the true, but unknown, value.

IMPUTATION TECHNIQUES

The idea of *nearest neighbor imputation* is that a record is located in the data set that resembles as much as possible the record in which a value is missing. Some kind of distance measure is defined to compare records on the basis of values of auxiliary variables that are available for all records.

If all auxiliary variables are of a quantitative nature, some kind of Euclidean distance may be used. Suppose there are p such variables. Let X_{kj} be the value of variable X_j for element k, for $k = 1, 2, \ldots, N$ and $j = 1, 2, \ldots, p$. Then, the distance between the records of two elements i and k could be defined by

$$D_{ik} = \sqrt{\sum_{j=1}^{p}(X_{ij} - X_{kj})^2}. \tag{8.8}$$

Let k be an element in the sample for which the value Y_k is missing. The imputed value is copied from the record of a sampled element i with the smallest distance D_{ki}, and for which the value of Y is available.

Ratio imputation assumes a relationship between the target variable Y (with missing values) and an auxiliary variable X (without missing values). If this relationship is (approximately) of the form $Y_k = B \times X_k$, for some constant B, then a missing value of Y for element k can be estimated by $B \times X_k$. If the value of B is not known, it can be estimated using the available data by

$$b = \frac{\sum_{k=1}^{N} a_k R_k Y_k}{\sum_{k=1}^{N} a_k R_k X_k}. \tag{8.9}$$

Let k be an element in the sample for which the value Y_k is missing. The imputed value is defined by

$$\hat{Y}_k = bX_k. \tag{8.10}$$

Ratio estimation is often used when the same variable is measured at two different moments in time in a longitudinal survey.

Regression imputation assumes a relationship between the target variable Y (with missing values) and an auxiliary variable X (without missing values). If this relationship is (approximately) of the form $Y_k = A + B \times X_k$, for some constants A and B, then a missing value of Y for element k can be estimated by $A + B \times X_k$.

If the values of A and B are not known, they can be estimated by applying ordinary least squares on the available data by

$$b = \frac{\sum_{k=1}^{N} a_k R_k (Y_k - \bar{y}_r)(X_k - \bar{x}_r)}{\sum_{k=1}^{N} a_k R_k (X_k - \bar{x}_r)^2} \tag{8.11}$$

and

$$a = \bar{y}_r - b\bar{x}_r. \tag{8.12}$$

Let k be an element in the sample for which the value Y_k is missing. The imputed value is defined by

$$\hat{Y}_k = a + bX_k. \tag{8.13}$$

The regression model above contains only one auxiliary variable. Of course it is possible to include more variables in the regression models. This will often increase the explanatory power of the model, and therefore imputed values will be closer to the true (but unknown) values.

At first sight, all single-imputation techniques mentioned above seem rather different. Nevertheless, almost all of them fit in a general model. Let k be an element in the sample for which the value Y_k is missing. The imputed value is defined by

$$\hat{Y}_k = B_0 + \sum_{j=1}^{p} B_j X_{kj} + E_k, \tag{8.14}$$

where X_{kj} denotes the value of auxiliary variable X_j for element k, $B_0, B_1, \ldots, B_p$ are regression coefficients, and E_k is a random term the nature of which is determined by the specific imputation technique.

By taking B_0 equal to the mean of the available values of Y, and setting the other coefficients B_j and E_k equal to 0, the model reduces to imputation of the mean.

If the auxiliary variable $X_1, X_2, \ldots, X_p$ are taken to be dummy variables that indicate to which group an element belongs ($X_{kj} = 1$ if element k is in group j, and otherwise $X_{kj} = 0$), $B_0 = 0$ and $E_k = 0$, then (8.14) is equal to imputation of the group mean.

Model (8.14) reduces to random imputation if the model for imputation of the mean is used, but a random term E_k is added. Its value is obtained by a random drawing from the set of values

$$\bar{y}_R - Y_k \tag{8.15}$$

for which $a_k = 1$ and $R_k = 1$.

Random imputation within groups is obtained by adding a random term E_k to the model for imputation of the group mean. The value of E_k is a random drawing from a set of values. These values are obtained by subtracting the available values from their respective group means.

It is clear that ratio imputation and regression imputation also are special cases of model (8.14). Nearest neighbor imputation does not fit in this model.

8.3.2 Properties of Single Imputation

There are many single-imputation techniques. So, the question may arise which technique to use in a practical situation. There are several aspects that may play a role in this decision. A number of these effects are discussed in this section.

IMPUTATION TECHNIQUES

The first aspect is the type of variable for which missing values have to be imputed. In principle all mentioned imputation techniques can be applied for quantitative variables. However, not every single-imputation technique can be used for qualitative variables. A potential problem is that the synthetic value produced by the imputation technique does not necessarily belong to the domain of valid values of the variable. For example, if the variable gender has to be imputed, mean imputation produces an impossible value (what is the mean gender?). Therefore, only some form of "donor imputation" is applicable for qualitative variables. These techniques always produce "real" values.

Single-imputation techniques can be divided into two groups. One contains deterministic imputation techniques and the other random imputation techniques. The random term E_k in model (8.14) is zero for deterministic techniques and not for random techniques.

For some deterministic imputation techniques (e.g., imputation of the mean), the mean of a variable before imputation is equal to the mean after imputation. This shows that not every imputation technique is capable of reducing a bias caused by missingness. For random imputation techniques, the mean before imputation is never equal to the mean after imputation. However, expected values before and after imputation may be equal.

Deterministic imputation may affect the distribution of a variable. It tends to produce synthetic values that are close to the center of the original distribution. The imputed distribution is more "peaked." This may have unwanted consequences. Estimates of standard errors may turn out to be too small. A researcher using the imputed data (not knowing that the data set contains imputed values) may get the impression that his estimates are very precise, while in reality this is not the case.

The possible effects of imputation on estimators are explored by analyzing two single-imputation techniques in some more detail: imputation of the mean and random imputation.

8.3.3 Effects of Imputation of the Mean

Imputation of the mean replaces missing values by the mean of the available values. Let $Y_1, Y_2, \ldots, Y_N$ be the values of the variable to be imputed. A sample of size n without replacement is denoted by the set of indicators $a_1, a_2, \ldots, a_N$. Missingness is indicated by the set of indicators $R_1, R_2, \ldots, R_N$, where $R_k = 1$ only if k is in the sample ($a_k = 1$) and the value Y_k is available. Of course, $R_k = 0$ if $a_k = 0$. A missing value for a sampled element k is indicated by $a_k = 1$ and $R_k = 0$. The number of available observations is denoted by

$$m = \sum_{k=1}^{N} a_k R_k. \tag{8.16}$$

The mean of the available observations is equal to

$$\bar{y}_R = \frac{1}{m} \sum_{k=1}^{N} a_k R_k Y_k. \tag{8.17}$$

In case of imputation of the mean, a missing observation is replaced by the mean of the available values. So, if $a_k = 1$ and $R_k = 0$ for some element k, the imputed value

$$\hat{Y}_k = \bar{y}_R \qquad (8.18)$$

is used. Now a new estimator for the population mean is obtained by taking the average of all "real" and all "synthetic" values of Y. Let

$$\bar{y}_I = \frac{1}{n-m} \sum_{k=1}^{N} a_k (1 - R_k) \hat{Y}_k \qquad (8.19)$$

denotes the mean of all imputed values. Consequently, the mean after imputation can be written as

$$\bar{y}_{IMP} = \frac{m\bar{y}_R + (n-m)\bar{y}_I}{n}. \qquad (8.20)$$

In case of imputation of the mean, expression (8.19) reduces to

$$\bar{y}_I = \frac{1}{n-m} \sum_{k=1}^{N} a_k (1 - R_k) \hat{Y}_k = \frac{1}{n-m} \sum_{k=1}^{N} a_k (1 - R_k) \bar{y}_R = \bar{y}_R \qquad (8.21)$$

and therefore,

$$\bar{y}_{IMP} = \frac{m\bar{y}_R + (n-m)\bar{y}_I}{n} = \frac{m\bar{y}_R + (n-m)\bar{y}_R}{n} = \bar{y}_R. \qquad (8.22)$$

The conclusion can be drawn that the mean after imputation is equal to the mean before imputation. Imputation does not affect the mean.

To determine the characteristics of an estimator after imputation, it should be realized that two different types of probability mechanisms may play a role. Of course, there always is the probability mechanism of the sample selection. An extra source of randomness may be introduced by the imputation technique, for example, if a randomly selected value is imputed. To take this into account, the expected value of an estimator (after imputation) is determined with the expression

$$E(\bar{y}_{IMP}) = E_S E_I(\bar{y}_{IMP}|S). \qquad (8.23)$$

E_I denotes the expectation over the imputation distribution and E_S the expectation over the sampling distribution S. This is applied to the case of imputation of the mean. Given the sample, the estimator is a constant. So taking the expectation over the imputation distribution results in the same constant. Hence, the expected values of the estimators before and after imputation are the same. Imputation of the mean will not be able to reduce a possibly existing bias due to missingness.

To compute the variance of the estimator after imputation of the mean, the expression

$$V(\bar{y}_{IMP}) = V_S E_I(\bar{y}_{IMP}|S) + E_S V_I(\bar{y}_{IMP}|S) \qquad (8.24)$$

is used. Given the sample, estimator (8.20) is a constant. This means the second term in expression (8.24) is equal to 0. The first term is equal to the variance of the estimator

IMPUTATION TECHNIQUES

before imputation. Consequently, imputation of the mean does not change the variance of the estimator.

Problems may arise when an unsuspecting researcher attempts to estimate the variance of estimators, for example, for constructing a confidence interval. To keep things simple, it is assumed the available observations can be seen as a simple random sample without replacement, that is, missingness does not cause a bias. Then the variance after imputation is equal to

$$V(\bar{y}_{\text{IMP}}) = V(\bar{y}_R) = \frac{1 - (m/N)}{m} S^2 \tag{8.25}$$

in which S^2 is the population variance.

It is a well-known result from sampling theory that in case of a simple random sample without replacement, the sample variance s^2 is an unbiased estimator of the population variance S^2. This also holds for the situation before imputation: the s^2 computed using the m available observations is an unbiased estimator of S^2.

What would happen if an attempt would be made to estimate S^2 using the complete data set, without knowing that some values have been imputed? The sample variance would be computed, and it would be assumed that this quantity is an unbiased estimator of the population variance. However, this is not the case. For the sample variance of the imputed data set, the following expression holds:

$$\begin{aligned} s_{\text{IMP}}^2 &= \frac{1}{n-1} \left(\sum_{k=1}^{N} a_k R_k (Y_k - \bar{y}_R)^2 + \sum_{k=1}^{N} a_k (1 - R_k)(\hat{Y}_k - \bar{y}_R)^2 \right) \\ &= \frac{1}{n-1} \left(\sum_{k=1}^{N} a_k R_k (Y_k - \bar{y}_R)^2 + 0 \right) = \frac{m-1}{n-1} s^2. \end{aligned} \tag{8.26}$$

Hence,

$$E(s_{\text{IMP}}^2) = \frac{m-1}{n-1} S^2. \tag{8.27}$$

This is not an unbiased estimator of the population variance. The population variance is underestimated. One gets the impression that estimators are very precise, whereas in reality this is not the case. So someone analyzing imputed data runs a substantial risk of drawing wrong conclusions from the data. This risk is larger as there are more imputed values.

Imputation has an impact also on the correlation between variables. Suppose the variable Y is imputed using imputation of the mean. And suppose the variable X is completely observed for the sample. It can be shown that in this case the correlation after imputation is equal to

$$r_{\text{IMP},X,Y} = \sqrt{\frac{m-1}{n-1}} r_{XY} \tag{8.28}$$

where r_{XY} is the correlation in the data set before imputation. So, the more observations are missing for Y, the smaller the correlation coefficient will be. Researchers not aware of their data set having been imputed will obtain the impression that relationships between variables are weaker than they are in reality. In addition, there is a risk of drawing wrong conclusions.

8.3.4 Effects of Random Imputation

The effects of imputation mentioned in the previous section are less severe when random imputation is applied. Random imputation means that a missing value is replaced by a value that is randomly chosen from the available values for the variable. Usually, synthetic values are selected by means of drawing values without replacement. This is, of course, impossible if there are more missing values than nonmissing values.

The estimator

$$\bar{y}_{\text{IMP}} = \frac{m\bar{y}_R + (n-m)\bar{y}_I}{n} \qquad (8.29)$$

is now composed of two random terms: the mean of the m "real" observations and the mean of the $n-m$ "synthetic" observations.

Given the sample, the expected value of the mean of the synthetic values over the imputation distribution is equal to the mean of the real values. If applied to expression (8.23), the conclusion can be drawn that the expected value of the mean after imputation is equal to the expected value of the mean before imputation. Imputation does not change the expected value of the estimator.

The computation of the variance of the estimator is now a little bit more complex, because the second term in expression (8.24) is not any more equal to zero. The variance turns out to be equal to

$$V(\bar{y}_{\text{IMP}}) = \frac{1-(m/N)}{m}S^2 + \frac{(n-m)(2m-n)}{n^2 m}S^2. \qquad (8.30)$$

Apparently, random imputation increases the variance of the estimator. The variance consists of two components: the first one is contributed by the sampling design and the second one is contributed by the imputation mechanism.

What happens if a researcher is unaware of the fact that random imputation has been carried out? He computes the sample variance s^2 using the complete data set, and he assumes this is an unbiased estimator of the population variance S^2. This assumption is wrong in case of imputation of the mean. In case of random imputation, it can be shown that

$$E(s^2) = S^2\left(1 + \frac{2m-n}{n(n-1)}\right). \qquad (8.31)$$

So, s^2 is not an unbiased estimator of S^2, but for large samples the bias will be small. Therefore, s^2 is an asymptotically unbiased estimator.

It should be noted that random imputation affects the value of correlation coefficients. These values will generally be too low when computed using the imputed data set. This is caused by the fact that imputed values are randomly selected without taking into account possibly existing relationships with other variables. This phenomenon can also be encountered when applying other single-imputation techniques.

8.3.5 Multiple Imputation

Single imputation can be seen as a technique that solves the missing data problem by filling the holes in the data set by plausible values. When analyzing data, one is not bothered any more by missing data. This is clearly an advantage. However, there are also disadvantages. When a single-imputation technique is applied in a naïve way, it may create more problems than it solves. It was shown in the previous section that single imputation may distort the distribution of an estimator. Therefore, there is a serious risk of drawing wrong conclusions from the data set. More details about this aspect of imputation can be found in, for example, Little and Rubin (1987).

To address the problems caused by single-imputation techniques, Rubin (1987) proposed *multiple imputation*, a technique in which each missing value is replaced by $m > 1$ synthetic values. Typically, m is small, say 3–10. This leads to m complete data sets. Each data set is analyzed by using standard analysis techniques. For each data set, an estimate of a population parameter of interest is obtained. The m estimates for a parameter are combined to produce estimates and confidence intervals that incorporate missing data uncertainty.

Rubin (1987) developed his multiple-imputation technique primarily for solving the missing data problem in large public use sample survey data files and censuses files. With the advent of new computational methods and software for multiple imputation, this technique has become increasingly attractive for researchers in other sciences confronted by missing data (see also Schafer, 1997).

Multiple imputation assumes some kind of model. This model is used to generate synthetic values. Let Y be the variable of which some values are missing, and let $X_1, X_2, \ldots, X_p$ be variables that have been observed completely. The imputation model for a quantitative variable Y will often be some regression model like

$$\hat{Y}_k = B_0 + \sum_{j=1}^{p} B_j X_{kj} + E_k. \tag{8.32}$$

A loglinear model can be used for qualitative variables. For the sake of convenience, this overview will consider only quantitative variables.

The effects of imputation depend on the missing data mechanism that has generated the missing values. The most convenient situation is missing completely at random. This means that missingness happens completely at random. It is not related to any factor, known or unknown. Missingness does not cause a bias in estimates for Y. In this case, multiple synthetic drawings can be generated by means of applying random imputation a number of times. It is also possible to use imputation of the mean if the variation is modeled properly. For example, this can be done by adding a random component to the mean that has been drawn from a normal distribution with the proper variance.

MCAR is usually not a very realistic assumption. The next best assumption is that data are missing at random. This means that missingness depends on one or more auxiliary variables, but these variables have been observed completely. A model such as (8.32) can be used in this case. Sets of synthetic values are generated using the regression model to predict the missing values. To give the imputed values the proper variance, usually a random component is added to the predicted value. This component is drawn from a distribution with the proper variance.

The worst case is the situation in which data is not missing at random. Then missingness depends on unobserved variables, and therefore no valid imputation model can be built using the available data. The distribution of the estimators cannot be repaired by applying multiple imputation. There still is a risk of drawing wrong conclusions from the analysis.

Rubin (1987) describes how estimates for the multiple data sets can be combined into one proper estimate. This is summarized here, concentrating on estimating the population mean.

Let $\bar{y}_j$ denote the estimator of data set j (for $j = 1, 2, \ldots, m$), and let $S(\bar{y}_j)$ be the associated standard error. The overall estimator for the population mean of Y is now defined by

$$\bar{y}_{MI} = \frac{1}{m} \sum_{j=1}^{m} \bar{y}_j. \qquad (8.33)$$

The variance of this estimator is equal to

$$V(\bar{y}_{MI}) = \frac{1}{m} \sum_{j=1}^{m} V(\bar{y}_j) + \left(1 + \frac{1}{m}\right) \frac{1}{m-1} \sum_{j=1}^{m} (\bar{y}_{MI} - \bar{y}_j)^2. \qquad (8.34)$$

The first term in expression (8.34) can be seen as the within imputation variance (the variation within the data sets) and the second one as the between imputation variance (the variation caused by differences in imputed values).

Rubin (1987) claims that the number of imputations per missing value should not exceed $m = 10$. He shows that the relative increase in variance of an estimator based on m imputations to the one based on an infinite number of imputations is approximately equal to

$$\left(1 + \frac{\lambda}{m}\right) \qquad (8.35)$$

where λ is the rate of missing information. For example, with 50% missing information ($\lambda = 0.5$), the relative increase in variance of an estimator based on $m = 5$ imputations equals 1.1. This means that the standard error will only be 5% larger.

Multiple imputation can be a useful tool for handling the problems caused by missing data, but if it is not done carefully, it is potentially dangerous. If an imputation does not model the missing data mechanism properly, analysis of the imputed data sets can be seriously flawed. This means that used models should always be checked.

8.4 DATA EDITING STRATEGIES

Data editing was mainly a manual activity in the days of traditional survey processing. Domain errors were identified by visually scanning the answers to the questions one at a time. Consistency errors were typically caught only when they involved a small number of questions on the same page or on adjacent pages. Route errors were found by following the route instructions and by noting deviations. In general, manual editing could identify only a limited number of problems in the data set.

The data editing process was greatly facilitated by the introduction of computers. Initially, these were mainframe computers, which permitted only batch-wise editing. Tailor-made editing programs, usually written in COBOL or FORTRAN, were designed for each survey. Later, general-purpose batch editing programs were developed and extensively used. These programs performed extensive checks on each record and generated printed lists of error reports by case ID. The error lists were then sent to subject-matter experts or clerical staff, who attempted to manually reconcile these errors. This staff then prepared correction forms, which were keyed to update the data file, and the process was repeated.

Batch computer editing of data sets improved the data editing process because it permitted a greater number and more complex error checks. Thus, more data errors could be identified. However, the cycle of batch-wise checking and manual correction was proved to be labor-intensive, time consuming, and costly.

Statistics Netherlands carried out a Data Editing Research Project in 1984 (see Bethlehem, 1997). A careful evaluation of data editing activities was conducted in a number of different surveys: large and small surveys, and social and economic surveys. Although considerable differences were observed between surveys, still some general characteristics could be identified. The traditional data editing process is summarized in Fig. 8.3.

After collection of the questionnaire forms, subject-matter specialists checked them for completeness. If necessary and possible, skipped questions were answered and obvious errors were corrected on the forms. Sometimes, forms were manually copied to a new form to allow the subsequent step of data entry. Next, the forms were transferred to the data entry department. Data typists entered the data in the computer at high speed without error checking. The computer was a dedicated system for data entry. After data entry, the files were transferred to the mainframe computer system. On the mainframe, an error detection program was run. Detected errors were printed on a list. The lists with errors were sent to the subject-matter department. Specialists investigated the error messages, consulted corresponding forms, and corrected errors on the lists. Lists with corrections were sent to the data entry department, and data typists entered the corrections in the data entry computer. The file with corrections was transferred to the mainframe computer. Corrected records and already present correct records were merged. The cycle of batch-wise error detection and manual correction was repeated until the number of detected errors was considered to be sufficiently small.

After the last step of the editing process, the result was a "clean" data set, which could be used for tabulation and analysis. Detailed investigation of this process for the

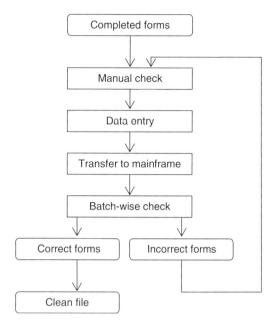

Figure 8.3 The traditional data editing process.

four selected surveys leads to a number of conclusions. These conclusions are summarized below.

First, various people from different departments were involved. Many people dealt with the information: respondents filled-in forms, subject-matter specialists checked forms and corrected errors, data typists entered data in the computer, and programmers from the computer department constructed editing programs. The transfer of material from one person/department to another could be a source of error, misunderstanding, and delay.

Second, different computer systems were involved. Most data entry was carried out on Philips P7000 minicomputer systems, and data editing programs ran on a CDC Cyber 855 mainframe. Furthermore, there was a variety of desktop (running under MS-DOS) and other systems. About 300 interviewers had been equipped with laptops running under CP/M. Transfer of files from one system to another caused delay, and incorrect specification and documentation could produce errors.

Third, not all activities were aimed at quality improvement. A lot of time was spent just on preparing forms for data entry, not on correcting errors. Subject-matter specialists had to clean up forms to avoid problems during data entry. The most striking example was manually assigning a code for "unknown" to unanswered questions.

Another characteristic of the process is that it was going through *macrocycles*. The whole batch of data was going through cycles: from one department to another, and from computer system to another. The cycle of data entry, automatic checking, and manual correction was in many cases repeated three times or more. Due to these macrocycles, data processing was very time consuming.

Finally, the nature of the data (i.e., the metadata) had to be specified in nearly every step of the data editing process. Although essentially the same, the "metadata language" was completely different for every department or computer system involved. The questionnaire itself was the first specification. The next one was with respect to data entry. Then, automatic checking program required another specification of the data. For tabulation and analysis, for example, using the statistical package SPSS, again another specification was needed. All specifications came down to a description of variables, valid answers, routing, and possibly valid relations.

With the emergence of microcomputers in the early 1980s, completely new methods of data editing became possible. One of these approaches has been called computer-assisted data input (CADI). The same process has also been called computer-assisted data entry (CADE). CADI provides an interactive and intelligent environment for combined data entry and data editing of paper forms by subject-matter specialists or clerical staff. Data can be processed in two ways: either in combination with data entry or as a separate step. In the first approach, the subject-matter employees process the survey forms with a microcomputer one by one. They enter the data "heads up," which means that they tend to watch the computer screen as they make entries. After completion of entry for a form, they activate the check options to test for all kinds of errors (omission, domain, consistency, and route errors). Detected errors are displayed and explained on the screen. Staff can then correct the errors by consulting the form or by contacting the supplier of the information. After the elimination of all visible errors, a "clean" record, that is, one that satisfies all check edit criteria, is written to file. If staff members do not succeed in producing a clean record, they can write it to a separate file of problem records. Specialists can later deal with these difficult cases with the same CADI system. This approach of combining capture and editing is efficient for surveys with relatively small samples but complex questionnaires.

In the second approach, clerical staff (data typists or entry specialists) enter data through the CADI system "heads down," that is, without much error checking. When this entry step is complete, the CADI system checks all the records in a batch run and flags the cases with errors. Then subject-matter specialists take over, examine the flagged records and fields one by one on the computer screen, and try to reconcile the detected errors. This approach works best for surveys with large samples and simple questionnaires.

The second advance in data editing occurred with the development of computer-assisted interviewing (CAI). It replaced the paper questionnaire with a computer program that was in control of the interviewing process. It began in the 1970s with computer-assisted telephone interviewing (CATI) using minicomputers. The emergence of small, portable computers in the 1980s made computer-assisted personal interviewing (CAPI) possible. Computer-assisted interviewing is being increasingly used in social and demographic surveys. CAI offers three major advantages over traditional paper and pencil interviewing (PAPI):

- Computer-assisted interviewing integrates three steps in the survey process: data collection, data entry, and data editing. Since interviewers use computers to record the answers to the questions, they take care of data entry during the

interview. Since most of the data editing is carried out during the interview, a separate data editing step has become superfluous in many surveys. After all interviewer files have been combined into one data file, the information is clean and therefore ready for further processing. Thus, computer-assisted interviewing reduces the length and the cost of the survey process.
- The interview software takes care of selecting the proper next question to ask and ensuring that entries are within their domains. Hence, routing and range errors are largely eliminated during data entry. This also reduces the burden on the interviewers, since they need not worry about routing from item to item and can concentrate on getting the answers to the questions.
- With CAI, it becomes possible to carry out consistency checking during the interview. Since both the interviewer and the respondent are available when data inconsistencies are detected, they can immediately reconcile them. In this way, computer-assisted interviewing should produce more consistent and accurate data, correcting errors in the survey office after the interview is over.

Computer-assisted interviewing has been shown to increase the efficiency of the survey operations and the quality of the results. For more information about these aspects, see Couper and Nicholls (1998).

The use of computer-assisted interviewing techniques makes it possible to move data editing to the front of the statistical process. The interviewers can take over many of the data editing tasks. This raises the question as to whether all data editing should be carried out during the interview, thereby avoiding a separate data editing step. There is much to say in favor of this approach. In his famous book on quality control, Deming (1986) strongly advises against dependence on mass inspection of the final product. It is ineffective and costly. Instead, quality control should be built into the production process and be carried out at the first opportunity. For computer-assisted interviewing, that first opportunity occurs in the interview itself. Powerful interviewing software, such as the Blaise system, can perform checks on data as entered and report any detected errors. Both the interviewer and the respondent are available, so together they can correct any problem. Experience has shown that many errors are detected and resolved in this way. Data editing during the interview has been shown to produce better data than editing after data collection.

Data editing during the interview has also some drawbacks. First, checks built into the interviewing program can be very complex, resulting in error message that are difficult for the interviewers and respondents to understand. Correction of some detected errors may prove to be a very difficult task. The developer of the interviewing program has to recognize that the interviewer is not a subject-matter specialist. Only errors that the interviewer can easily handle should be made part of the interview.

Second, having many checks in the interviewing program will increase the length of the interview as the interviewer is stopped to correct each detected error. Interviews should be kept as short as possible. If longer interviews, the respondent may lose interest, with a possible loss of data quality, and this may offset quality gains from additional editing.

Third, not all forms of data editing are possible during the interview. When comparisons of entered data are necessary with information from other sources, such as large database systems, laptop computers used in CAPI may not have sufficient facilities and capacities to use them.

Fourth, improperly specified checks can completely obstruct the completion of an interview. The interviewing software will not accept entries that violate programmed checks, and if the respondent claims that the answers are correct, there can be an impasse. Fortunately, most interviewing software has ways to avoid these deadlocks. One solution is to permit both hard checks and soft checks.

Hard checks designate errors that must be corrected. The interviewer is permitted to continue the interview until changes have been made that no longer violate the check. Soft checks result in warnings of situations that are highly unlikely although possible. If the respondent insists that the answer is correct, the interviewer can accept the answer and continue. Soft checks must be used wherever there is a risk of an impasse in the interview. It is also possible to combine soft and hard checks. A soft check with somewhat relaxed conditions is used to detect suspicious cases, whereas the same type of check with more strict conditions is specified as a hard check to detect the real errors.

Despite the drawbacks mentioned, the possible extra burden on the interviewer, and the limitations imposed by hardware, there are time, money, and quality considerations generally encouraging as much data editing as possible in the interview. Only editing not possible during the interview should be carried out after data collection. This requires a careful thought during the design stage in the development of both the interview and the postinterview editing instruments.

Performing data editing during a computer-assisted interview is greatly facilitated when the interviewing software allows specification of powerful checks in an easy and user-friendly way. Although checks can be hard coded for each survey in standard programming languages, this is a costly, time-consuming, and error-prone task. Many CAI software packages now offer very powerful tools for microediting, permitting easy specification for a large number of checks, including those involving complex relationships among many questions. Editing during CAI is now extensively used both in government and private sector surveys.

Whether microediting is carried out during or after the interview, the entire process may have major disadvantages, especially when carried to extremes. Little and Smith (1987) have mentioned the risk of overediting. Powerful editing software offers ample means for almost any check one can think of, and it is sometimes assumed that the more checks one carries out, the more errors one will correct. But there are risks and costs.

First, the use of too many checks may cause problems in interviewing or postinterview data correction, especially if the checks are not carefully designed and thoroughly tested prior to use. Contradictory checks may cause virtually all records to be rejected, defeating the purpose of editing. Redundant checks may produce duplicate or superfluous error messages slowing the work. And checks for data errors that have little impact on the quality of published estimates may generate work that does not contribute to the quality of the finished product.

Second, since data editing activities make up a large part of the total survey costs, their cost effectiveness has to be carefully evaluated at a time when many survey agencies face budget reductions. Large numbers of microedits that require individual corrections will increase the costs of a survey. Every attempt should be made to minimize data editing activities so that they do not affect the quality of the survey results.

Third, it must be recognized that not all data problems can be detected and repaired with microediting. One such problem is that of outliers. An *outlier* is a value of a variable that is within the domain of valid answers to a question but is highly unusual or improbable when compared with the distribution of all valid values. An outlier can be detected only if the distribution of all values is available. Macroediting is required for this.

The remaining sections of this chapter describe three alternative approaches to editing that address some of the limitations of traditional microediting. In some situations, they could replace microediting. In other situations, they could be carried out in combination with traditional microediting or with each other. They are

- *Automatic editing* attempts to automate microediting. Since human intervention is eliminated, costs are reduced and timeliness is increased.
- *Selective editing* attempts to minimize the number of edits in microediting. Only edits having an impact on the survey results are performed.
- *Macroediting* offers a top–down approach. Edits are carried out on aggregated cases rather than on individual records. Microediting of individual records is invoked only if problems are identified by macroedits.

A more detailed description of these data editing approaches can be found in Bethlehem and Van de Pol (1998).

8.4.1 Automatic Editing

Automatic editing is a process in which records are checked and corrected automatically by a software package. Since no human activities are involved, this approach is fast and cheap. For automatic editing, the usual two stages of editing, error detection and error correction, are expanded to three:

- *Error Detection*. As usual, the software detects errors or inconsistencies by reviewing each case using the prespecified edit rules.
- *Determining the Variables Causing the Error*. If an edit detects an error that involves several variables, the system must next determine which variable caused the error. Several strategies have been developed and implemented to solve this problem.
- *Error Correction*. Once the variable causing the error has been identified, its value must be changed so that the new value no longer causes an error message.

There is no straightforward way to determine which of the several variables causes a consistency error. One obvious criterion is the number of inconsistencies in which that variable is involved. If variable A is related to three other variables B, C, and D, an erroneous value of A may generate three inconsistencies with B, C, and D. If B, C, and D are involved in no other edit failures, A seems the likely culprit. However, it could be noted that no other edit rules have been specified for B, C, and D. Then, also B, C, and D could be candidates for correction.

Edit rules have to be specified to be able to detect errors. Such rules are mathematical expressions that describe relationships between variables. For quantitative variables, such relationships usually take the form of equalities or inequalities. Suppose, three variables are measured in a business survey: turnover, costs, and profit. By definition, the first two variables only assume nonnegative values. The third variable may be negative. The following edit rules may apply to these variables:

```
Profit + Costs = Turnover
Costs > 0.6 x Turnover
```

Edit rules for qualitative variables often take the form of IF-THEN-ELSE constructions. An example for two variables, age and marital status, is

```
IF Age < 15 THEN MarStat = Unmarried
```

If a record satisfies all specified edit rules, it is considered correct. If at least one edit rule is violated, it is considered incorrect and will need further treatment. As an example, Table 8.2 contains two records that have to be checked using the quantitative edit rules mentioned above.

The variables in record 1 satisfy both edit rules. Therefore, the record is considered correct. There is something wrong with record 2 as profit and costs do not add up to turnover. Note that the edit rules would be satisfied if the value 755 is replaced with 75. So, the error may have been caused by a typing error.

The Fellegi–Holt methodology takes a more sophisticated approach (Fellegi and Holt, 1976; United Nations, 1994). To reduce dependence on the number of checks defined, the Fellegi–Holt methodology performs for each variable an analysis of the pertinent edit checks. Logically superfluous checks are removed and all implied checks that can be logically derived from the checks in question are added. Records are then processed as a whole, and not on a variable-by-variable basis, with all consistency

Table 8.2 Examples of Two Records to be Checked

Record	Profit	Costs	Turnover
1	30	70	100
2	755	125	200

checks in place to avoid the introduction of new errors as identified ones are resolved. The smallest possible set of imputable fields is located, with which a record can be made consistent with all checks.

In the Fellegi–Holt methodology, erroneous values are often corrected with hot-deck imputation. Hot-deck imputation employs values copied from a similar donor record (another case) not violating any edit rules. When the definition of "similar" is very strict or when the receptor record is unique, it may be impossible to find a similar donor record. In this situation, a simple default imputation procedure is applied instead.

The Fellegi–Holt methodology has been programmed and put into practice by several survey agencies. For an overview of software and algorithms, see Bethlehem (1997). All these programs identify values that are likely to be incorrect and impute new values. In practical applications, many ties occur, that is, several variables are equally likely to be in error. With one check and two inconsistent values, there is a 50% chance that the wrong variable will be changed, an undesirably high percentage of erroneous corrections. Ties are less frequent when more edit rules are specified, but the Fellegi–Holt methodology makes more checks costly as more computing resources are required. When checks are interrelated, there can be hundreds of thousands of implied checks, using a vast amount of computing time for their calculation. Nevertheless, a large number of original checks are advisable to avoid ties.

The Fellegi–Holt methodology is based on the idea that usually the number of errors in a record will be very small. Consequently, as few changes as possible should be made in a record to remove errors. So, if a record can be made to satisfy all edit rules by making either small changes in two values or a large change in one field, the latter should be preferred. This also ensures that large errors will be detected and corrected.

Suppose, the Fellegi–Holt methodology is applied to record 2 in Table 8.2. To question is whether the data can be made to satisfy all edit rules by changing just one value. There are three possibilities:

- *Change the Value of the Variable Costs*. To satisfy the first edit rule, the value of Costs must be equal to −555. However, this is not possible as it would violate the rule that the value of Costs must be nonnegative.
- *Change the Value of the Variable Turnover*. To satisfy the first edit rule, the value of Turnover must be made equal to 880. However, this is not possible, as it would violate the second edit rule that costs must be larger than 60% of the turnover.
- *Change the Value of the Variable Profit*. To satisfy the first edit rule, the value of Profit must be made equal to 75. This does not affect the second rule, which was already satisfied. So, this change will lead to a situation in which all edit rules are satisfied. This is the preferred correction.

The most likely correct value could be computed in the example above. The situation is not always that simple in practice. Usually, some imputation technique is used to compute a new value. See Section 8.3 for an overview of some imputation techniques. It may happen that after imputation (e.g., imputation of the mean) the records still do not satisfy all edit rules. This calls for another round of changes. One should be careful to change only imputed values and "real" values.

Algorithms for automatic imputation can be very complex. This is because they have to solve two problems simultaneously. First, the algorithm must see to it that a situation is created in which all edit rules are satisfied. Second, this must be accomplished with as few changes as possible.

8.4.2 Selective Editing

The implicit assumption of microediting is that every record receives the same treatment and the same effort. This approach may not be appropriate or cost-effective in business surveys, since not every record has the same effect on computed estimates of the population. Some large firms may make substantially larger contributions to the value of estimates than others.

Instead of conserving editing resources by fully automating the process, they may be conserved by focusing the process on correcting only the most necessary errors. Necessary errors are those that have a noticeable effect on published figures. This approach is called *selective editing*.

To establish the effect of data editing on population estimates, one can compare estimates based on unedited data with estimates based on edited data. Boucher (1991) and Lindell (1997) did this and found that for each variable studied, 50–80% of the corrections had virtually no effect on the estimate of the grand total. Similar results were obtained in an investigation carried out by Van de Pol and Molenaar (1995) on the effects of editing on the Dutch Annual Construction Survey. Research in this area shows that only a few edits have a substantial impact on the final figures. Therefore, data editing efforts can be reduced by identifying those edits. One way to implement this approach is to use a criterion that splits the data set into a critical and a noncritical stream. The *critical stream* contains records that have a high risk of containing influential errors and therefore requires thorough microediting. Records in the *noncritical stream* could remain unedited or could be limited to automatic editing.

The basic question of selective editing is: is it possible to find a criterion to split the data set into a critical and a noncritical stream? At first sight, one might suggest that only records of large firms will contain influential errors. However, Van de Pol and Molenaar (1995) show that this is not the case. Both large and small firms can generate influential errors. A more sophisticated criterion is needed than just the size of the firm. For selective editing to be effective and efficient, powerful and yet practical criteria must be available. This involves taking account of inclusion probabilities, nonresponse adjustments, size of relevant subpopulations, relative importance of record, and most important of all, a benchmark to determine whether an observed quantity may be in error. Examples of such benchmarks could be deviations from the sample mean (or median) for that quantity.

Hidiroglou and Berthelot (1986) probably were the first to use a score function to select records with influential errors in business surveys. Their approach was followed by Lindell (1997) and Engström (1995). Van de Pol and Molenaar (1995) use a somewhat modified approach. They concentrate on edits based on ratios. Let

$$R_{ijk} = Y_{ik}/Y_{ij} \qquad (8.36)$$

be the ratio of the values of two variables j and k for firm i. This ratio is compared with the median value M_{jk} of all the ratios, by computing the distance

$$D_{ijk} = \text{Max}\left\{\frac{R_{ijk}}{M_{jk}}, \frac{M_{jk}}{R_{ijk}}\right\} \qquad (8.37)$$

or, equivalently,

$$D_{ijk} = e^{|\log(R_{ijk}) - \log(M_{jk})|}. \qquad (8.38)$$

A cutoff criterion may be used to set D_{ijk} to zero when it is not suspiciously high. Next, a risk index is computed as a weighted sum of the distances for all edits in a record:

$$\text{RI}_i = \frac{I_i}{\pi_i}\left[\sum e^{W_{jk}\log(D_{ijk})} - Q\right]. \qquad (8.39)$$

The number of ratios involved is denoted Q. The quantity I_i denotes the relative importance of firm i. It may be included to ensure that more important firms get a higher edit priority than small firms. The inclusion probability π_i is determined by the sampling design. The weight W_{jk} is the reciprocal of the estimated standard deviation of the $\log(D_{ijk})$.

This risk index can be transformed into an *OK index* by carrying out the transformation

$$\text{OK}_i = 100 - \frac{100\,\text{RI}_i}{\text{Med}(\text{RI}_i) + \text{RI}_i}. \qquad (8.40)$$

Low values of the OK index indicate a record is not OK and is in need for further treatment. The transformation causes the values of the OK index to be more or less uniformly distributed over the interval [0, 100]. This has the advantage of a simple relationship between the criterion value and the amount of work to be done: the decision to microedit records with an OK index value below a certain value c means that approximately $c\%$ of the records are in the critical stream.

The OK index can be used to order the records from the lowest OK index value to the highest. If microediting is carried out in this sequence, the most influential errors will be taken care of first. The question arises when to stop editing records. Latouche and Berthelot (1992), Engström (1995), and Van de Pol and Molenaar (1995) discuss several stop criteria. Van de Pol and Molenaar (1995) suggest that editing records with an OK index value under 50 would have little effect on the quality of estimates. Hence, it would generally be sufficient to edit only half of the records.

8.4.3 Macroediting

Macroediting provides a solution to some of the data problems left unsolved by microediting. It can address data problems at the aggregate level. The types of edit rules employed by macroediting are similar to those of microediting, but the difference is that macroedit checks involve aggregated quantities. Two general methods of macroediting are described here.

The first method is sometimes called the *aggregation method* (see Granquist, 1990; United Nations, 1994). It formalizes and systematizes what statistical agencies routinely do before publishing statistical tables. They compare the current figures with those of the previous periods to see if they appear plausible. Only when an unusual value is observed at the aggregate level, the individual records contributing to the unusual quantity are edited at the microlevel. The advantage of this form of editing is that it concentrates on editing activities at those points that have an impact on the final results of the survey. No superfluous microediting activities are carried out on records that do not produce unusual values at the aggregate level. A disadvantage is that results are bent in the direction of one's expectations. There is also a risk that undetected errors may introduce undetected biases.

A second method of macroediting is called the *distribution method*. The distribution of variables is computed using the available data, and the individual values are compared with the distribution. Measures of location, spread, and covariation are computed. Records containing values that appear unusual or atypical in their distributions are candidates for further inspection and possible editing.

Many macroediting techniques analyze the behavior of a single observation in the distribution of all observations. *Exploratory data analysis* (EDA) is a field of statistics for analyzing distributions of variables. Tukey (1977) advocates using graphical techniques as they provide more insight into the behavior of variables than numerical techniques do. Many of these techniques can be applied directly to macroediting and are capable of revealing unusual and unexpected properties that might not be discovered through numerical inspection and analysis. There are two main groups of techniques. The first group analyzes the distribution of a single variable and concentrates on detection of outliers. This can be done by means of *one-way scatter plots*, *histograms*, and *box plots*.

There are also numerical ways to characterize the distribution and to search for outliers. Obvious quantities to compute are the mean $\bar{y}$ and standard deviation s of the observations. If the underlying distribution is normal, approximately 95% of the values must lie in the interval

$$(\bar{y} - 1.96 \times s, \quad \bar{y} + 1.96 \times s). \tag{8.41}$$

Outliers can now be defined as values outside one of these intervals. This simple technique has two important drawbacks:

- The assumptions are only satisfied if the underlying distribution is approximately normal. Hence, this technique should be used only if a graphical method has shown that this model assumption is not unrealistic.
- Traditional statistical analysis is very sensitive to outliers. A single outlier can have a large effect on the values of mean and standard deviation, and may therefore obscure the detection of outliers. Hence, this numeric technique should be used only after graphical methods have justified assumptions of normality.

Numeric techniques based on the median and quartiles of the distribution are less vulnerable to extreme values. For example, the box plot could be applied in a numerical way. Values smaller than the lower adjacent value or larger than the upper adjacent value can be identified as outliers. See also the description of the box plot in Section 12.4.

The second group of macroediting techniques analyzes the relationship between two variables and tries to find records with unusual combinations of values. The obvious graphical technique to use is the *two-dimensional scatter plot*. If points in a scatter plot show a clear pattern, this indicates a certain relationship between the variables. The simplest form of relationship is a linear relationship. In this case, all points will lie approximately on a straight line. When such a relationship seems present, it is important to look for points not following the pattern. They may indicate errors in the data.

EXERCISES

8.1 Which of the sources of error below does not belong to the category of observation errors?
   a. Measurement error
   b. Overcoverage
   c. Undercoverage
   d. Processing error

8.2 Memory effects occur if respondents forget to report certain events or when they make errors about the date of occurrence of events. To which source of errors do these memory effects belong?
   a. Estimation error
   b. Undercoverage
   c. Measurement error
   d. Nonobservation error

8.3 A CADI system reports many errors for a form. It turns out that one variable is involved in all these errors. What kind of action should be undertaken?
   a. The corresponding record should be removed from the data file.
   b. The variable should be removed from the data file.
   c. Correct the value of this variable in such a way that the error messages disappear.
   d. Impute a value for the variable that makes the error message disappear.

8.4 Which of the sources of error below belongs to the category "sampling error?"
   a. Selection error
   b. Overcoverage
   c. Nonresponse
   d. Undercoverage

EXERCISES

8.5 The general imputation model can be written as

$$\hat{Y}_i = B_0 + \sum_{j=1}^{p} B_j X_{ij} + E_j.$$

Which values of the parameters have to be used to obtain imputation of the mean as special case?

a. Take all B_j and E_j equal to 0, except B_0.
b. Take all B_j equal to 0, except B_0.
c. Take all B_j equal to 0.
d. Just take all X_{ij} equal to 0.

8.6 The ministry of agriculture in a country wants to have more information about the manure production by pigs on pig farms. The target variable is the yearly manure production per farm. Among other variables recorded are the number of pigs per farm and the region of the country (north or south). The table below contains part of the data:

Farm	Manure Production	Region	Number of Pigs
1	295,260	North	220
2	259,935	North	195
3	294,593	North	221
4	253,604	North	188
5	?	North	208
6	520,534	South	398
7	?	South	435
8	559,855	South	375
9	574,528	South	416
10	561,865	South	405

The value of manure production is missing in two records due to item nonresponse. Describe six imputation techniques for replacing the missing value by a synthetic value. Compute for each technique which values are obtained. Explain the consequences of estimating the mean manure production per region. Indicate whether these consequences are acceptable.

8.7 It is assumed that there is a relationship between the energy consumption of a house and its total floor space. A simple random sample (with equal probabilities and without replacement) of four houses has been selected. The table below contains the collected data:

House	Floor space (m^2)	Gas Consumption (m^3)	Electricity Consumption (kWh)
1	116	1200	1715
2	81	950	1465
3	73	650	1020
4	99	1050	–

It is a known fact that the mean floor space of all houses in target population is equal to 103.7 m².

a. Compute the value of the ratio estimator for the mean gas consumption in the population.

b. Apply regression imputation to compute the missing value of electricity consumption of house 4.

8.8 Four variables are measured in a business survey: income (I), personnel costs (PC), other costs (OC), and profit (P). The following there rules are checked in the data editing process:

- $P = I - PC - OC$
- $PC > 1.5 \times OC$
- $150 < I < 250$

Check the four records in the table below. Determine which rules are satisfied and which are not. Correct the records using the Fellegi and Holt principle.

Record	I	PK	OK	W
1	260	110	70	50
2	180	80	50	50
3	210	160	20	30
4	240	50	40	30

8.9 A town council has carried out an income survey among its inhabitants. A simple random sample of 1000 persons has been selected. The total size of the population is 19,000. All selected persons have been asked to reveal their net monthly income. From previous research it has become clear that the standard deviation of the net monthly income is always equal to 600.

a. Assuming all sampled persons cooperate and provide their income data, compute the standard error of the sample mean.

b. Suppose 10% of the sampled persons do not provide their income data. The researcher solves this problem by imputing the mean.

Compute the standard error of the mean after imputation.

c. Suppose the sample standard deviation before imputation happens to be equal to the population standard deviation (600). The imputed survey data set is made available to a researcher. He does not know that imputation has been carried out.

If he computes the estimated standard error, what value would he get?

d. Which conclusion can be drawn from comparing the results of (a) and (c)?

CHAPTER 9

The Nonresponse Problem

9.1 NONRESPONSE

Nonresponse occurs when elements in the selected sample that are also eligible for the survey do not provide the requested information or that the provided information is not usable. The problem of nonresponse is that the researcher does not have control any more over the sample selection mechanism. Therefore, it becomes impossible to compute unbiased estimates of population characteristics. Validity of inference about the population is at stake.

This chapter gives a general introduction of the phenomenon of nonresponse as one of the factors affecting the quality of survey based estimates. It is shown that nonresponse has become an ever more serious problem in course of time. Attention is paid to two approaches that provide insight in the possible consequences of nonresponse: the *follow-up survey* and the *basic question approach*. These techniques can also be successful in reducing a possible bias of estimates.

Adjustment weighting is one of the most important nonresponse correction techniques. Chapter 10 will be devoted to adjustment weighting.

There are two types of nonresponse: unit nonresponse and item nonresponse. *Unit nonresponse* occurs when a selected element does not provide any information at all, that is, the questionnaire form remains empty. *Item nonresponse* occurs when some questions have been answered but no answer is obtained for some other, possibly sensitive, questions. So, the questionnaire form has been partially completed.

In case of unit nonresponse, the realized sample size will be smaller than planned. This will lead to increased variances of estimates and thus will lead to a lower precision of estimates. Valid estimates can still be obtained because computed confidence intervals still have the proper confidence level.

To avoid the realized sample of being too small, the initial sample size should be taken larger. For example, if a sample of 1000 elements is required and the expected

---

Applied Survey Methods: A Statistical Perspective, Jelke Bethlehem
Copyright © 2009 John Wiley & Sons, Inc.

response rate is in the order of 60%, the initial sample size should be approximately $1000/0.6 = 1667$.

The main problem of nonresponse is that estimates of population characteristics may be biased. This situation occurs if, due to nonresponse, some groups in the population are over- or underrepresented in the sample, and these groups behave differently with respect to the characteristics to be investigated. Then, nonresponse is said to be *selective*.

It is likely that survey estimates are biased unless very convincing evidence to the contrary is provided. Bethlehem and Kersten (1985) mention a number of Dutch surveys were nonresponse was selective:

- A follow-up study of the Dutch Victimization Survey showed that people who are afraid to be alone at home during night are less inclined to participate in the survey.
- In the Dutch Housing Demand Survey, it turned out that people who refused to participate have lesser housing demands than people who responded.
- For the Survey of Mobility of the Dutch Population, it was obvious that the more mobile people were underrepresented among the respondents.

It will be shown in Section 9.3 that the amount of nonresponse is one of the factors determining magnitude of the bias of estimates. The higher the nonresponse rate, the larger will be the bias.

The effect of nonresponse is shown using a somewhat simplified example that uses data from the Dutch Housing Demand Survey. Statistics Netherlands carried out this survey in 1981. The initial sample size was 82,849. The number of respondents was 58,972, which comes down to a response rate of 71.2%.

To obtain more insight in the nonresponse, a follow-up survey was carried out among the nonrespondents. Among other things they were also asked whether they intended to move within 2 years. The results are summarized in Table 9.1.

Based on the response, the percentage of people with the intention to move within 2 years is 29.7%. However, for the complete sample (response and nonresponse) a much lower percentage of 24.8% is obtained. The reason is clear: there is a substantial difference between respondents and nonrespondents with respect to the intention to move within 2 years. For nonrespondents, this is only 12.8%.

Nonresponse can have many causes. It is important to distinguish these causes. To reduce nonresponse in the field, it is important to know what caused it. Moreover,

Table 9.1 Nonresponse in the Dutch Housing Demand Survey 1981

Do You Intend to Move Within 2 Years?	Response	Nonresponse	Total
Yes	17,515	3056	20,571
No	41,457	20,821	62,278
Total	58,972	23,877	82,849

NONRESPONSE

different types of nonresponse can have different effects on estimates and therefore may require different treatment.

There are no unique ways to classify nonresponse by its cause. This makes it difficult to compare the nonresponse for different surveys. Unfortunately, no standardized classification exits. There have been some attempts. The American Association for Public Opinion Research (AAPOR) has published a report with a comprehensive list of definitions of possible survey outcomes (see AAPOR, 2000). However, these definitions apply only to household surveys with one respondent per household and samples selected by means of Random Digit Dialing (RDD). Lynn et al. (2002) have proposed a more general classification. This classification will be used here.

The classification follows the possible courses of events when selected elements are approached in an attempt to get cooperation in a survey (see Fig. 9.1).

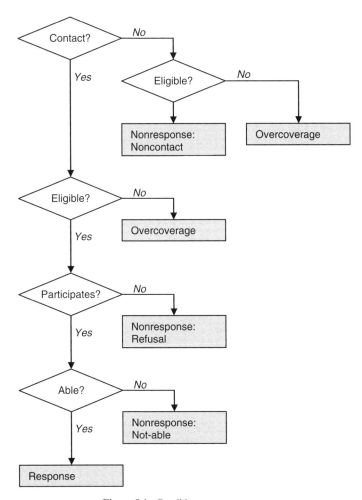

Figure 9.1 Possible survey outcomes.

First, contact must be established with the selected element. If this is not successful, there are two possibilities:

- If the selected element belongs to the target population (i.e., it is eligible), it should have been part of the sample. So, this is nonresponse due to *noncontact*.
- If the selected element does not belong to the target population (i.e., it is not eligible), it should not be included in the sample. This is an example of *overcoverage*, and therefore it can be excluded from the survey.

In practical situations, it is often impossible to determine whether a noncontact belongs to the target population or not. This makes it difficult to calculate response rates.

If there is contact with a selected element, the next step is to establish whether it belongs to the target population or not. If not, it can be dismissed as a case of *overcoverage*.

In the case of contact with an eligible element, its cooperation is required to get the answers to the questions. If the selected element refuses to cooperate, this is nonresponse due to *refusal*.

If there is an eligible element, and it cooperates, there may be still problems if this element is not able to provide the required information. Reasons for this may be, for example, illness or language problems. This is a case of nonresponse due to *not-able*.

Finally, if an eligible element wants to cooperate and is able to provide information, then the result is *response*.

Figure 9.1 shows that there are three main causes for nonresponse: noncontact, refusal, and not-able. Nonresponse need not be permanent. In case of a noncontact, another contact attempt may be tried at some other moment. Some surveys may undertake six contact attempts before the case is closed as a noncontact. Also, a refusal may be temporary. If an interviewer calls at an inconvenient moment, it may be possible to make an appointment for some other date. However, many refusals turn out to be permanent. In case someone is not able to participate because of illness, an interviewer may be successful after the patient has recovered.

9.2 RESPONSE RATES

Due to the negative impact nonresponse may have on the quality of survey results, the *response rate* is considered to be an important indicator of the quality of a survey. Response rates are frequently used to compare the quality of surveys and also to explore the quality of a survey that is repeated over time.

Unfortunately, there is no standard definition of a response rate. Here a definition is used that is similar to the one introduced by Lynn et al. (2002): The response rate is defined as the proportion of eligible elements in the sample for which a questionnaire has been completed.

RESPONSE RATES

The initial sample size n_I can be written as

$$n_I = n_{NC} + n_{OC} + n_{RF} + n_{NA} + n_R, \quad (9.1)$$

where n_{NC} denotes the number of noncontacts, n_{OC} the number of noneligible elements among the contacts (i.e., cases of overcoverage), n_{RF} the number of refusers, n_{NA} the number of not-able elements, and n_R the number of respondents.

The response rate is defined as the number of respondents divided by the number of n_E eligible elements in the sample:

$$\text{Response rate} = \frac{n_R}{n_E}. \quad (9.2)$$

There is a problem in computing the number of eligible elements. This problem arises because the noncontacts consist of eligible noncontacts and noneligible noncontacts. It is not known how many of these noncontacts are eligible. If it is assumed that all noncontacts are eligible, then $n_E = n_{NC} + n_{RF} + n_{NA} + n_R$. Consequently, the response rate is given as follows:

$$\text{Response rate} = \frac{n_R}{n_{NC} + n_{RF} + n_{NA} + n_R}. \quad (9.3)$$

This might not be a realistic assumption. Another assumption is that the proportion of eligibles among the noncontacts is equal to the proportion of eligibles among the contacts. Then, the response rate would be equal to

$$\text{Response rate} = \frac{n_R}{n_{NC}[(n_{RF} + n_{NA} + n_R)/(n_{OC} + n_{RF} + n_{NA} + n_R)] + n_{RF} + n_{NA} + n_R}. \quad (9.4)$$

Response rate definitions like (9.3) or (9.4) can be used in a straightforward way for surveys in which one person per household is selected. The situation is more complicated when the survey population consists of households for which several or all of its members have to provide information. Then, partial response may also occur. It is possible to introduce response rates for households and for persons, and these response rates would be different.

Another complication concerns self-administered surveys. These are surveys in which there are no interviewers, like a mail survey or a Web survey. For such surveys, it is very difficult to distinguish between different sources of nonresponse and also very difficult to determine eligibility. The questionnaire is either returned or not returned. The response rate simplifies to

$$\text{Response rate} = \frac{n_R}{n_R + n_{NR}}. \quad (9.5)$$

The computation of the response rate is illustrated using data from the Survey on Well-being of the Population. The results are listed in Table 9.2.

The category "not able" contains nonresponse because of illness, handicap, or language problems. The extra nonresponse category "other nonresponse" contains

Table 9.2 Fieldwork Results of the Survey on Well-Being of the Population in 1998

Outcome	Frequency
Overcoverage	129
Response	24,008
Noncontact	2,093
Refusal	8,918
Not-able	1,151
Other nonresponse	3,132
Total	39,431

cases that are not processed by interviewers due to workload problems. Also, people who had moved and could not be found any more are included in this category.

If it is assumed that all noncontacts are eligible, the response rate of this survey is

$$100 \times \frac{24,008}{24,008 + 2093 + 8918 + 1151 + 3132} = 61.09\%.$$

If it is assumed that the proportion of eligibles among contacts and noncontacts is the same, the response rate is equal to

$$100 \times \frac{24,008}{24,008 + 2093 \times [(39,431 - 2093 - 129)/(39,431 - 2093)] + 8918 + 1151 + 3132}$$

$$= 61.11\%.$$

The differences in response rates are very small. This is due to small amount of overcoverage.

Another aspect making the definition of response rate difficult is the use of sampling designs with unequal selection probabilities. If, on the one hand, the response rate is used as an indicator of the quality of survey outcomes, the sizes of the various outcome categories should reflect the structure of the population. Consequently, observation should be weighted with inverse selection probabilities. This leads to a so-called *weighted response rate*. If, on the other hand, the response rate is used as an indicator of the quality of the fieldwork, and more specifically the performance of interviewers, an *unweighted response rates* may be more appropriate.

Response rates have declined over time in many countries. Table 9.2 contains (unweighted) response rates for a number of surveys of Statistics Netherlands. The definition of response rates is more or less the same for each survey. It is not easy to explain differences in response rates between surveys. Response rates are determined by a large number of factors such as the topic of the survey, the target population, the time period, the length of the questionnaire, the quality of the interviewers, and the organization of the fieldwork.

It is clear from Table 9.3 that nonresponse is a considerable problem. The problem has become more serious over the years. It also has an impact on the costs of the survey. It

Table 9.3 Response Rates of Some Survey of Statistics Netherlands

Year	Labor Force Survey	Consumer Sentiments Survey	Survey on Well-Being of the Population	Mobility Survey	Holiday Survey
1972		71			
1973	88	77			
1974		75	72		
1975	86	78			86
1976		72	77^a		87
1977	88	69	70		81
1978		64		67	78
1979	81	63	65^b	69	74
1980		61	61	68	74
1981	83	65		68	74
1982		60	64^a	66	71
1983	81	63	58	66	74
1984		65^c		64	69
1985	77	69		61	68
1986		71	59	59	66
1987	60^c	71		59	
1988	59	68		55	
1989	61	68	44	58	
1990	61	68	47	55	
1991	60	69	46	57	
1992	58	69	45	57	
1993	58	72	46	56	
1994	59	70	52^c	55	
1995	60	67	54	54	
1996	58	67	52	52	
1997	56	57	63	50	
1998	54	64	60		
1999	56	62	60		
2000	56	61	57		
2001	58	64	60		
2002	58	65			
2003	59	65	62		

^{a} Young only.
^{b} Elderly only.
^{c} Change in survey design.

takes more and more effort to obtain estimates with the precision as specified in the survey design.

The Labor Force Survey is probably one of the most important surveys of Statistics Netherlands. We will denote it by its Dutch acronym EBB (Enquête Beroepsbevolking). It has been exposed to many redesigns, the most important one taking place in 1987. Then, several things were changed in the design:

- Before 1986, data collection was carried out by means of a paper questionnaire form (PAPI). In 1987, Statistics Netherlands changed to computer-assisted personal interviewing (CAPI). The Blaise System was developed for this. The EBB became a CAPI survey. Each month, approximately 400 interviewers equipped with laptops visited 12,000 addresses.
- Until 1987, the fieldwork for the EBB was carried out by the municipal employees. So, they were not professional interviewers. From 1987 onward, the fieldwork was done by the professional interviewers.
- In 1987, the questionnaire of the EBB was completely redesigned.

Another important survey of Statistics Netherlands is the Survey of Well-being of the Population, denoted by its Dutch acronym POLS (Permanent Onderzoek Leefsituatie). It is a continuous survey in which every month a sample of 3000 persons is selected. The survey has a modular structure. There is a base module with questions for all sampled persons and in addition there are a number of modules about specific themes (such as employment situation, health, and justice). The sampled persons are selected for one of the thematic modules; the base module is answered by everyone. POLS exists only since 1997, before that all the modules were separate surveys.

The Consumer Sentiments Survey (denoted by CCO) measures consumer confidence (for instance in the economic situation). Since April 1986, it is performed monthly by means of computer-assisted telephone interviewing (CATI). Before 1984, the interview was conducted by pen and paper (PAPI). Every month 1500 households are selected in a simple random sample. The Dutch telephone company (KPN) adds telephone numbers to the selected addresses. Only listed numbers of fixed-line telephones can be added. This is possible for about two-third of the addresses. These phone numbers are then passed through to the CATI interviewers. Only one person in every household is interviewed.

The response rates of these three major surveys are also graphically presented in Fig. 9.2. From 1972 to 1983, response percentages of CCO and POLS show a similar, falling trend. After 1983, the response percentage for CCO stabilized whereas for POLS it kept on falling. It seems as though both rates start to converge in 1993 and show a similar pattern in the last 6 years. The two breakpoints coincide with redesigns these surveys (CCO in 1984 and POLS in 1997). The redesign of CCO in 1984 caused a temporary increase in response rates. The same is true for the redesign of POLS in 1997.

The response percentage of the EBB was initially higher than that of the other two surveys, but during 1983–1984 it decreased and reached the same level as the rates of CCO and POLS. From 1987, response shows a more or less stable pattern. As mentioned before, there was a comprehensive redesign of EBB in 1987.

Table 9.4 shows an international comparison of response rates. Stoop (2005) used data from the European Social Surveys (ESS) for this. The ESS is a biannual survey of values, attitudes, beliefs, and behavioral patterns in the context of a changing Europe. Its major aim is to provide data to social scientists for both substantive and

RESPONSE RATES 217

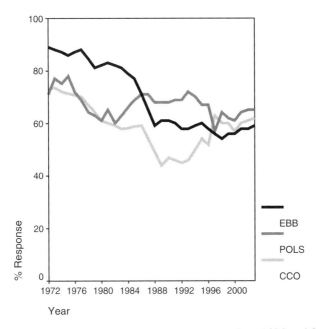

Figure 9.2 Response percentages for three Dutch surveys (EBB, POLS, and CCO).

Table 9.4 Response and Nonresponse Rates in the European Social Survey

Country	Response Rate	Noncontact Rate	Refusal Rate	Not-Able Rate
Austria	60	12	26	2
Belgium	59	8	25	8
Czech Republic	43	12	20	5
Denmark	68	4	24	5
Finland	73	4	19	4
France	43	15	39	4
Germany	57	8	26	8
Greece	80	3	16	1
Hungary	70	7	14	9
Ireland	64	10	20	5
Israel	71	6	22	1
Italy	44	4	44	8
Luxemburg	44	11	45	0
The Netherlands	68	3	24	3
Norway	65	3	25	7
Poland	73	2	20	5
Portugal	69	3	26	1
Slovenia	71	5	17	5
Spain	53	11	32	3
Sweden	69	4	21	6
Switzerland	33	3	55	9
UK	56	5	33	5

Reprinted by permission of Ineke Stoop (2005), The Hunt for the Last Respondent.

methodological studies and analyses. The first round of the ESS took place in 2002–2003. Data were collected in 22 European countries.

To improve comparability between countries, there was one centrally specified sampling design for all participating countries. Furthermore, the response target was 70% and the target for the noncontact rate was 3%. Central fieldwork specifications saw to it that variations due to different procedures in the field were minimized. Table 9.4 is taken from Stoop (2005) and shows the differences in response rates.

There are large differences in response rates. Switzerland has a very low response rate (33%), followed by the Czech Republic (43%), Italy, and Luxemburg (both 44%). The highest response rate was obtained in Greece (80%), followed by Finland (73%), Israel, and Slovenia (both 71%). Note that many countries were not able to reach the target of 70% response.

The noncontact rates differ substantially across countries. The rates vary from 2% in Poland to 15% in France.

The refusal rates vary from 14% in Hungary to 55% in Switzerland. The large difference may partly be due to the differences in procedures for dealing with refusers. For example, refusers were reapproached in Switzerland, United Kingdom, The Netherlands, Finland, and Greece. This hardly ever happened in Luxemburg, Hungary, and Italy.

The not-able rates vary from 0% in Luxemburg to 9% in Switzerland. These figures seem to indicate that difference may be caused by differences in reporting than differences in fieldwork results.

9.3 MODELS FOR NONRESPONSE

To be able to investigate the possible impact of nonresponse on estimators of population characteristics, this phenomenon should be incorporated in sampling theory. Two approaches are described. One is the *random response model* and the other is the *fixed response model*. Both approaches are discussed in Lindström et al. (1979), Kalsbeek (1980), Cassel et al. (1983), and Bethlehem and Kersten (1986). Both models give insight in conditions under which nonresponse causes estimators to be biased.

It is also explored in this chapter what the effect of biased estimators will be on the validity of confidence intervals.

9.3.1 The Fixed Response Model

The *fixed response model* assumes the population to consist of two mutually exclusive and exhaustive strata: the response stratum and the nonresponse stratum. If selected in the sample, elements in the response stratum will participate in the survey with certainty and elements in the nonresponse stratum will not participate with certainty.

MODELS FOR NONRESPONSE

A set of response indicators

$$R_1, R_2, \ldots, R_N \quad (9.6)$$

is introduced, where $R_k = 1$ if the corresponding element k is part of the response stratum and $R_k = 0$ if element k belongs to the nonresponse stratum. So, if selected, $R_k = 1$ means response and $R_k = 0$ means nonresponse.

The size of the response stratum can be denoted by

$$N_R = \sum_{k=1}^{N} R_k \quad (9.7)$$

and the size of the nonresponse stratum can be denoted by

$$N_{NR} = \sum_{k=1}^{N} (1 - R_k), \quad (9.8)$$

where $N = N_R + N_{NR}$. The mean of the target variable Y in the response stratum is equal to

$$\bar{Y}_R = \frac{1}{N_R} \sum_{k=1}^{N} R_k Y_k. \quad (9.9)$$

Likewise, the mean of the target variable in the nonresponse stratum can be written as

$$\bar{Y}_{NR} = \frac{1}{N_{NR}} \sum_{k=1}^{N} (1 - R_k) Y_k. \quad (9.10)$$

The *contrast* K is introduced as the difference between the means of the target variable in response stratum and the nonresponse stratum:

$$K = \bar{Y}_R - \bar{Y}_{NR}. \quad (9.11)$$

It as an indicator of the extent to which respondents and nonrespondents differ on average.

Now suppose a simple random sample without replacement of size n is selected from this population. This sample is denoted by the set of indicators $a_1, a_2, \ldots, a_N$, where $a_k = 1$ means that element k is selected in the sample and otherwise $a_k = 0$. It is not known beforehand to which of the two strata selected elements belong. There will be

$$n_R = \sum_{k=1}^{N} a_k R_k \quad (9.12)$$

elements from the response stratum and

$$n_{NR} = \sum_{k=1}^{N} a_k (1 - R_k), \quad (9.13)$$

where $n = n_R + n_{NR}$.

Only the values for the n_R selected elements in the response stratum become available. The mean of these values is denoted by

$$\bar{y}_R = \frac{1}{n_R} \sum_{k=1}^{N} a_k R_k Y_k. \qquad (9.14)$$

Theoretically, it is possible that no observations at all become available. This occurs when all sample elements happen to fall in the nonresponse stratum. In practical situations, this event has a very small probability of happening. Therefore, it will be ignored. Then, it can be shown that the expected value of the response mean is equal to

$$E(\bar{y}_R) = \bar{Y}_R. \qquad (9.15)$$

This is not surprising since the responding elements can be seen as a simple random sample without replacement from the response stratum.

Of course, it is not the objective of the survey to estimate the mean of the response stratum but the mean in the population. If both means have equal values, there is no problem, but this is generally not the case. Therefore, estimator (9.14) will be biased and this bias is given as

$$B(\bar{y}_R) = \bar{Y}_R - \bar{Y} = \frac{N_{NR}}{N}(\bar{Y}_R - \bar{Y}_{NR}) = QK, \qquad (9.16)$$

where K is the contrast and $Q = N_{NR}/N$ is the relative size of the nonresponse stratum. From expression (9.16), it is clear that the bias is determined by two factors:

- The amount to which respondents and nonrespondents differ, on average, with respect to the target variable. The more they differ, the larger the bias will be.
- The relative size of the nonresponse stratum. The bigger the group of nonrespondents is, the larger the bias will be.

The fixed response model is applied to data from the Dutch Housing Demand Survey. Statistics Netherlands carried out this survey in 1981. The sample size was 82,849. The number of respondents was 58,972, which comes down to a response rate of 71.2%. One of the target variables was whether one had the intention to move within 2 years. The population characteristic to be estimated was the percentage of people with the intention to move within 2 years.

To obtain more insight in the nonresponse, a follow-up survey was carried out among the nonrespondents. One of the questions asked was the intention to move within 2 years. The results are summarized in Table 9.5.

The percentage of potential movers in the response stratum can be estimated using the response data. The estimate is equal to $100 \times 17{,}517/58{,}972 = 29.7\%$. The percentage of potential movers in the nonresponse stratum can be estimated using the data in the follow-up survey. The estimate is equal to $100 \times 3056/23{,}877 = 12.8\%$. Hence, the contrast K is equal to $29.7 - 12.8 = 16.9\%$. Apparently, the intention to move is much higher among respondents than under nonrespondents.

MODELS FOR NONRESPONSE

Table 9.5 Nonresponse in the 1981 Dutch Housing Demand Survey

Do you Intend to Move Within 2 Years?	Response	Nonresponse	Total
Yes	17,515	3056	20,571
No	41,457	20,821	62,278
Total	58,972	23,877	82,849

The relative size of the nonresponse stratum is estimated by $23,877/82,849 = 0.288$. Therefore, the bias of the estimator just based on the response data is equal to $16.9 \times 0.288 = 4.9\%$.

9.3.2 The Random Response Model

The *random response model* assumes every element k in the population to have (an unknown) response probability ρ_k. If element k is selected in the sample, a random mechanism is activated that results with probability ρ_k in response and with probability $1 - \rho_k$ in nonresponse. Under this model, a set of response indicators

$$R_1, R_2, \ldots, R_N \tag{9.17}$$

is introduced, where $R_k = 1$ if the corresponding element k responds; $R_k = 0$, otherwise. So, $P(R_k = 1) = \rho_k$ and $P(R_k = 0) = 1 - \rho_k$.

Now, suppose a simple random sample without replacement of size n is selected from this population. This sample is denoted by the set of indicators $a_1, a_2, \ldots, a_N$, where $a_k = 1$ means that element k is selected in the sample, and otherwise $a_k = 0$. The response only consists of those elements k for which $a_k = 1$ and $R_k = 1$. Hence, the number of available cases is equal to

$$n_R = \sum_{k=1}^{N} a_k R_k. \tag{9.18}$$

Note that this realized sample size is a random variable. The number of non-respondents is equal to

$$n_{NR} = \sum_{k=1}^{N} a_k (1 - R_k), \tag{9.19}$$

where $n = n_R + n_{NR}$.

The values of the target variable become available only for the n_R responding elements. The mean of these values is denoted by

$$\bar{y}_R = \frac{1}{n_R} \sum_{k=1}^{N} a_k R_k. \tag{9.20}$$

Theoretically, it is possible that no observations at all become available. This happens when all sample elements do not respond. In practical situations, this event has a very small probability of happening. Therefore, we will ignore it. It can be shown (see Bethlehem, 1988) that the expected value of the response mean is approximately equal to

$$E(\bar{y}_R) \approx \tilde{Y}, \tag{9.21}$$

where

$$\tilde{Y} = \frac{1}{N} \sum_{k=1}^{N} \frac{\rho_k}{\bar{\rho}} Y_k \tag{9.22}$$

and

$$\bar{\rho} = \frac{1}{N} \sum_{k=1}^{N} \rho_k \tag{9.23}$$

is the mean of all response probabilities in the population. From expression (9.21), it is clear that, generally, the expected value of the response mean is unequal to the population mean to be estimated. Therefore, this estimator is biased. This bias is approximately equal to

$$B(\bar{y}_R) = \tilde{Y} - \bar{Y} = \frac{S_{\rho Y}}{\bar{\rho}} = \frac{R_{\rho Y} S_\rho S_Y}{\bar{\rho}}, \tag{9.24}$$

where $S_{\rho Y}$ is the covariance between the values of the target variable and the response probabilities, $R_{\rho Y}$ is the corresponding correlation coefficient, S_Y is the standard deviation of the variable Y, and S_ρ is the standard deviation of the response probabilities. From this expression of the bias a number of conclusions can be drawn:

- The bias vanishes if there is no relationship between the target variable and response behavior. Then $R_{\rho Y} = 0$. The stronger the relationship between target variable and response behavior, the larger the bias will be.
- The bias vanishes if all response probabilities are equal. Then $S_\rho = 0$. Indeed, in this situation the nonresponse is not selective. It just leads to a reduced sample size.
- The magnitude of the bias increases as the mean of the response probabilities decreases. Translated in practical terms, this means that lower response rates will lead to larger biases.

The effect of nonresponse is shown by means of a simulation experiment. From the working population of the small country of Samplonia 1000 samples of size 40 were selected. For each sample, the mean income was computed as an estimate of the mean income in the population. The distribution of these 1000 estimates is displayed in Fig. 9.3. The sampling distribution is symmetric around the population value to be estimated (indicated by the vertical line). Therefore, the estimator is unbiased.

Now the experiment is repeated, but also nonresponse is generated. Response probabilities are taken linearly related to income. People with the lowest income have a response probability of 0.95 and people with the highest income have a response probability of 0.05. So the higher the income is, the lower is the probability of response.

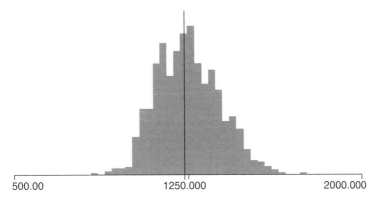

Figure 9.3 The distribution of the mean income in 1000 samples from the working population of Samplonia. There is no nonresponse.

Again 1000 samples of (initial) size 40 are generated. The resulting sampling distribution of the estimator is displayed in Fig. 9.4.

The distribution has shifted to the left. Apparently, people with lower incomes are overrepresented and people with high incomes are underrepresented. The vertical line representing the population mean is not in the center of the distribution any more. The average of the sample mean of all 1000 samples is equal to 970 whereas the population mean is equal to 1234. Clearly, the estimator has a substantial bias.

9.3.3 The Effect of Nonresponse on the Confidence Interval

The precision of an estimator is usually quantified by computing the 95% confidence interval. Suppose, for the time being, that all sampled elements cooperate. Then, the sample mean can be computed. This is an unbiased estimator for the population

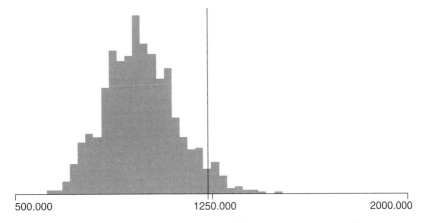

Figure 9.4 The distribution of the mean income in 1000 samples from the working population of Samplonia. Nonresponse increases with income.

mean. Since the sample mean has (approximately) a normal distribution, the 95% confidence interval for the population mean is equal to

$$I = (\bar{y} - 1.96 \times S(\bar{y}); \bar{y} + 1.96 \times S(\bar{y})), \qquad (9.25)$$

where $S(\bar{y})$ is the standard error of the sample mean. The probability that this interval contains the true value is, by definition (approximately), equal to

$$P(\bar{Y} \in I) = 0.95. \qquad (9.26)$$

In case of nonresponse, only the response mean $\bar{y}_R$ can be used to compute the confidence interval. This confidence interval is denoted by I_R. It can be shown that

$$P(\bar{Y} \in I_R) = \Phi\left(1.96 - \frac{B(\bar{y}_R)}{S(\bar{y}_R)}\right) - \Phi\left(-1.96 - \frac{B(\bar{y}_R)}{S(\bar{y}_R)}\right), \qquad (9.27)$$

in which Φ is the standard normal distribution function. Table 9.6 presents values of this probability as a function of the *relative bias*, which is defined as the bias divided by the standard error.

It is clear that the confidence level can be much lower than expected. If the bias is equal to the standard error, that is, the relative bias is 1, the confidence level is only 0.83. As the relative bias increases, the situation becomes worse. The conclusion is that due to nonresponse the interpretation of the confidence interval is not correct any more.

The effect of nonresponse on the confidence interval can also be shown by means of a simulation experiment. From the working population of Samplonia, samples of size 40 were selected. Again nonresponse was generated. People with the lowest income had a response probability of 0.95. Nonresponse increased with income. People with the highest income had a response probability of 0.05.

For each sample, the 95% confidence interval was computed. Figure 9.5 shows the result of the first 30 samples. Each confidence interval is indicated by a horizontal line. The vertical line denotes the true population mean to be estimated. Note that only 10

Table 9.6 The Confidence Level of the 95% Confidence Interval as a Function of the Relative Bias

$|B(\bar{y}_R)/S(\bar{y}_R)|$	$P(\bar{Y} \in I_R)$
0.0	0.95
0.2	0.95
0.4	0.93
0.6	0.91
0.8	0.87
1.0	0.83
1.2	0.78
1.4	0.71
1.6	0.64
1.8	0.56
2.0	0.48

ANALYSIS OF NONRESPONSE

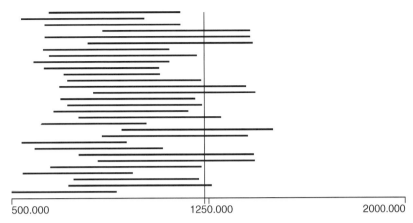

Figure 9.5 Confidence intervals for the mean income in the working population of Samplonia. Nonresponse increases with income.

out of 30 confidence intervals contain the population mean. This suggests a confidence level of 33.3% instead of 95%.

9.4 ANALYSIS OF NONRESPONSE

One should always be aware of the potential negative effects of nonresponse. It is therefore important that a nonresponse analysis is carried out on the data that have been collected in a survey. Such an analysis should make clear whether or not response is selective, and if so, which technique should be applied to correct for a possible bias.

This chapter gives an example of such a nonresponse analysis. Data used here are the data from the Integrated Survey on Household Living Conditions (POLS) that has been conducted by Statistics Netherlands in 1998.

9.4.1 How to Detect a Bias?

How can one detect that the nonresponse is selective? The available data with respect to the target variables will not be of much use. There are data only for the respondents and not for the nonrespondents. So, it is not possible to establish whether respondents and nonrespondents differ with respect to these variables. The way out for this problem is to use auxiliary variables (see Fig. 9.6).

An auxiliary variable in this context is a variable that has been measured in the survey and for which the distribution in the population (or in the complete sample) is available. So, it is possible to establish a relationship between this variable and the response behavior.

Three different response mechanisms were introduced in Chapter 8. The first one is *missing completely at random* (MCAR). The occurrence of nonresponse (R) is

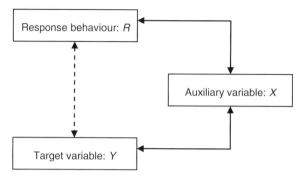

Figure 9.6 Relationships between target variable, response behavior, and auxiliary variable.

completely independent of both the target variable (Y) and the auxiliary variable (X). The response is not selective. Estimates are not biased. There is no problem.

In case of MCAR, the response behavior (R) and any auxiliary variable (X) are unrelated. If it is also known that there is a strong relationship between the target variable (Y) and the auxiliary variable (X), this is an indication that there is no strong relationship between target variable (Y) and response behavior (R) and thus the estimators do not have a severe bias.

It should be noted that if there is no strong relationship between the auxiliary variable (X) and the target variable (Y), analysis of the relationship between the auxiliary variable (X) and the response behavior will provide no information about a possible bias of estimates.

The second response mechanism is *missing at random* (MAR). This situation occurs when there is no direct relationship between the target variable (Y) and the response behavior (R), but there is a relationship between the auxiliary variable (X) and the response behavior (R). The response will be selective, but this can be cured by applying a weighting technique using the auxiliary variable. Chapter 10 is devoted to such weighting techniques.

In case of MAR, response behavior (R) and the corresponding auxiliary variable (X) will turn out to be related. If it is also known that there is a strong relationship between the target variable (Y) and the auxiliary variable (X), this is an indication there is (an indirect) relationship between target variable (Y) and response behavior (R), and thus the estimators may be biased.

The third response mechanism is *not missing at random* (NMAR). There is a direct relationship between the target variable (Y) and the response behavior (R) and this relationship cannot be accounted for by an auxiliary variable. Estimators are biased. Correction techniques based on use of auxiliary variables will be able to reduce such a bias.

All this indicates that the relationship between auxiliary variables and response behavior should be analyzed. If such a relationship exists and it is known that there is also a relationship between the target variables and auxiliary variables, there is a serious the risk of biased estimates. So, application of nonresponse correction techniques should be considered.

9.4.2 Where to Find Auxiliary Variables?

To be able to analyze the effects of nonresponse, auxiliary variables are needed. Those variables have to be measured in the survey and moreover information about the distribution of these variables in the population (or in the complete sample) must be available.

One obvious source of auxiliary information is the sampling frame itself. For example, if the sample is selected from a population register, variables such as age (computed from date of birth), gender, marital status, household composition, and geographical location (e.g., neighborhood) are available. The values of these variables can be recorded for both respondents and nonrespondents.

The sample for the 1998 Integrated Survey on Household Living Conditions (POLS) of Statistics Netherlands was selected from the population register. It was a stratified two-stage sample. In the first stage municipalities were selected within regional strata. In the second stage, a sample was drawn in each selected municipality. Sampling frames were the population registers of the municipalities. These registers contain, among other variables, marital status. So, marital status is known for both respondents and nonrespondents. Figure 9.7 shows the response behavior for the various categories of marital status.

Married people have the highest response rates (62.6%). Response is also reasonably high for unmarried people (61.5%), but response is much lower for divorced people (51.0%) and widowed people (53.4%).

It is also possible to collect auxiliary information about respondents and nonrespondents by letting interviewers record observations about the location of the selected persons. Examples are the neighborhood, type of house, and age of house.

Figure 9.8 shows the relationship between response behavior and the building period of the house. Response is worse in houses that have been built between the two world wars. Of course, there is no causal relationship between the building period of a house and the response behavior of its inhabitants. The differences in response rates are probably caused by different socioeconomic characteristics of the people living in the house. This calls for more analysis.

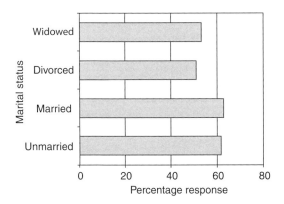

Figure 9.7 Response by marital status in POLS.

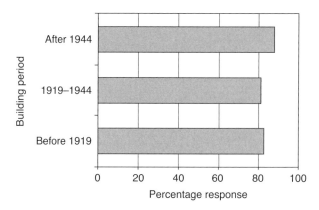

Figure 9.8 Response by building period of house in the Housing Demand Survey 1977–1978.

National statistical institutes and related agencies are a third source of auxiliary information. The publications (on paper or electronic) of these institutes often contain population distributions of auxiliary variables.

By comparing the town of residence of respondents in the 1998 Integrated Survey on Household Living Conditions (POLS) of Statistics Netherlands with the population distribution over towns, the relation between response behavior and town size can be explored (see Fig. 9.9).

A well-known phenomenon can be observed in this figure: getting response in big towns is much harder than getting it in small towns. The response is high in rural areas (60.3%), but low in urbanized areas (41.1%). Getting response is particularly difficult in the three big cities in The Netherlands: Amsterdam, Rotterdam, and The Hague.

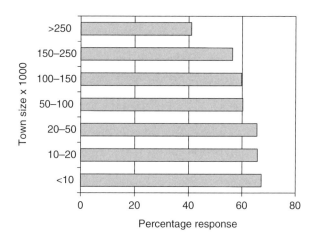

Figure 9.9 Response by town size in POLS.

9.4.3 Nonresponse Analysis of POLS 1998

As an example, nonresponse in the 1998 Integrated Survey on Household Living Conditions (POLS) is analyzed in this section. A lot of auxiliary information was available for this survey. POLS is a large continuous survey of Statistics Netherlands. Every month, a sample is selected. The survey consists of a number of thematic modules. Persons are selected by means of a stratified two-stage sample. In the first stage, municipalities are selected within regional strata with probabilities proportional to the number of inhabitants. In the second stage, an equal probability sample is drawn in each selected municipality. Sampling frames are the population registers of the municipalities.

The fieldwork of POLS 1998 covered a period of two months. In the first month, selected persons where approached with CAPI. For persons who could not be contacted or refused and who had a listed phone number, a second attempt was made in the second month using CATI. Table 9.7 contains the fieldwork results.

The sample size mentioned in Table 9.7 is the final sample size. The initial sample size was larger. It consisted of 39,431 persons. In 129 cases, persons did not belong to the target population of the survey. So, they were removed from the sample (overcoverage).

Ultimately, about 61% of the sampled persons responded. Note that almost 60% of these respondents (14,275 out of 24,008) refused one or more times before they cooperated.

The composition of the nonresponse is displayed in Fig. 9.10. By far, it is clear that refusal is the largest cause of nonresponse (58%). In 16% of the cases, no contact could be established with the sampled persons. Also, 16% of the cases were not processed in the field. Reasons for this type of nonresponse are lack of capacity (high workload of the interviewer) and interviewer not available (illness, holiday). Only 8% of the

Table 9.7 The Fieldwork Results of POLS 1998

Result	Frequency	Percentage
Sample size	39,302	100.0
Response	24,008	61.1
Immediate response	9,718	24.7
Converted refusers	14,275	36.3
Other response	15	0.0
Nonresponse	15,294	38.9
Unprocessed cases	2,514	6.4
Non contact (not-at-home)	2,093	5.3
Non contact (moved)	376	1.0
Not-able (illness, handicap)	735	1.9
Not-able (language problem)	416	1.1
Refusal	8,918	22.7
Other nonresponse	242	0.6

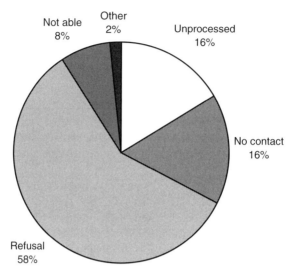

Figure 9.10 The composition of the nonresponse in POLS.

nonresponse is caused by the people who are not able to answer the questions due to illness, handicap, or language problems.

In the early nineties of the last century, Statistics Netherlands started the development of an integrated system of social statistics. This system is called the *Social Statistics Database* (SSD). The SSD will ultimately contain a wide range of characteristics on each individual in The Netherlands. There will be data on demography, geography, income, labor, education, health, and social protection. These data are obtained by combining data from registers and administrative data sources. Moreover, data from surveys are included. These data relate to attitude, behavior, and so on. For more information about the SSD, see Everaers and Van Der Laan (2001).

SSD records can be linked to the survey data records using internal personal identification numbers. This can be done both for respondents and nonrespondents. Thus, demographic variables such as sex, age, province of residence, and ethnicity became available for all sampled persons and also socioeconomic variables such as employment and various types of social security benefits.

The Netherlands is divided in approximately 420,000 postal code areas. A postal code area contains, on average, 17 addresses. These areas are homogeneous with respect to social and economic characteristics of its inhabitants. Using information from the population register, Statistics Netherlands has computed some demographic characteristics for these postal code areas. Since postal codes are included in the survey data file for both respondents and nonrespondents, these characteristics can be linked to the survey data file. Among the variables used in this analysis are degree of urbanization, town size, and percentage of people with a foreign background (nonnatives). From another source also the average house value was included.

During the fieldwork period, interviewers kept record of all contact attempts. For each attempt, its contact result was recorded (contact, or not). In case contact was established, the result of the cooperation request was recorded (response or nonresponse, and in case of nonresponse the reason of nonresponse). Also other information was included, like the mode of the fieldwork attempt (CAPI or CATI), and whether there was contact with the person to be interviewed or another member of the household. All this fieldwork information was included in the analysis data file.

Two other variables were included in the survey data file. The first one was the interviewer's district code. Thus, for every respondent and nonrespondent, it is known which interviewer made the contact attempts. The second variable was an indicator whether a selected person has a listed telephone number or not.

In the nonresponse analysis, possible relationships between auxiliary variables and response behavior were explored. The most interesting results are presented here.

Figure 9.11 shows the relationship between response behavior and age. Response is high for the people younger than the age of 20. Response is much lower for those between 20 and 30 years of age. There are relatively many unprocessed cases and noncontacts. Over the years, response rates tend to increase, but they drop again for the elderly. The group of not-able persons is particularly large here.

Figure 9.12 shows the possible effects of marital status on the fieldwork results. The response rate is highest for married people. Both the groups of noncontacts and not-ables are small. This is a common phenomenon. These are often young or middle-aged people with a family. Making contact is relatively easy.

There is a larger number of unprocessed cases and noncontacts for unmarried people. This group may coincide at least partially with the young people in the previous graph. For divorced people, it is apparently difficult to make contact. Also, the number of unprocessed cases is large. Among the widowed people, the group of not-able is large. This group probably coincides with the elderly in Fig. 9.11.

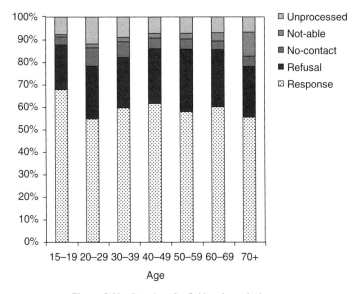

Figure 9.11 Bar chart for fieldwork results by age.

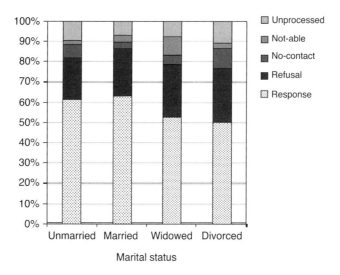

Figure 9.12 Bar chart for fieldwork results by marital status.

Figure 9.13 shows the relationship between the fieldwork result and the size of the household. There is a clear trend: response rates increase with the size of the household. Not surprisingly, nonresponse due to noncontact is less likely as the household size increases. Also, the refusal rate and the number of unprocessed cases are smaller for larger households.

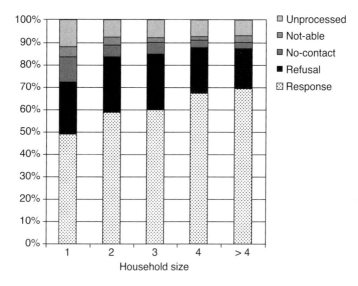

Figure 9.13 Bar chart for fieldwork results by household size.

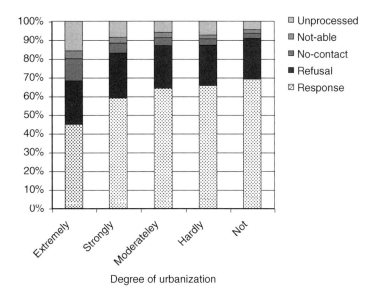

Figure 9.14 Bar chart for fieldwork results by degree of urbanization.

The results of this graph confirm an earlier conclusion that it is relatively easy to obtain response from families with children.

The results in the following four graphs explore relationships between response behavior and characteristics of the neighborhood in which people live. The first variable is degree of urbanization. Figure 9.14 shows its relationship with the fieldwork result.

Response rates are very low in the extremely urbanized areas. These are the four largest towns in The Netherlands (Amsterdam, Rotterdam, The Hague, and Utrecht). Also, note the high number of unprocessed cases here. Furthermore, the noncontact rate is high in densely populated areas. Response rates are high in rural areas. Note that there is not much variation in refusal rates.

The Netherlands is divided into 12 provinces. Figure 9.15 shows how the fieldwork results differ by province. Response rates are low in three provinces: Utrecht, Noord-Holland, and Zuid-Holland. These are the three most densely populated provinces. The four largest cities lie in these provinces. So, this confirms the pattern found in Fig. 9.14 that it is difficult to get a high response rate in big cities.

The Netherlands is divided into approximately 420,000 postal code areas. Each area contains around 17 houses. The average house value is available in each area. Since the postal code of each sampled person is known, the relationship between response behavior and the average house value in the area can be explored. Figure 9.16 shows the result.

The graph shows a clear pattern: response is low in areas with cheap housing. Note that nonresponse is particularly caused by a high noncontact rate and a large number of unprocessed cases. Refusal rates are somewhat lower here.

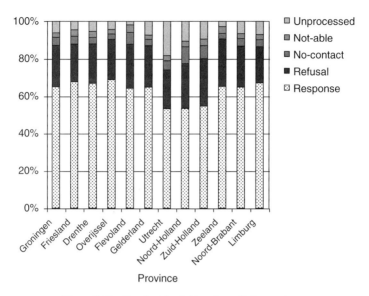

Figure 9.15 Bar chart for fieldwork results by province.

Approximately 3 million of the total 16 million inhabitants of The Netherlands have a foreign background. There are substantial ethnic minority groups from Turkey, Morocco, and the former colonies in the West Indies and South America (Surinam). The percentage of nonnative people in each of the approximately 420,000 postal code areas is available. So, a possible relationship between response behavior and nonnative background can be analyzed. The results are displayed in Fig. 9.17.

A clear, almost linear, pattern can be observed: Response rates decrease as the percentage of nonnatives in the neighborhood increases. In areas with more than 50% nonnatives, response rate drops to 40%. The high number of unprocessed cases is a major cause of nonresponse. Also, the noncontact rate is high. The high percentage of not-able cases is caused by language problems.

It is remarkable that the refusal rate is very low among nonnatives. This seems to contradict the believe of many natives that nonnatives refuse to integrate in the population.

Together, Figs 9.16 and 9.17 seem to suggest that response rates are low in areas with a low socioeconomic status.

One more variable turned out to be interesting. This variable indicates whether a selected person has a listed phone number or not. For every person selected in the sample, it is known whether he or she has a listed phone number or not. The telephone company provides phone numbers, but only for those people with a fixed-line phone that is listed in the directory.

From Fig. 9.18, it becomes clear that people with a listed phone number have a much higher response rate. People without such a number tend to refuse more and are much harder to contact. Also, there are a larger number of unprocessed cases. Figure 9.18 seems to confirm the hypothesis sometimes found in the literature that social isolation may be a factor contributing to nonresponse.

ANALYSIS OF NONRESPONSE   **235**

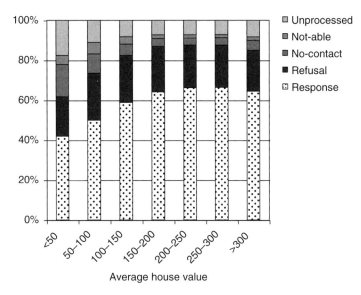

Figure 9.16 Bar chart for fieldwork results by average house value.

Analysis of the POLS data shows that additional auxiliary variables help to explain what is going on with respect to response and nonresponse. Not only demographic and socioeconomic variables are useful in this respect but also fieldwork variables that describe various contact attempts. Traditionally, fieldwork reports are made to monitor fieldwork and interviewer performance. Use of this type of information in a nonresponse analysis requires this information to be recorded in a more systematic way in

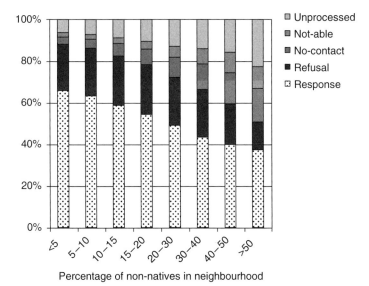

Figure 9.17 Bar chart for fieldwork results by percentages of nonnatives.

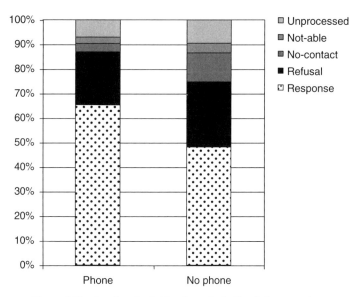

Figure 9.18 Bar chart for fieldwork results by listed phone number.

the survey data file. Also, it is important that fieldwork information becomes a standard part of this file.

It is a good idea to split the response mechanism in two sequential phases. The first phase is that of the contact attempt. The second phase is that of the cooperation attempt once contact has been established. Analysis of both phenomena may require different models and different auxiliary variables. However, in practical survey situations it is not easy to separate both mechanisms. Future survey design should attempt to take care of this in a better way. Of course, it also remains important to distinguish other groups of nonrespondents.

9.5 NONRESPONSE CORRECTION TECHNIQUES

There is ample evidence that nonresponse often causes population estimates to be biased. This means that something has to be done to prevent wrong conclusions to be drawn from the survey data. There are several correction approaches possible.

A frequently used correction technique is *adjustment weighting*. It assigns weights to the observed elements. These weights are computed in such a way that overrepresented groups get a smaller weight than underrepresented groups. Adjustment weighting has many aspects. Chapter 10 is completely dedicated to this approach.

In the remainder of this chapter, two other approaches are described: the *follow-up survey* and the *basic question approach*. To be able to assess whether nonresponse causes estimators to be biased, information about nonrespondents is needed. This is difficult to achieve as nonrespondents by definition do not provide information. The follow-up survey and the basic question approach attempt to at least partially solve this problem.

9.5.1 The Follow-Up Survey

Hansen and Hurwitz (1946) were among the first to recognize that nonresponse can lead to biased estimates of population parameters. They proposed investigating nonresponse in mail surveys by taking a sample of nonrespondents and trying to obtain the required information by means of a face-to-face interview. If the information collected in the second phase is representative for all nonrespondents, an indication can be obtained of the differences between respondents and nonrespondents. Furthermore, it is possible to correct for a nonresponse bias.

The basic idea of Hansen Hurvitz was to conduct a *follow-up survey* among nonrespondents. Such a follow-up survey is also possible if data collection in the main survey is carried out by means of face-to-face interviews instead of through mail questionnaires. Then, specially trained interviewers can reapproach the nonrespondents. Of course, this substantially increases the survey costs.

The follow-up survey is described under the fixed response model. Then, the target population consists of two strata: a stratum of respondents and a stratum of nonrespondents. Suppose a simple random sample of size n is selected without replacement from this population. The sample is denoted by the set of indicators $a_1, a_2, \ldots, a_N$, where $a_k = 1$ means that element k is selected in the sample, and otherwise $a_k = 0$. The number of elements selected in the response stratum is denoted by

$$n_R = \sum_{k=1}^{N} a_k R_k \tag{9.28}$$

and the number of selected elements in the nonresponse stratum is denoted by

$$n_{NR} = \sum_{k=1}^{N} a_k (1 - R_k), \tag{9.29}$$

where $n = n_R + n_{NR}$.

Only the values of Y of the n_R selected elements in the response stratum are available for estimation purposes. The mean of these values is denoted by

$$\bar{y}_R = \frac{1}{n_R} \sum_{k=1}^{N} a_k R_k Y_k. \tag{9.30}$$

The bias of this estimator is equal to

$$B(\bar{y}_R) = \bar{Y}_R - \bar{Y} = \frac{N_{NR}}{N} (\bar{Y}_R - \bar{Y}_{NR}) = QK, \tag{9.31}$$

where K is the contrast and $Q = N_{NR}/N$ is the relative size of the nonresponse stratum.

For the follow-up survey, a simple random sample is selected from the nonrespondents in the main survey. This comes down to drawing a simple random sample from the nonresponse stratum.

Formally, this approach contradicts the assumptions underlying the fixed response model. This model assumes the existence of a subpopulation consisting of elements who would never respond in a survey. However, the fixed response model should be

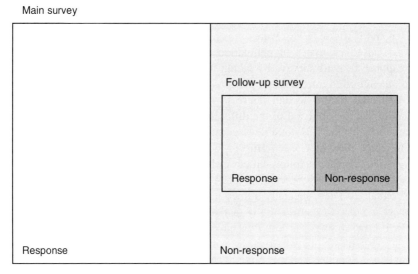

Figure 9.19 The follow-up survey.

seen as conditional on the survey design. The design of a follow-up survey assumes a different population (the main survey nonrespondents) with different response and nonresponse strata. The situation is depicted in Fig. 9.19.

If everybody responds in the follow-up survey, or if the nonresponse in the follow-up survey is ignorable (i.e., there is no direct correlation with target variables), it is possible to compute unbiased estimates of the parameters of the nonresponse stratum. However, one may wonder whether this condition is always fulfilled in practical situations. It is, for example, possible that the population consists of three strata: respondents, soft respondents (they cooperate in the follow-up), and hard nonrespondents (they never cooperate).

Suppose a sample of size m is selected for the follow-up survey. The number of respondents in this survey is denoted by m_R.

Let $\bar{y}_{NR}$ denote the mean of the m_R values of the responding elements. It is assumed that these m_R observations constitute a simple random sample from the nonresponse stratum of the main survey. This mean is an unbiased estimator of the mean of the target variable in the nonresponse stratum of the main survey. Consequently,

$$\frac{N_R}{N}\bar{y}_R + \frac{N_{NR}}{N}\bar{y}_{NR} \tag{9.32}$$

is an unbiased estimator of the population mean of Y. Unfortunately, the sizes of the response stratum and the nonresponse stratum are unknown. Therefore, the quantities N_R/N and N_{NR}/N are replaced by their unbiased estimates n_R/n and n_{NR}/n. This results in the estimator

$$\frac{n_R}{n}\bar{y}_R + \frac{n_{NR}}{n}\bar{y}_{NR} \tag{9.33}$$

Under the condition mentioned, this is an unbiased estimator.

From July 2005 to December 2005, Statistics Netherlands conducted a large-scale follow-up of nonrespondents in the Dutch Labor Force Survey (LFS). For a detailed description of this study, see Schouten (2007).

A sample of 775 LFS nonrespondents was approached once more by a small number of selected interviewers. The interviewers had received additional training in doorstep interaction, they could offer incentives and they could earn a bonus based on their response rate. The households that were eligible for the follow-up survey were former refusals, noncontacts, and nonprocessed households.

An additional response of 43% was obtained, leading to a weighted overall response rate of 77%. It turned out that the respondents in the follow-up survey differed from the LFS respondents with respect to geographical variables, having a listed landline telephone and ethnicity. The follow-up survey respondents more often lived in the more urbanized, western parts of The Netherlands. Furthermore, households that did not have a listed landline telephone were overrepresented as were Moroccan and non-Western households other than Moroccan and Turkish households.

Furthermore, the follow-up respondents resembled the follow-up nonrespondents with respect to demographic and socioeconomic characteristics. So, they were a good representation of the nonresponse in the LFS.

Besides the differences in background characteristics between follow-up respondents and the LFS respondents, there was no significant difference in job and employment status. This implies that the survey estimates of employment were unaffected by the addition of the follow-up response.

9.5.2 The Basic Question Approach

A follow-up survey such as proposed by Hansen and Hurwitz (1946) will be expensive. The costs per interview in the follow-up survey will be much higher than that in the main survey. It requires a lot of travel for a relative small amount of interviews. Another factor is timeliness. Sampling and interviewing nonrespondents will substantially increase the duration of the fieldwork period.

Kersten and Bethlehem (1984) have proposed the basic question approach as an alternative to the follow-up survey. It can be applied in situations where a follow-up survey cannot be carried out due to time and money constraints.

The *basic question approach* assumes that many survey questionnaires are composed around a few basic questions. Answers to these questions are required to be able to formulate the most important conclusions of the survey. If interviewers face problems in getting cooperation during the fieldwork, they can change their strategy and attempt to obtain only answers to a few basic questions "with the foot in the door." Another approach could be to carry out the basic question approach afterward by means of telephone or mail follow-up (for example, for the not-at-homes).

One way to apply the basic question approach is to let the interviewers attempt the basic questions straight away after they have been confronted with a refusal for the main questionnaire. This may lead to higher nonresponse rates for the main questionnaire. Therefore, it is better, but also more expensive, to reapproach the refusers after a short while with different interviewers.

The main goal of the basic question approach is to gain insight in possible differences between respondents and nonrespondents with respect to the most important variables of the survey. If such differences are detected, the approach also provides information for correcting estimates for other variables.

The basic question approach was born from the observation that people who refuse to participate, can often be persuaded to answer a few basic questions. Many surveys have basic questions. Only these questions are asked when it is clear that the further attempts to get the questionnaire completed will be useless.

It is stressed that these questions can only be a limited approximation of the set of research variables that have to be measured. Often the values of important research variables depend on the answers to several questions. If not all these questions can be asked, the value of the research variable cannot be derived.

Here it is supposed that there is just one basic question. Of course, it is possible to ask more basic questions. It is advisable to keep the number of basic questions as small as possible. The more questions are asked, the higher the risk of getting no information at all. Here are some examples of basic questions that have been used in surveys of Statistics Netherlands:

- *Housing Demand Survey*. Do you intend to move within 2 years?
- *Labor Force Survey*. How many people in this household have a paid job?
- *Holiday Survey*. Have you been on holiday during the last 12 months?
- *Family Planning Survey*. Taking into account your present circumstances and your expectations of the future, how many children do you think to get from this moment on?

The basic question approach helps to get as much answers as possible to the important questions of the survey. The approach seems to have worked well in several specific surveys. People who refuse to cooperate can be persuaded to answer just a few questions if the interviewer states "OK, I accept your refusal, but please help me to fix my administration," or "OK, I will not persist any more, but at least answer this question." Even for refusal in a telephone survey the basic question approach may work.

Considerable insight into the characteristics of nonrespondents can be obtained in situations where especially the name of the survey causes nonresponse. This may occur when people think that the survey does not apply to them, that is, if they do not intend to move (in a housing demand survey), they do not have a job (in a labor force survey), or they do not visit a doctor (in a health survey).

The large-scale follow-up of nonrespondents to the Dutch LFS also comprised the basic question approach. A sample out of the nonresponse was reapproached using a basic questionnaire. The regular LFS was face-to-face. The second wave (with basic questions) was by telephone for those addresses where a listed landline telephone was available. If no telephone number was available, households were asked to fill in either a paper or web questionnaire. The questionnaires used were a strongly condensed form

Table 9.8 Results of the Basic Question Approach in the Dutch LFS

Group	Mode	Sample Size	Response Rate (%)
Nonrespondents	Telephone	564	50
	Paper or Web	378	23
Control group	Telephone	667	80
	Paper or Web	333	25

of the regular questionnaire. The condensed questionnaire contained a maximum of 10 questions and took between 1 and 3 min to answer. For analytical purposes also a fresh control group received the same treatment. Table 9.8 shows the response rates of the various groups.

A response rate of 50% could be obtained by reapproaching nonrespondents by telephone with basic questions. This is, of course, lower than the 80% of the control group. This should come as no surprise as the control group also contains households that would have participated in the regular LFS questionnaire.

To be able to compare respondents and nonrespondents, the basic question must be answered by both respondents and nonrespondents. To avoid all kinds of interviewing effects, questionnaire effects and mode effects, the basic question must be presented to respondents in a situation that resembles the nonresponse situation as much as possible. Therefore, the basic questions should be among the first questions in the questionnaire. It is also important that the answer to the basic question is not changed when the answers to subsequent questions indicate that the answer to the basic question may be wrong.

The estimation procedure for the basic question approach is described under the fixed response model. Then, the population can be divided in a stratum U_R of N_R (potential) respondents and a stratum U_{NR} of N_{NR} nonrespondents, with $N = N_R + N_{NR}$.

Suppose, there is some target variable Y with values $Y_1, Y_2, \ldots, Y_N$, and a basic question variable Z with values $Z_1, Z_2, \ldots, Z_N$. A simple random sample of size n is selected without replacement. There are n_R respondents and n_{NR} nonrespondents. Not every nonrespondent answers the basic question. The number of nonrespondents who answer the basic question is denoted by m_{NR}.

The response means $\bar{y}_R$ and $\bar{z}_R$ are unbiased estimators of the response stratum means of the target variable and the basic question variable, respectively.

Estimating the mean of the basic question variable in the nonresponse stratum is not so simple. The fundamental question is: May nonrespondents who answer the basic question be regarded as a simple random sample from all nonrespondents? It is assumed that this is the case.

Let $\bar{z}_{NR}$ be the mean of the available m_{NR} values of the basic question variable in the nonresponse stratum. Then, this mean is an unbiased estimator of the mean of the basic question variable in nonresponse stratum. Consequently,

$$\frac{N_R}{N}\bar{z}_R + \frac{N_{NR}}{N}\bar{z}_{NR} \qquad (9.34)$$

is an unbiased estimator. Unfortunately, the sizes of the response stratum and the nonresponse stratum are not known. Therefore, the quantities N_R/N and N_{NR}/N are replaced by their unbiased estimates n_R/n and n_{NR}/n. This results in the estimator

$$\bar{z}_{BQ} = \frac{n_R}{n}\bar{z}_R + \frac{n_{NR}}{n}\bar{z}_{NR}. \tag{9.35}$$

This is an unbiased estimator, provided the nonrespondents answering the basic question are a simple random sample from all nonrespondents.

The basic question approach has been tested in the Dutch Housing Demand Survey 1981 (see Kersten and Bethlehem, 1984). Excluding overcoverage, the sample size of this face-to-face survey was 82,849. The number of respondents was 58,972 and this amounts to a response percentage of 71%. When contacted people refused to cooperate, the basic question approach was tried at the door. In total, 8383 refusers could be persuaded to answer the basic question. This implies that $58,972 + 8383 = 67,355$ people answered the basic questions, which comes down to a response percentage of 81% (for this variable).

The basic question in the survey was: "Do you intend to move within two years?" Table 9.9 shows the results of these questions for two groups: the initial respondents and the refusers who answered the basic question. It is clear that there is a difference between the two groups. Initial respondents are much more inclined to move than refusers answering the basic question.

In the Dutch Housing Demand Survey 1981, also a second wave of fieldwork was carried out. Callbacks were made for a sample of nonrespondents including those who answered the basic question. This provided a means to check the answers to the basic question. For the 1638 refusers in the first wave, both their answer to the basic question in the first wave and their answer in the complete interview in the second wave became available. The results are presented in Table 9.10.

The same answer was given in $8.6 + 74.1 = 82.7\%$ of the cases. So, there is a reasonable amount (but not complete) of consistency. Note that there was a time lag of 3 months between the two waves. It is not unlikely that at least some people may have changed their mind in this period.

It is clear that estimation for basic variables can be improved. But what about the other variables in the survey? It is also possible to improve estimation for these

Table 9.9 Results of the Basic Question Approach in the Dutch Housing Demand Survey 1981

Do you Intend to Move Within 2 Years?	Initial Respondents (%)	Refusers Answering the Basic Question (%)
Yes	29.7	12.8
No	70.3	87.2
Total	100.0	100.0

Table 9.10 Checking the Basic Question Approach in the Dutch Housing Demand Survey 1981

	Second Wave	
First Wave	Intends to Move (%)	Does not Intend to Move (%)
Intends to move	8.6	5.1
Does not intend to move	12.1	74.1

variables. To that end, the basic question variable is treated as an auxiliary variable. If the basic question variable is a qualitative variable, the poststratification estimator can be used and if the basic question variable is a quantitative variable, the ratio estimator or regression estimator can be used.

First, the case of a qualitative basic question variable is considered. The expression for the poststratification estimator for a target variable Y other than a basic question is

$$\bar{y}_{PS} = \frac{1}{N} \sum_{h=1}^{L} N_h \bar{y}^{(h)}. \qquad (9.36)$$

To be able to apply poststratification, the numbers N_h of population elements in the strata corresponding to the categories of the basic question variable must be available. This is not the case, but they can be estimate using the answers to the basic question (both for the initial respondents and the refusers). Next, the average of the target variable is computed for every category of the basic question variable. Only the data for the initial respondents can be used for this. By substituting these means in expression 9.36, an estimate is obtained, that is, hopefully, less biased. It has been shown already that poststratification works better as target variable and auxiliary variable have a stronger relationship. Since the basic question variable is also a target variable of the survey and target variables are often correlated, it is not unlikely that the basic question approach produces better estimates.

Now the case of a quantitative basic question variable is considered. One way to improve the estimate for a target variable is to use a ratio estimator in which the basic question variable is used as an auxiliary variable. In the case, the ratio estimator would take the form

$$\bar{y}_{RAT} = \bar{y}_R \frac{\bar{z}_{BQ}}{\bar{z}_R}. \qquad (9.37)$$

The more the values of Y and Z are proportional, the more effective the ratio estimator is.

An even better approach is to use the regression estimator in which the basic question variable plays the role of auxiliary variable. This estimator would take the form

$$\bar{y}_{REG} = \bar{y}_R - b(\bar{z}_R - \bar{z}_{BQ}). \qquad (9.38)$$

Table 9.11 Results of the Follow-Up Survey and the Basic Question Approach in an Election Survey

Result	Cases	Percentage
Response in first wave	508	51.1
Response in basic question approach	196	19.7
Response in callback approach	224	22.5
Final nonresponse	67	6.7
Total	995	100.0

This estimator is effective if there is a more or less linear relationship between the values of the target variable and the basic question variable.

Voogt (2004) presents an interesting example of a survey in which both a follow-up survey and the basic question approach were applied. His research focused on nonresponse bias in election research. He selected a simple random sample of 995 voters from the election register of the town of Zaanstad in The Netherlands. There were two basic questions in this survey:

- Did you vote in the parliamentary election on Wednesday May 6, 1998?
- Are you interested in politics, fairly interested or not interested?

In the first wave of the survey, people were contacted by phone if a phone number was available. If not, they were send a questionnaire by mail. The basic question approach was applied in a separate follow-up. All refusers were offered the possibility to answer just the two basic questions (by phone or mail). The follow-up approach was applied to those who refused to cooperate in the basic question approach. This time the refusers were visited at home by interviewers. The results of the fieldwork are summarized in Table 9.11.

One conclusion that can be drawn from this table is that the situation need not be hopeless if the response is low in the first wave. With additional measures, response rates can be increased substantially.

Because the researcher had access to the voting register of the town, he could establish with certainty whether all 995 people in the survey had voted or not. In this group, 72.9% had voted. The voting behavior for the various groups is listed in Table 9.12.

Table 9.12 Voting Behavior in the Follow-Up Survey and the Basic Question Approach of an Election Survey

Group	% Voters
Response in first wave	85.4
Response in basic question approach	66.3
Response in follow-up approach	55.8
Final nonresponse	53.7

The groups are ordered in growing reluctance to participate. There seems to be a relationship between this reluctance and voting behavior: the more reluctant the group, the lower is the percentage of voters. It can be concluded that the response in the basic question approach is not representative for all nonresponse after the first wave. Applying estimator (9.35) leads to an estimate of

$$\frac{508}{995} \times 85.4 + \frac{487}{995} 66.3 = 76.4. \qquad (9.39)$$

This value is much better than the 85.4% for the initial response, but it is still too high.

EXERCISES

9.1 A survey is usually carried out to measure the state of a target population at a specific reference date. The survey outcomes are supposed to describe the status of the population at that point in time. Ideally, the fieldwork of the survey should take place at that date. This is not possible in practice, so interviewing usually takes place in a period of a number of days or weeks around the reference date.

Suppose, a business survey is carried out. A sample of companies is selected from the sampling frame (the business register) 2 weeks before the reference date. Interviewing takes place in the period of 4 weeks: the 2 weeks between sample selection and reference date and the 2 weeks after the reference date.

For each of the situations described below, explain whether there is a problem and if so, explain what kind of problem it is: nonresponse, undercoverage, overcoverage, or an error in the sampling frame (a frame error).

a. The contact attempt takes place between the sample selection date and the reference date. It turns out the company that went bankrupt (and thus it does not exist any more) before the sample selection date.

b. The contact attempt takes place between the sample selection date and the reference date. It turns out the owner who went out of business (and thus the company does not exist any more) after the sample selection date.

c. The contact attempt takes place after the reference date. It turns out the company has moved to a different country before the sample selection date.

d. The contact attempt takes place after the reference date. It turns out the company went bankrupt (and thus it does not exist any more) between the sample selection date and the reference date.

e. The contact attempt takes place after the reference date. It turns out the company that was destroyed by a fire (and thus the company does not exist any more) after the reference date.

9.2 A town council wants to do something about the traffic problems in its town center. There is a plan to turn it into a pedestrian area. So, cars will not be able to access the center any more. The town council wants to know what companies think of this plan. A simple random sample of 1000 companies is selected. Each selected company is invited to participate in the survey. They are asked whether they are in favor of the plan, or not. Furthermore, the location of the company is recorded (town center or suburb). The results of the survey are summarized in the table below:

	Suburbs	Town Center
In favor	120	80
Not in favor	40	240

  a. Compute the response percentage.
  b. Compute the percentage of respondents in favor of the plan.
  c. Compute a lower bound and an upper bound for the percentage in favor in the complete sample.

9.3 A survey is carried to measure how much money people spend on health care. The target population consists of 24,000 people. A sample of 800 persons is selected. Only 600 people respond. Among the respondents, the average amount spent on health care per year is €1240. Suppose it is known that the health care costs of nonrespondents are on average 10% higher. Using this information, compute a better estimate of average health care costs.

9.4 A researcher wants to find out whether inhabitants of a town are interested in local politics. To that end, he carries out a survey. Unfortunately, the survey is affected by nonresponse. The total population of potential voters consists of 38,000 people. Suppose, the fixed response model applies and we have the following distribution for interest in local politics over response and nonresponse stratum:

	Interest in Politics	
Response	Yes	No
Yes	15,200	3,800
No	7,600	11,400

  a. Compute the expected fraction of interested people if a simple random sample is selected and the response mechanism works as described in the above table.
  b. Compute the value of the contrast K.
  c. Compute the bias of the estimator.

EXERCISES

9.5 A researcher carries out a time budget survey. Among the things, he wants to know is the time spent (per week) on surfing the Internet. A simple random sample of 20 households is selected. Each selected household is asked for its number of members and the numbers of hours spent on the Internet. The results are in the table below:

Household	Members	Internet Hours	Household	Members	Internet Hours
1	1	6	11	1	–
2	2	–	12	2	9
3	3	17	13	4	–
4	4	–	14	5	27
5	4	–	15	6	28
6	1	–	16	1	–
7	2	–	17	3	–
8	4	18	18	4	20
9	5	23	19	5	–
10	6	32	20	7	35

The survey suffers from nonresponse. Therefore, it is not possible to record time spent on the Internet for some households. The household size can be retrieved from the sampling frame.

a. Assuming the households for which the Internet variable is available form a simple random sample, estimate the average hours spent on the Internet.

b. Looking at the available data, explain why the nonresponse will probably be selective with respect to hours spent on the Internet.

c. Use the ratio estimator to computer a better estimate. Use household size as auxiliary variable.

d. Compare the outcomes under (a) and (c). Explain why (or why not) the ratio estimator produces better estimates.

9.6 A simple random sample of size 2000 is selected from the population of 20,000 potential voters in the town of Harewood. Objective of this opinion poll is to estimate the percentage of voters for the new political party "Forza Harewood." Only 50% of the selected voters wants to participate in the survey. Among those, 300 say that they will vote for the new party.

a. Assuming the response is a simple random sample from the population, compute an estimate, and also the 95% confidence interval, for the percentage of "Forza Harewood."

b. A simple random sample of size 100 is selected from the nonrespondents. With a lot of extra efforts and specially trained interviewers, these nonrespondents are reapproached. It turns out that they all want to cooperate in this follow-up survey and 10 people say they will vote for the new party.

Use all available information to compute a better estimator.
   c. Assuming the margin of the confidence interval computed under (a) is not affected by nonresponse, what can be said about the confidence level of the interval computed under (a)?

9.7 The basic question approach can be used to reduce the negative effects of nonresponse. Assuming there is only one basic question, what should its position be in the questionnaire for the respondents?
   a. At the beginning of the questionnaire.
   b. At the end of the questionnaire.
   c. The location of the question is not relevant.
   d. The question need not be included in the questionnaire.

CHAPTER 10

Weighting Adjustment

10.1 INTRODUCTION

There is ample evidence that nonresponse often causes estimates to be biased. This means that something has to be done to correct this bias. A frequently used technique is *adjustment weighting*. Adjustment weighting is typically applied in case of unit nonresponse. Different correction techniques are available for item nonresponse (see Chapter 8).

Adjustment weighting is based on the use of *auxiliary information*. Auxiliary information is defined in this context as a set of variables that have been measured in the survey and for which information on the population (or the complete sample) distribution is available. By comparing the population distribution of an auxiliary variable with its response distribution, it can be assessed whether or not the response is representative for the population (with respect to this variable). If these distributions differ considerably, one must conclude that nonresponse has resulted in a selective sample.

As a next step, this auxiliary information can be used to compute *adjustment weights*. Weights are assigned to all observed records of observations. Estimates of population characteristics can now be obtained by using the weighted values instead of the unweighted values. The weights are defined in such a way that population characteristics for the auxiliary variables can be computed without error. Then the weighted sample is said to be *representative* with respect to the auxiliary variables used.

Suppose, the *inclusion weight* $c_i = 1/\pi_i$ is introduced as one over the first-order inclusion probability of selected element i, for $i = 1, 2, \ldots, n$. Consequently, the Horvitz–Thompson estimator can be written as

$$\bar{y}_{HT} = \frac{1}{N} \sum_{i=1}^{n} c_i y_i. \qquad (10.1)$$

Applied Survey Methods: A Statistical Perspective, Jelke Bethlehem
Copyright © 2009 John Wiley & Sons, Inc.

Adjustment weighting replaces this estimator by a new estimator

$$\bar{y}_W = \frac{1}{N} \sum_{i=1}^{n} w_i y_i, \qquad (10.2)$$

where the weight w_i is equal to

$$w_i = c_i \times d_i \qquad (10.3)$$

and d_i is a *correction weight* produced by an weighting adjustment technique.

If the response can be made representative with respect to several auxiliary variables, and if all these variables have a strong relationship with the phenomena to be investigated, then the (weighted) sample will also be (approximately) representative with respect to these phenomena, and hence estimates of population characteristics will be more accurate.

Several weighting techniques will be described in this chapter. It starts with the simplest and most commonly used one: *poststratification*. Next *linear weighting* is described. It is more general than poststratification. This technique can be applied in situations where the auxiliary information is inadequate for poststratification. Then *multiplicative weighting* is discussed as an alternative for linear weighting. Furthermore, an introduction into *calibration* is provided. This can be seen as an even more general theoretical framework for adjustment weighting that includes linear weighting and multiplication as special cases. Finally, an overview of propensity weighting is given.

10.2 POSTSTRATIFICATION

Poststratification is a well-known and often used weighting method. Note that poststratification has already been introduced in Chapter 6 as an estimation technique that can lead to more precise estimators. In this chapter, it is shown that poststratification can also be effective in reducing nonresponse bias. First, the case of complete response is considered.

To be able to carry out poststratification, one or more qualitative auxiliary variables are needed. Suppose, there is an auxiliary variable X having L categories. So it divides the population U into L strata $U_1, U_2, \ldots, U_L$. The number of population elements in stratum U_h is denoted by N_h, for $h = 1, 2, \ldots, L$. So $N = N_1 + N_2 + \cdots + N_L$.

A sample of size n is selected from the population. If n_h denotes the number of sample elements in stratum U_h (for $h = 1, 2, \ldots, L$), then $n = n_1 + n_2 + \cdots + n_L$. Note that the values of the n_h are the result of a random selection process. So, they are random variables.

Poststratification assigns identical adjustment weights to all elements in the same stratum. In case of simple random sampling without replacement, the correction weight d_i for an observed element i in stratum U_h is equal to

$$d_i = \frac{N_h/N}{n_h/n}. \qquad (10.4)$$

POSTSTRATIFICATION

If the values of the inclusion probabilities ($c_i = n/N$) and correction weights (10.4) are substituted in expression (10.2), the result is the *poststratification estimator*

$$\bar{y}_{PS} = \frac{1}{N} \sum_{h=1}^{L} N_h \bar{y}^{(h)}, \qquad (10.5)$$

where $\bar{y}^{(h)}$ is the mean of the observed elements in stratum h. So, the poststratification estimator is equal to a weighted sum of sample stratum means.

The computation of adjustment weights is shown in an example. A sample of size 100 is selected from the Samplonian population of size 1000. There are two auxiliary variables: *Sex* (with two categories *male* and *female*), and *AgeClass* (with three categories *Young*, *Middle*, and *Old*). Table 10.1 contains the population and sample distribution of these variables.

The sample is not representative for the population. For example, the percentage of young females in the population is 20.9%, whereas the corresponding sample percentage is 15.0%. The sample contains too few young females.

The correction weights in Table 10.1 have been computed by means of expression (10.4). For example, the weight for young female is equal to $(209/1000)/(15/100) = 1.393$. Young females are underrepresented in the sample and therefore get a weight larger than 1. People in overrepresented strata get a weight less than 1.

The adjustment weights w_i are obtained by multiplying the correction weights d_i by the inclusion weights c_i. Here, all inclusion weights are equal to $N/n = 10$.

Table 10.1 Computation of Adjustment Weights in Case of Poststratification

	Population		
	Male	Female	Total
Young	226	209	435
Middle	152	144	296
Elderly	133	136	269
Total	511	480	1000
	Sample		
	Male	Female	Total
Young	23	15	38
Middle	16	17	33
Elderly	13	16	29
Total	52	48	100
	Weights		
	Male	Female	
Young	0.983	1.393	
Middle	0.950	0.847	
Elderly	1.023	0.850	

Suppose, these weights are used to estimate the number of young females in the population. The weighted estimate would be $15 \times 10 \times 1.393 = 209$, and this is exactly the population total. Thus, application of weights to the auxiliary variables results in perfect estimates. If there is a strong relationship between the auxiliary variable and the target variable, estimates for the target variable will be improved if these weights are used.

Now suppose the sample is affected by nonresponse. Then the poststratification estimator takes the form

$$\bar{y}_{R,PS} = \frac{1}{N} \sum_{h=1}^{L} N_h \bar{y}_R^{(h)}, \qquad (10.6)$$

where $\bar{y}_R^{(h)}$ denotes the mean of the responding elements in stratum h. It can be shown that the bias of this estimator is equal to

$$B(\bar{y}_{R,PS}) = \frac{1}{N} \sum_{h=1}^{L} N_h B(\bar{y}_R^{(h)}). \qquad (10.7)$$

Apparently, the bias of this estimator is the weighted sum of the biases of the stratum estimators. By applying the random response model, this bias can be written as

$$B(\bar{y}_{R,PS}) = \frac{1}{N} \sum_{h=1}^{L} N_h (\bar{Y}^{(h)} - \tilde{Y}^{(h)}), \qquad (10.8)$$

where $\bar{Y}^{(h)}$ is the mean of the target variable in stratum h, and

$$\tilde{Y}^{(h)} = \frac{1}{N_h} \sum_{k=1}^{N_h} \frac{\rho_k^{(h)}}{\bar{\rho}^{(h)}} Y_k^{(h)}. \qquad (10.9)$$

Here, $Y_k^{(h)}$ denotes value of the target value of element k in stratum, $\rho_k^{(h)}$ is the corresponding response probability, and

$$\bar{\rho}^{(h)} = \frac{1}{N_h} \sum_{k=1}^{N_h} \rho_k^{(h)} \qquad (10.10)$$

is the mean of the response probabilities in stratum h. In a fashion similar to expression (9.24) in Chapter 9, the bias can be rewritten as

$$B(\bar{y}_{PS,R}) = \frac{1}{N} \sum_{h=1}^{L} N_h \frac{R_{\rho Y}^{(h)} S_\rho^{(h)} S_Y^{(h)}}{\bar{\rho}^{(h)}}, \qquad (10.11)$$

where $R_{\rho Y}^{(h)}$ is the correlation between the Y and ρ in stratum h. $S_\rho^{(h)}$ and $S_Y^{(h)}$ are the standard errors of ρ and Y in stratum h, respectively.

The bias of the poststratification estimator is small if the biases within strata are small. A stratum bias is small in the following situations:

- If there is little or no relationship between the target variable and the response behavior within all strata, then their correlations are small.

- If response probabilities within a stratum are more or less equal, then their standard errors are small.
- If values of the target variable within a stratum are more or less equal, then their standard errors are small.

These conclusions give some guidance with respect to the construction of strata. Preferably, strata should be used that are homogeneous with respect to the target variable, response probabilities, or both. The more the elements resemble each within strata, the smaller the bias will be.

Two variables were used for weighting in Table 10.1: *AgeClass* and *Sex*. Strata were formed by crossing these two variables. Therefore, this weighting model is denoted by

$$AgeClass \times Sex.$$

The idea of crossing variables can be extended to more than two variables. As long as the table with population frequencies is available, and all response frequencies are greater than 0, weights can be computed. However, if there are no observations in a stratum, the corresponding weight cannot be computed. This leads to incorrect estimates. If the sample frequencies in the strata are very small, say less than 5, weights can be computed, but estimates will be unstable.

As more variables are used in a weighting model, there will be more strata. Therefore, the risk of empty strata or strata with too few observations will be larger. There are two solutions for this problem. One is to use less auxiliary variables, but then a lot of auxiliary information is thrown away. Another is to use *collapse strata*. This means merging a stratum having too few observations with another stratum. It is important to combine strata that resemble each other as much as possible. Collapsing strata is not a simple job, particularly if the number of auxiliary variables and strata is large. It is often a manual job.

Another problem with the use of several auxiliary variables is the lack of a sufficient amount of population information. This is shown in Table 10.2. The population distributions of the two variables *AgeClass* and *Sex* are known separately, but the distribution in the cross-classification is not known. In this case, the poststratification *AgeClass* × *Sex* cannot be carried out because weights cannot be computed for the strata in the cross-classification.

One way to solve this problem is to use only one variable, but this would mean ignoring all information with respect to the other variable. What is needed is a weighting technique that uses both marginal frequency distributions simultaneously. There are two weighting techniques that can do this: linear weighting and multiplicative weighting. These two techniques are described in the next two sections.

10.3 LINEAR WEIGHTING

The technique of linear weighting is based on the theory of general regression estimation. The regression estimator was already introduced in Chapter 6. It uses

Table 10.2 Lack of Population Information

	Population		
	Male	Female	Total
Young	?	?	435
Middle	?	?	296
Elderly	?	?	269
Total	511	480	1000

	Sample		
	Male	Female	Total
Young	23	15	38
Middle	16	17	33
Elderly	13	16	29
Total	52	48	100

	Weights	
	Male	Female
Young	?	?
Middle	?	?
Elderly	?	?

an auxiliary variable to produce more precise estimates. This estimator is extended here to the *generalized regression estimator*. It is shown that this estimator can also help to reduce a bias due to nonresponse. The theory of linear weighting is described assuming that data have been collected by means of simple random sampling without replacement. The theory can easily be generalized (Bethlehem, 1988).

First, the case of full response is considered. Suppose there are p auxiliary variables available. The p vector of values of these variables for element k is denoted by

$$X_k = (X_{k1}, X_{k2}, \ldots, X_{kp})'. \tag{10.12}$$

The symbol $'$ denotes transposition of a matrix or vector. Let Y be the N vector of all values of the target variable, and let X be the $N \times p$ matrix of all values of the auxiliary variables. The vector of population means of the p auxiliary variables is defined by

$$\bar{X} = (\bar{X}_1, \bar{X}_2, \ldots, \bar{X}_p)'. \tag{10.13}$$

If the auxiliary variables are correlated with the target variable, then for a suitably chosen vector $B = (B_1, B_2, \ldots, B_p)'$ of regression coefficients for a best fit of Y on X, the residuals $E = (E_1, E_2, \ldots, E_N)'$ defined by

$$E = Y - XB \tag{10.14}$$

LINEAR WEIGHTING

vary less than the values of the target variable itself. Application of ordinary least squares results in

$$B = (X'X)^{-1}XY' = \left(\sum_{k=1}^{N} X_k X_k'\right)^{-1} \left(\sum_{k=1}^{N} X_k Y_k\right). \quad (10.15)$$

For a simple random sample without replacement, the vector B can be estimated by

$$b = \left(\sum_{k=1}^{N} a_k X_k X_k'\right)^{-1} \left(\sum_{k=1}^{N} a_k X_k Y_k\right) = \left(\sum_{i=1}^{n} x_i x_i'\right)^{-1} \left(\sum_{i=1}^{n} x_i y_i\right), \quad (10.16)$$

where $x_i = (x_{i1}, x_{i2}, \ldots, x_{ip})'$ denotes the p vector of values of the p auxiliary variables for sample element i (for $i = 1, 2, \ldots, n$). The estimator b is an asymptotically design unbiased (ADU) estimator of B. It means the bias vanishes for large samples. The *generalized regression estimator* is now defined by

$$\bar{y}_{GR} = \bar{y} + (\bar{X} - \bar{x})'b, \quad (10.17)$$

where $\bar{x}$ is the vector of sample means of the auxiliary variables.

The generalized regression estimator is an ADU estimator of the population mean of the target variable. If there exists a p vector c of fixed numbers such that $Xc = I$, where I is a vector consisting of 1's, the generalized regression estimator can also be written as

$$\bar{y}_{GR} = \bar{X}'b. \quad (10.18)$$

It can be shown that the variance of the generalized regression estimator can be approximated by

$$V(\bar{y}_{GR}) = \frac{1-f}{n} S_E^2, \quad (10.19)$$

where S_E^2 is the population variance of the residuals $E_1, E_2, \ldots, E_N$.

Expression (10.19) is identical to the variance of the simple sample mean, but with the values Y_k replaced by the residuals E_k. This variance will be small if the residual values E_k are small. Hence, the use of auxiliary variables that can explain the behavior of the target variable will result in a precise estimator.

In case of nonresponse, the following modified version of the general regression estimator is introduced:

$$\bar{y}_{GR,R} = \bar{y}_R + (\bar{X} - \bar{x}_R)'b_R = \bar{X}'b_R, \quad (10.20)$$

in which b_R is defined by

$$b_R = \left(\sum_{k=1}^{N} a_k R_k X_k X_k'\right)^{-1} \left(\sum_{k=1}^{N} a_k R_k X_k Y_k\right). \quad (10.21)$$

So b_R is the analogue of b, but just based on the response data. Bethlehem (1988) shows that the bias of estimator (10.20) is approximately equal to

$$B(\bar{y}_{\text{GR,R}}) = \bar{X}B_R - \bar{Y}, \qquad (10.22)$$

where B_R is defined by

$$B_R = \left(\sum_{k=1}^{N} \rho_k X_k X'_k\right)^{-1} \left(\sum_{k=1}^{N} \rho_k X_k Y_k\right). \qquad (10.23)$$

The bias of this estimator vanishes if $B_R = B$. Thus, the regression estimator will be unbiased if nonresponse does not affect the regression coefficients. Practical experience (at least in The Netherlands) shows that nonresponse often seriously affects estimators, such as means and totals, but less often causes estimates of relationships to be biased. Particularly, if relationships are strong (the regression line fits the data well), the risk of finding wrong relationships is small.

By writing

$$B_R = B + \left(\sum_{k=1}^{N} \rho_k X_k X'_k\right)^{-1} \left(\sum_{k=1}^{N} \rho_k X_k E_k\right), \qquad (10.24)$$

the conclusion can be drawn that the bias will be small if the residuals are small.

This theory shows that use of the generalized regression estimator has the potential of improving the precision and reducing the bias in case of ignorable nonresponse. Therefore, it forms the basis for linear weighting adjustment techniques.

Bethlehem and Keller (1987) have shown that the generalized regression estimator (10.17) can be rewritten in the form of weighted estimator (10.2). The adjustment weight w_i for observed element i is equal to $w_i = v'X_i$, and v is a vector of weight coefficients that is equal to

$$v = n \left(\sum_{i=1}^{n} x_i x'_i\right)^{-1} \bar{X}. \qquad (10.25)$$

Poststratification is a special case of linear weighting, where the auxiliary variables are qualitative variables. To show this, qualitative auxiliary variables are replaced by sets of dummy variables. Suppose there is one auxiliary variable with L categories. Then L dummy variables $X_1, X_2, \ldots, X_L$ can be introduced. For an observation in a certain stratum h, the corresponding dummy variable X_h is assigned the value 1, and all other dummy variables are set to 0. Consequently, the vector of population means of these dummy variables is equal to

$$\bar{X} = \left(\frac{N_1}{N}, \frac{N_2}{N}, \ldots, \frac{N_L}{N}\right), \qquad (10.26)$$

and

$$v = \frac{n}{N}\left(\frac{N_1}{n_1}, \frac{N_2}{n_2}, \ldots, \frac{N_L}{n_L}\right)'. \qquad (10.27)$$

LINEAR WEIGHTING

Table 10.3 Weighting by Crossing the Variables Sex and AgeClass

Sex	AgeClass	X_1	X_2	X_3	X_4	X_5	X_6
Male	Young	1	0	0	0	0	0
Male	Middle	0	1	0	0	0	0
Male	Elderly	0	0	1	0	0	0
Female	Young	0	0	0	1	0	0
Female	Middle	0	0	0	0	1	0
Female	Elderly	0	0	0	0	0	1
Population means		0.226	0.152	0.133	0.209	0.144	0.136
Weight coefficients		0.983	0.950	1.023	1.393	0.847	0.850

If this form of v is used to compute $w_i = v'X_i$ and the result is substituted in expression (10.2) of the weighted estimator, the poststratification estimator is obtained.

Suppose there are two qualitative auxiliary variables: *Sex* and *AgeClass* (in three categories). Crossing these two variables produces a table with $2 \times 3 = 6$ cells. A dummy variable is introduced for each cell. So, there are six dummy variables. The possible values of these dummy variables are shown in Table 10.3.

The table also contains the vector of population means of the auxiliary variables. These values are equal to the population fractions in the cells of the population table.

The weight coefficients in the vector v are given in the bottom row of the table. These weight coefficients are used to compute the adjustment weights for the observed elements. For example, the weight for a young male is equal to 0.983.

Linear weighting can address the problem of the lack of sufficient population information. It offers a possibility to include variables in the weighting scheme without having to know the population frequencies in the cells obtained by cross-tabulating all variables. The trick is to use a different set of dummy variables. Instead of defining one set of dummy variables corresponding to the cells in the table, a set of dummy variables is defined for each variable separately. This approach allows the use of all marginal frequency distributions simultaneously. Of course, the amount of information used is less than that for a complete poststratification. However, still information about all auxiliary variables is used.

Continuing the example in Table 10.3, it is now shown how to use just the marginal distributions of *Sex* and *AgeClass*. Two sets of dummy variables are introduced: one set of two dummy variables for the categories of *Sex*, and another set of three dummy variables for the categories of *AgeClass*. Then there are $2 + 3 = 5$ dummy variables. In each set, always one dummy has the value 1, whereas all other dummies are 0. The possible values of the dummy variables are shown in Table 10.4.

The first dummy variable X_1 represents the constant term in the regression model. It always has the value 1. The second and third dummy variables relate to the two sex categories, and the last three dummies represent the three age categories. The vector of population means is equal to the fractions for all dummy variables separately. Note that in this weighting model always three dummies in a row have the value 1.

Table 10.4 Weighting by Using the Marginal Distributions of Sex and AgeClass

Sex	AgeClass	X_1	X_2	X_3	X_4	X_5	X_6
Male	Young	1	1	0	1	0	0
Male	Middle	1	1	0	0	1	0
Male	Elderly	1	1	0	0	0	1
Female	Young	1	0	1	1	0	0
Female	Middle	1	0	1	0	1	0
Female	Elderly	1	0	1	0	0	1
Population means		1.000	0.511	0.489	0.435	0.296	0.269
Weight coefficients		0.991	−0.033	0.033	0.161	−0.095	−0.066

The weight for an observed element is now obtained by summing the appropriate elements of this vector. The first value corresponds to the dummy X_1, which always has the value 1. So there is always a contribution of 0.991 to the weight. The next two values correspond to the categories of *Sex*. Note that their sum equals zero. For males, an amount 0.033 is subtracted, and for females, the same amount is added. The final three values correspond to the categories of *AgeClass*. Depending on the age category a contribution is added or subtracted. For example, the weight for a young male is now equal to $0.991 - 0.033 + 0.161 = 1.119$.

No information is used about crossing *Sex* by *AgeClass* here, but only the marginal distributions. Therefore, a different notation is introduced. This weighting model is denoted by

$$Sex + AgeClass.$$

Owing to the special structure of the auxiliary variables, the computation of the weight coefficients *v* cannot be carried out without imposing extra conditions. Here, for every qualitative variable the condition is imposed that the sum of the weight coefficients for the corresponding dummy variables must equal zero.

The weights obtained by using the model *Sex* + *AgeClass* are not equal to the weights obtained by complete poststratification. This is not surprising since the model *Sex* + *AgeClass* uses less information than the model *Sex* × *AgeClass*.

The examples in Tables 10.3 and 10.4 use only two auxiliary variables. More variables can be used in a weighting model. This makes it possible to define various weighting models with these variables. Suppose there are three auxiliary variables *Sex*, *AgeClass*, and *MarStat* (marital status). If the complete population distribution on the crossing of all three variables is available, then the weighting model

$$Sex \times AgeClass \times MarStat$$

can be applied. If only the bivariate population distributions of every crossing of two variables are available, the following weighting scheme could be applied:

$$(Sex \times AgeClass) + (AgeClass \times MarStat) + (Sex \times MarStat).$$

LINEAR WEIGHTING

Note that in this scheme three poststratifications are carried out simultaneously. If only marginal frequency distributions are available, the model

$$Sex + AgeClass + MarStat$$

could be considered. More details about the theory of linear weighting can be found in Bethlehem and Keller (1987).

Until now only linear weighting with qualitative auxiliary variables was described. It is also possible to apply linear weighting with quantitative auxiliary variables, or a combination of qualitative and quantitative variables.

If there is only one quantitative variable, linear weighting comes down to applying the simple regression estimator that was described in Chapter 6, and if there are more quantitative auxiliary variables, the generalized regression estimator can be used. It is also possible to combine qualitative and quantitative auxiliary variables. This is shown in an example with one quantitative variable *Age* and one qualitative variable *Sex*. Three different weighting models are described. Table 10.5 contains the first few cases.

The first weighting model only uses the variable *Age*. If a weighting model contains quantitative variables, always a column of constants must be included in the model. The matrix X for this model consists of the two columns X_1 and X_2 in the table.

Note that table also contains the population means. The value 34.369 denotes the mean age in the population. The row indicated by "weight coefficients 1" contains the weight coefficients for this model. There are only two coefficients: one corresponding to the constant term and other for *Age*. The second weight coefficient is negative (-0.003). This implies that weight decreases with age: younger people have higher weight than older people. Apparently, young people are underrepresented in the survey, while older people are overrepresented.

Table 10.5 Weighting by Using a Quantitative and Qualitative Variable

Age	Sex	X_1	X_2	X_3	X_4	X_5	X_6
65	Male	1	65	1	0	65	0
36	Male	1	36	1	0	36	0
73	Female	1	73	0	1	0	73
6	Male	1	6	1	0	6	0
33	Female	1	33	0	1	0	33
82	Female	1	82	0	1	0	82
2	Male	1	2	1	0	2	0
32	Male	1	32	1	0	32	0
66	Female	1	66	0	1	0	66
2	Female	1	2	0	1	0	2
Population means		1.000	34.369	0.511	0.489	33.509	35.268
Weight coefficients 1		1.101	−0.003				
Weight coefficients 2		1.101	−0.003	−0.032	0.032		
Weight coefficients 3		1.087				−0.001	−0.004

The second weighting model uses both variables *Age* and *Sex*. The weighting model is denoted by *Age* + *Sex*. This model uses columns X_1 (the constant term), X_2 (age), X_3 (dummy for male), and X_4 (dummy for female). The row marked "weight coefficients 2" contains the weight coefficients for this model. There are four coefficients: one for the constant term, one for age, one for males, and one for females. The second weight coefficient is negative. This implies that again weight decreases with age: for males, an extra amount is subtracted from the weight and for females, the same amount is added. A look at the adjustment weights would reveal that, for example, young females are underrepresented and that old males are overrepresented. The model *Age* + *Sex* should be a better model than the one just containing age since more population information is used.

The third example of a weighting model is a model in which the qualitative variable *Sex* and the quantitative variable *Age* are crossed. This weighting model is denoted by *Age* × *Sex*. The theory allows only one quantitative variable to be crossed with a number of qualitative variables in each term of the model. Crossing a quantitative variable with qualitative variables means that no longer the relative sizes of the strata in the population are required to be known, but rather the population means of the quantitative variables in the strata. Hence, for the model *Age* × *Sex* mean ages of males and females are required. For this model, columns X_1, X_5, and X_6 are used. Note that the ages for males are set to zero in the column for females, and vice versa the ages of females are set to zero in the column for males. The resulting weights can be found in the row "weight coefficients 3". There are three coefficients: one for the constant term, one for the age of males, and one for the age of females. The weight coefficients for both strata are negative. This means that for both males and females weights decrease with age. The weight coefficient for females is more negative than that for males. So females get a lower weight than males.

10.4 MULTIPLICATIVE WEIGHTING

If linear weighting is applied, correction weights are obtained that are computed as the sum of a number of *weight coefficients*. It is also possible to compute correction weights in a different way, namely, as the product of a number of *weight factors*. This weighting technique is usually called *raking* or *iterative proportional fitting*. Here, it is denoted by *multiplicative weighting* because weights are obtained as the product of a number of factors contributed by various auxiliary variables.

Multiplicative weighting can be applied in the same situations as linear weighting as long as only qualitative variables are used. It computes correction weights by means of an iterative procedure. The resulting weights are the product of factors contributed by all cross-classifications.

The iterative proportional fitting technique was already described by Deming and Stephan (1940). Skinner (1991) discusses application of this technique in multiple frame surveys. Little and Wu (1991) describe the theoretical framework and show that this technique comes down to fitting a loglinear model for the probabilities of getting observations in strata of the complete cross-classification given the probabilities for

MULTIPLICATIVE WEIGHTING

marginal distributions. To compute the weight factors, the following scheme must be carried out:

Step 1. Introduce a weight factor for each stratum in each cross-classification term. Set the initial values of all factors to 1.

Step 2. Adjust the weight factors for the first cross-classification term so that the weighted sample becomes representative with respect to the auxiliary variables included in this cross-classification.

Step 3. Adjust the weight factors for the next cross-classification term so that the weighted sample is representative for the variables involved. Generally, this will disturb representativeness with respect to the other cross-classification terms in the model.

Step 4. Repeat this adjustment process until all cross-classification terms have been dealt with.

Step 5. Repeat steps 2, 3, and 4 until the weight factors do not change anymore.

Use of multiplicative weighting is illustrated using the same data as in Tables 10.3 and 10.4. The weighting model contains the two qualitative auxiliary variables *Sex* and *AgeClass*.

Suppose only the population distribution of *Sex* (two categories) and *AgeClass* (three categories) are separately available and not the cross-classification. Table 10.6 contains the starting situation. The upper left part of the table contains the weighted relative frequencies in the sample for each combination of *AgeClass* and *Sex*.

The row and column denoted by "weight factor" contain the initial values of the weight factors (1.000). The values in the row and column denoted by "weighted sum" are obtained by first computing the weight for each sample cell (by multiplying the relevant row and column factors) and then summing the weighted cell fractions. Since the initial values of all factors are equal to 1, the weighted sums in Table 10.6 are equal to the unweighted sample sums. The row and the column denoted by "population distribution" contain the fractions for *AgeClass* and *Sex* in the population.

Table 10.6 Multiplicative Weighting, Starting Situation

	Starting Situation				
	Male	Female	Weight Factor	Weighted Sum	Population Distribution
Young	0.230	0.150	1.000	0.380	0.435
Middle	0.160	0.170	1.000	0.330	0.296
Elderly	0.130	0.160	1.000	0.290	0.269
Weight factor	1.000	1.000			
Weighted sum	0.520	0.480		1.000	
Population distribution	0.511	0.489			1.000

Table 10.7 Multiplicative Weighting, Situation after Adjusting the Rows

	Situation after Adjusting for AgeClass				
	Male	Female	Weight Factor	Weighted Sum	Population Distribution
Young	0.230	0.150	1.145	0.435	0.435
Middle	0.160	0.170	0.897	0.296	0.296
Elderly	0.130	0.160	0.928	0.269	0.269
Weight factor	1.000	1.000			
Weighted sum	0.527	0.473		1.000	
Population distribution	0.511	0.489			1.000

The iterative process must result in row and column factors with such values that the weighted sums match the population distribution. This is clearly not the case in the starting situation. First, the weight factors for the rows are adjusted. This leads to weight factors 1.145, 0.897, and 0.925 for *Young, Middle,* and *Elderly* (see Table 10.7). The weighted sums for the rows are now correct, but the weighted sums for the columns are 0.527 and 0.473 and thus still show a discrepancy.

The next step will be to adjust the weight factors for the columns such that the weighted column sums match the corresponding population frequencies. Note that this adjustment for *Sex* will disturb the adjustment for *AgeClass*. The weighted sums for the age categories no longer match the relative population frequencies. However, the discrepancy is much less than that in the initial situation.

The process of adjusting for *AgeClass* and *Sex* is repeated until the weight factors do not change anymore. The final situation is reached after a few iterations. Table 10.8 shows the final results.

The adjustment weight for a specific sample element is now obtained by multiplying the relevant weight factors. For example, the weight for a young male is equal to

Table 10.8 Multiplicative Weighting, Situation after Convergence

	Situation after Convergence				
	Male	Female	Weight Factor	Weighted Sum	Population Distribution
Young	0.230	0.150	1.151	0.435	0.435
Middle	0.160	0.170	0.895	0.296	0.296
Elderly	0.130	0.160	0.923	0.269	0.269
Weight factor	0.968	1.035			
Weighted sum	0.511	0.489		1.000	
Population distribution	0.511	0.489			1.000

$1.151 \times 0.968 = 1.114$. For this example, the adjustment weights differ only slightly from those obtained by linear weighting as described in the previous section.

10.5 CALIBRATION ESTIMATION

Deville and Särndal (1992) and Deville et al. (1993) have created a general framework for weighting, of which linear weighting and multiplicative weighting are special cases. Assuming simple random sampling, their starting point is that adjustment weights have to satisfy two conditions:

- The adjustment weights w_k have to be as close as possible to 1.
- The weighted sample distribution of the auxiliary variables has to match the population distribution, that is,

$$\bar{x}_W = \frac{1}{N} \sum_{i=1}^{n} w_i x_i = \bar{X}. \qquad (10.28)$$

The first condition sees to it that resulting estimators are unbiased, or almost unbiased, and the second condition guarantees that the weighted sample is representative with respect to the auxiliary variables used.

Deville and Särndal (1992) introduce a distance measure $D(w_i, 1)$ measuring the difference between w_i and 1 in some way. The problem is now to minimize

$$\sum_{i=1}^{n} D(w_i, 1) \qquad (10.29)$$

under the condition (10.28). This problem can be solved by using the method of Lagrange. By choosing the proper distance function, linear and multiplicative weighting can be obtained as special cases of this general approach. For linear weighting, the distance function is defined by

$$D(w_i, 1) = (w_i - 1)^2, \qquad (10.30)$$

and for multiplicative weighting the distance

$$D(w_i, 1) = w_i \log(w_i) - w_i + 1 \qquad (10.31)$$

must be used.

Deville and Särndal (1992) and Deville et al. (1993) only consider the full response situation. They show that estimators based on weights computed within their framework have asymptotically the same properties. This means that for large samples it does not matter whether linear or multiplicative weighting is applied. Estimators based on both weighting techniques will behave approximately the same way. Note that although the estimators behave in the same way, the individual weights computed by means of linear or multiplicative weighting may differ substantially.

The situation is different under nonresponse. Generally, the asymptotic properties of linear and multiplicative weighting will not be equal under nonresponse. The extent to which the chosen weighting technique is able to reduce the nonresponse bias depends on how well the corresponding underlying model can be estimated using the observed data. Linear weighting assumes a linear model to hold with the target variable as dependent variable and the auxiliary variables as explanatory variables. Multiplicative weighting assumes a loglinear model for the cell frequencies. An attempt to use a correction technique for which the underlying model does not hold will not help to reduce the bias.

10.6 OTHER WEIGHTING ISSUES

There are several reasons why survey researchers may want to have some control over the values of the adjustment weights. One reason is that extremely large weights are generally considered undesirable. Large weights usually correspond to population elements with rare characteristics. Use of such weights may lead to unstable estimates of population parameters. To reduce the impact of large weights on estimators, a weighting method is required that is able to keep adjustment weights within pre-specified boundaries and that at the same time enables valid inference.

Another reason to have some control over the values of the adjustment weights is that application of linear weighting might produce negative weights. Although theory does not require weights to be positive, negative weights should be avoided, since they are counterintuitive, they cause problems in subsequent analyses, and they are an indication that the regression model does not fit the data well.

Negative weights can be avoided by using a better regression model. However, it is not always possible to find such models. Another solution is to use the current model and force weights within certain limits. Several techniques have been proposed for this. A technique developed by Deville et al. (1993) comes down to repeating the (linear) weighting process a number of times. First, a lower bound L and an upper bound U are specified. After the first run, weights smaller than L are set to L and weights larger than U are set to U. Then, the weighting process is repeated, but records from the strata with the fixed weights L and U are excluded. Again, weights may be produced not satisfying the conditions. These weights are also set to the value either L or U. The weighting process is repeated until all computed weights fall within the specified limits. Convergence of this iterative process is not guaranteed. Particularly, if the lower bound L and upper bound U are not far apart, the process may not converge.

Huang and Fuller (1978) use a different approach. Their algorithm produces weights that are a smooth, continuous, monotone increasing function of the original weights computed from the linear model. The algorithm is iterative. At each step, the weights are checked against a user supplied criterion value M. This value M is the maximum fraction of the mean weight by which any weight may deviate from the mean weight. For example, if M is set to 0.75, then all weights are forced into the interval with lower bound equal to 0.25 times the mean weight and upper bound equal to 1.75 times the mean weight. Setting the value to 1 implies that all weights are forced to be positive. Huang and Fuller (1978) proved that the asymptotic properties of the regression estimator constructed with their algorithm are asymptotically the same as those of the generalized regression estimator. So, restricting the weights has (at least asymptotically) no effect on the properties of population estimates computed with these weights.

Another issue is the computation of weights that are consistent for persons and households. Some statistical surveys have complex sample designs. One example of such a complex design is cluster sampling. Many household surveys are based on cluster samples. First, a sample of households is selected. Next, all persons in the selected households are interviewed. The collected information can be used to make estimates for two populations: the population consisting of all households and the population consisting of all individual persons. In both situations, weighting can be carried out to correct for nonresponse. This results in two weights assigned to each record: one for the household and other for the individual. Having two weights in each record complicates further analysis.

If the aim of the survey is to make inference on the population of all individual persons, the process is fairly straightforward. The unit of measurement is the individual person. The data file must be approached as a file of records with data on persons. Available population information on the distribution of personal characteristics can be used to compute adjustment weights, and these weights are assigned to the individual records.

For making inference on the population of households, the same approach can be used. However, there is a problem. In The Netherlands, for example, there is no or limited information available on the population distribution of household variables. Even information on simple variables, such as size of the household and household composition, is lacking. This makes it impossible to carry out an efficient weighting procedure.

Since it is possible to compute weights for the members of the household, one may wonder whether it is possible to use the individual weights to compute household weights. Possible approaches could be to take (1) the weight of head of the household, (2) the weight of a randomly selected household member, or (3) to compute some kind of average weight of the household members. Whatever approach is used, there are always problems. If household weights are applied to members of the households, weighted estimates of individual characteristics will not match known population frequencies. This discrepancy will not occur if the individual weights are used. Furthermore, inconsistencies may turn up. For example, an estimate of the total income through the households will not be equal to an estimate based on the individual persons.

Generalized regression estimation offers a solution to these problems. The trick is to sum the dummy variables corresponding to the qualitative auxiliary variables for the individuals over the household. Thus, quantitative auxiliary variables are created at the household level.

The resulting weights are assigned to the households. Furthermore, all elements within a household are assigned the same weight, and this weight is equal to the household weight. This approach forces estimates computed using element weights to be consistent with estimates based on cluster weights. For an application of consistent weighting, see Nieuwenbroek (1993).

10.7 USE OF PROPENSITY SCORES

The weighting techniques described in the previous sections were all based on the principle of assigning correction weights to observations. These weights were computed in such a way that weighted estimators have better properties than unweighted estimators. The technique of *propensity scores* implements a slightly different approach. It concentrates on first estimating response probabilities. Then, the estimated probabilities are used to improve estimators.

The use of propensity scores is described under the *random response model*. It is assumed that whether or not an individual responds is the result of some random process, where each individual k has a certain, unknown probability ρ_k of responding when selected, for $k = 1, 2, \ldots, N$. Let R denote an indicator variable, where $R_k = 1$ if individual k responds, and where $R_k = 0$ otherwise. Then $P(R_k = 1) = \rho_k$.

Only those values Y_k become available in the survey for which individual k is selected in the sample ($a_k = 1$) and responds ($R_k = 1$). Therefore, the first-order inclusion probability for element k is equal to $\pi_k \rho_k$. To obtain an unbiased estimator, the Horvitz–Thompson estimator is replaced by an adapted Horvitz–Thompson estimator

$$\bar{y}'_{HT} = \frac{1}{N} \sum_{k=1}^{N} \frac{a_k R_k Y_k}{\pi_k \rho_k}. \tag{10.32}$$

The response probabilities ρ_k are unknown quantities. Therefore, they are estimated using the available data. Rosenbaum and Rubin (1983) introduced the technique of propensity scores to achieve this.

The propensity score $\rho(X)$ is the conditional probability that an individual with observed characteristics X responds in a survey when invited to do so ($R = 1$):

$$\rho(X) = P(R = 1|X). \tag{10.33}$$

It is assumed that within subpopulations defined by values of the observed characteristics X, all individuals have the same response probability. This is the missing at random (MAR) assumption that was introduced in Chapter 8. Both linear weighting and the propensity score method rely on this assumption.

Often, the propensity score is modeled by means of a logit model:

$$\log\left(\frac{\rho(X_k)}{1 - \rho(X_k)}\right) = \alpha + \beta' X_k. \tag{10.34}$$

Other models can be used too, but, Dehija and Wahba (1999) for example, conclude that different models often produce similar results.

The model is fitted with maximum likelihood estimation. The resulting model is used to predict the propensity scores. These predicted scores can be used in various ways: the first approach is called *propensity score weighting*. The response probabilities ρ_k in the adapted Horvitz–Thompson estimator (10.32) are replaced by their estimates $\rho(X_k)$ from the logit model. Cobben and Bethlehem (2005) show that this approach does not always performs very well. In an example, estimates of population parameters turn out to be unstable. This might be caused by the fact that estimates highly depend on the model used for the propensity scores.

A second approach is *propensity score stratification*. This is a form of poststratification where strata are formed on the basis of the propensity scores. Suppose the sample is stratified into L strata by means of the estimated propensity score. The poststratification estimator is defined by

$$\bar{y}_{\text{PS}} = \frac{1}{N}\sum_{h=1}^{L} N_h \bar{y}^{(h)}, \tag{10.35}$$

where N_h is the number of elements in stratum h and $\bar{y}^{(h)}$ is the mean of the available observations in stratum h. Bethlehem (1988) shows that the bias of the poststratified Horvitz–Thompson estimator can be written as

$$B(\bar{y}_{\text{PS,R}}) = \frac{1}{N}\sum_{h=1}^{L} N_h \frac{R_{\rho Y}^{(h)} S_\rho^{(h)} S_Y^{(h)}}{\bar{\rho}^{(h)}}, \tag{10.36}$$

where $R_{\rho Y}^{(h)}$ is the correlation between Y and ρ in stratum h. $S_\rho^{(h)}$ and $S_Y^{(h)}$ are the standard errors of ρ and Y in stratum h, respectively. $\bar{\rho}_h$ is the average of the response probabilities in stratum h.

This bias is small if the variation in response probabilities is small. So it makes sense to construct strata in such a way that most variation of these probabilities is between strata and not within strata. Cochran (1968) suggests that as much as five strata may be sufficient to remove a large part of the bias.

Cobben and Bethlehem (2005) have tested this approach. It turned out that values of estimates move in the right direction but are often still far away from the full sample estimates. So, stratification based on just propensity scores was not able to completely correct for the bias. They also explored the effects of using a different number of strata. Estimates based on 25 propensity score strata performed slightly better. This is not surprising because the strata will be more homogeneous with respect to the values of the propensity scores.

A third approach to using response propensities is *linear weighting with adjusted inclusion probabilities*. In its most general form, the generalized regression estimator is defined by

$$\bar{y}_{\text{GR}} = \bar{y}_{\text{HT}} + (\bar{X} - \bar{x}_{\text{HT}})'b, \tag{10.37}$$

where $\bar{X}$ is the vector of population means of a set of auxiliary variables, $\bar{x}_{\text{HT}}$ is the vector of Horvitz–Thompson estimators for the auxiliary variables, and b is a vector of regression coefficients defined by

$$b = \left(\sum_{k=1}^{N} \frac{a_k X_k X'_k}{\pi_k}\right)^{-1} \left(\sum_{k=1}^{N} \frac{a_k X_k Y_k}{\pi_k}\right). \tag{10.38}$$

Linear weighting produces only consistent estimates if the proper inclusion probabilities are used. Therefore, in case of nonresponse, the π_k in (10.38) should by replaced by $\pi_k \rho_k$. Unfortunately, the ρ_k are unknown, so they are replaced by their estimates $\rho(X_k)$.

Cobben and Bethlehem (2005) showed that in their example estimates based on this approach performed better than those based on propensity score stratification. Estimates were closer to the true values. This could be expected because now the adjusted inclusion probabilities have been used.

A final approach is *linear weighting including propensity score strata*. This comes down to using a normal weighting model but including in it a categorical propensity score variable.

Cobben and Bethlehem (2005) tested this approach with two versions of a propensity score variable, one with 5 categories and other with 25 categories. The weight model with the second variable performed better than the model with the five-category variable. Again, this is no surprise. Including a categorical propensity score variable in the model pays.

10.8 A PRACTICAL EXAMPLE

Since 1995, Statistics Netherlands has an integrated system of social surveys. It is known under its Dutch acronym POLS (Permanent Onderzoek Leefsituatie). POLS is a continuous survey. Every month a sample is selected. The target population consists of people of age 12 and older. The samples are stratified two-stage samples. In the first stage, municipalities are selected within regional strata with probabilities proportional to the number of inhabitants. In the second stage, an equal probability sample is drawn in each selected municipality. Sampling frames are the population registers of the municipalities. The samples are self-weighted samples; that is, all individuals have the same probability of being selected in the sample.

In this example, the effect of weighting on one social participation variable is studied. This is the variable recording whether or not a person is doing any volunteer work. It is to be expected that there is a relationship between social participation and response behavior: people participating more in social activities will also be more inclined to respond. The sample size of the thematic module on social participation was 6672, with a response percentage of 56.6%.

For this example, the only population information used was taken from the Statistical Yearbook of Statistics Netherlands, see Statistics Netherlands (1998). It contains frequency distributions with respect to five variables: gender, age, marital status, province of residence, and degree of urbanization of the area of residence.

The ideal situation would be to have the complete crossing of these five variables. However, the Statistical Yearbook contains only information with respect to partial crossings. Table 10.9 contains counts for the distribution of gender by age by marital status.

With respect to the two variables, province and degree of urbanization, only the crossing with age is available as displayed in Tables 10.10 and 10.11, respectively.

A PRACTICAL EXAMPLE

Table 10.9 Population Distribution of Age × Gender × Marital Status (×1000)

	Male				Female			
Age	Unmarried	Married	Widowed	Divorced	Unmarried	Married	Widowed	Divorced
12–19	752.4	0.4	0.0	0.0	716.5	3.5	0.0	0.0
20–29	981.5	185.7	0.2	10.2	785.0	330.6	0.7	22.7
30–39	445.4	795.1	1.9	72.1	283.5	879.3	5.7	93.8
40–49	164.7	899.0	6.9	113.9	103.1	882.9	21.5	138.4
50–59	67.3	732.9	15.8	86.3	44.4	675.9	56.1	98.8
60–69	42.0	519.2	31.7	42.6	41.4	458.9	140.0	51.5
70–79	21.4	308.4	52.5	16.6	43.0	239.9	254.3	27.9
80+	8.0	84.0	50.4	4.0	35.0	49.6	243.9	12.4

Note that in Tables 10.10 and 10.11, the age variable only has five categories, whereas in Table 10.9 it has eight categories. In a linear weighting model, this causes no problems. It is possible to use both age variables simultaneously.

Note that poststratification can only be applied if one of these three tables is used. And even Table 10.9 cannot be used as it is because it contains four empty cells. This problem could be solved by *collapsing strata* with other strata, see, for example, Tremblay (1986), Kalton and Maligalig (1991), Little (1993), and Gelman and Carlin (2000). For example, the two age categories 12–19 and 20–29 could be merged into one new age category 12–29.

To select a weighting model, the general guideline could be applied to use as much population information as possible. The more auxiliary variables are used, the better the regression model will be able to explain the behavior of the target variables, and so

Table 10.10 Population Distribution of Province × Age (×1000)

	Age				
Province	12–19	20–44	45–64	65–79	80+
Groningen	49.3	222.1	127.8	48.4	20.8
Friesland	61.5	225.7	144.0	65.7	21.6
Drenthe	43.7	165.9	114.3	53.9	15.3
Overijssel	106.2	404.2	240.2	110.8	31.9
Flevoland	33.8	115.7	53.0	21.8	4.2
Gelderland	183.4	720.6	443.4	195.7	56.9
Utrecht	103.7	439.4	238.6	102.2	32.5
Noord-Holland	220.5	999.9	574.4	254.3	82.1
Zuid-Holland	316.2	1307.9	759.5	347.1	117.7
Zeeland	35.0	130.1	89.2	44.1	15.6
Noord-Brabant	218.7	898.7	562.4	225.2	57.9
Limburg	100.8	429.5	289.8	127.1	30.8

Table 10.11 Population Distribution of Degree of Urbanization × Age (×1000)

Urbanization	Age				
	12–19	20–44	45–64	65–79	80+
Very strong	223.2	1196.6	565.3	293.4	113.0
Strong	336.5	1468.4	839.0	389.8	114.6
Moderate	317.1	1223.7	766.3	321.7	90.5
Little	333.7	1226.3	820.6	328.8	93.4
None	262.3	944.7	645.4	262.6	75.8

the smaller the remaining bias will be. On the contrary, auxiliary variables having no relationship with the target variables will not help to reduce the bias. Moreover, use of many auxiliary variables may inflate variance estimates. Therefore, it is a good idea to compare population and response distributions for each auxiliary variable (Table 10.12).

For the variable age, nonresponse is highest for people between 20 and 30 years of age (mainly not at home) and elderly people (mainly refusal). A look at marital status shows a relatively high response for married people. Divorced people tend to respond less than average. Response is particularly high in the provinces Gelderland and Noord-Brabant. There is a lot of nonresponse in the more densely populated and more urbanized provinces of Noord-Holland and Zuid-Holland. This phenomenon is also reflected in the variable degree of urbanization. Quite striking is the low response rate in the very strongly urbanized areas.

This analysis indicates that at least the variables marital status, province, and degree of urbanization should be included in the weighting model. Note that the last two variables are partially but not completely confounded.

A number of different weighting models have been computed for this example. The computations were carried out with the software package Bascula, developed by Statistics Netherlands (Bethlehem, 1996). Weights obtained in this way have been used to estimate the percentage of people doing some kind of volunteer work. Since it is expected that people doing this kind of work are overrepresented, the estimated percentage should decrease as more effective weighting models are applied. Table 10.13 contains the results of the computations.

A clear pattern can be distinguished: the more auxiliary information is used, the lower the estimate of the percentage of people doing volunteer work. Of course, the effectiveness of a model cannot be judged by just looking at the deviation from the unadjusted estimate. However, use of more information also leads to a decrease in standard error, and this is an indication of better fitting models. Hence, it is not unlikely that in this example the number volunteers are overrepresented in the response, and the weighting models correct for this. Standard errors were computed using the method of balanced half-samples (Renssen et al., 1997).

If only one auxiliary variable is used for weighting, it turns out that variable gender has no effect. Weighting using the variable marital status, degree of urbanization, or age (in eight categories) reduces the estimate by 0.5–0.6%.

A PRACTICAL EXAMPLE

Table 10.12 Population and Response Distributions of the Auxiliary Variables (%)

Age	Response	Population	Difference
12–19	12.8	11.1	1.7
20–29	15.9	17.5	−1.6
39–39	20.5	19.4	1.1
40–49	17.9	17.6	0.3
50–59	14.0	13.4	0.6
60–69	10.0	10.0	0.0
70–79	6.5	7.3	−0.8
80+	2.5	3.7	−1.2

Private	Response	Population	Difference
Groningen	2.7	3.5	−0.8
Friesland	4.3	3.9	0.4
Drenthe	2.3	3.0	−0.7
Overijssel	6.8	6.7	0.1
Flevoland	1.8	1.7	0.1
Gelderland	15.4	12.1	3.3
Utrecht	5.4	6.9	−1.5
N-Holland	14.0	16.1	−2.1
Z-Holland	18.0	21.5	−3.5
Zeeland	2.7	2.4	0.3
N-Brabant	17.6	14.8	2.8
Limburg	9.1	7.4	1.7

Marital Status	Response	Population	Difference
Unmarried	32.7	34.2	−1.5
Married	57.2	53.2	4.0
Widowed	5.2	6.0	−0.8
Divorced	4.9	6.7	−2.8

Urbanization	Response	Population	Difference
Very strong	11.8	18.0	−6.2
Strong	24.0	23.8	0.2
Moderate	23.2	20.5	2.7
Little	23.3	21.1	2.2
None	17.7	16.5	1.2

Gender	Response	Population	Difference
Male	48.6	49.1	−0.5
Female	51.4	50.9	0.5

Models containing the degree of urbanization produce the largest shift in the estimate. This suggests that this variable is more effective than the other auxiliary variables.

Poststratification by gender, marital status, and age is not possible due to empty cells. But even merging the age categories 12–19 and 20–29 would produce cells with less than five observations, which could produce unstable weights. Therefore, instead

Table 10.13 Estimates of the Percentage of People Doing Volunteer Work, Based on Various Weighting Models

	Weighting Model	Number of Parameters	Estimate	Standard Error
1	No weighting	0	43.4	1.2
2	Gender	2	43.4	1.2
3	Province	12	43.3	1.2
4	MarStat	4	42.9	1.2
5	Urban	5	42.9	1.0
6	Age8	8	42.8	1.2
7	Age5 × Province	60	42.9	1.2
8	(Gender × Age8) + (Gender × MarStat)	22	42.3	1.1
9	Age5 × Urban	25	42.5	1.0
10	Gender + Age8 + MarStat + Urban + Province	23	42.1	1.0
11	(Gender × Age8) + (Gender × MarStat) + (Age5 × Urban) + Province	53	42.0	0.9

of attempting to carry out the poststratification *Gender × MarStat × Age8*, the linear weighting model (*Gender × Age8*) + (*Gender × MarStat*) was used. Application of this model produces a decrease in the estimate of 1.1%.

Model 11 in Table 10.13 contains the maximum possible weighting model. The population information required for the term (*Gender × Age8*) + (*Gender × MarStat*) is taken from Table 10.9 and that for the term *Age5 × Urban* from Table 10.10. Note that only the term *Province* is used and not *Age5 × Province* because the response table contains cells with too few observations. Population counts for *Province* are taken from Table 10.11. Application of this maximum weighting model shows the greatest decrease in the estimate, from 43.3 to 42.0.

It is also interesting to look at the result of model 10 in Table 10.9. In this model, all auxiliary variables are used, but only their marginal distributions. This is a much smaller model, which can be seen by looking at the number of model parameters in Table 10.9 (53 for model 11 and 23 for model 10). Still the simpler model 10 performs almost as well as the maximum model 11. This is an indication that in this example the main effects of the auxiliary variables play an important role in reducing the bias of the estimates, whereas all kinds of interaction effects are less important.

EXERCISES

10.1 Which property of an auxiliary variable makes is useful for including in a weighting adjustment model?

  a. The response distribution of the variable is approximately equal to its population distribution.

  b. The sample distribution of the variable is approximately equal to its population distribution.

EXERCISES 273

c. The response distribution of the variable differs considerably from its population distribution.

d. The response distribution of the variable is approximately equal to its sample distribution.

10.2 A large company has 2500 employees. The management has installed coffee machines everywhere in the building. After a while, the management wants to know whether or not the employees are satisfied with the coffee machines.

a. Determine the sample size under the condition that the margin of the 95% confidence interval may not exceed 4%.

b. It is decided to draw a simple random sample without replacement of 500 employees. It turns out that 380 employees complete the questionnaire form. Of them, 310 are satisfied with the coffee machines.

Compute the 95% confidence interval of the percentage of the percentage of employees in the company who are satisfied with the coffee machines.

c. Only 380 out of 500 selected employees responded. So there is a nonresponse problem.

Compute a lower bound and an upper bound for the percentage of employees in the sample who are satisfied with the coffee machines.

d. Previous research has showed that employees with a higher level of education are less satisfied with the coffee facilities. The management knows the level of education of each employee in the company: 21% has a high education and 79% has a low education. The table below shows the relationship between coffee machine satisfaction and level of education for the 380 respondents:

	Low Education	High Education	Total
Satisfied	306	4	310
Not satisfied	40	30	70
Total	346	34	380

A weighting adjustment procedure is carried out to reduce the nonresponse bias.

Compute weights for low- and high-educated employees.

e. Compute the weighted estimate of the percentage of employees in the company satisfied with the coffee facilities.

10.3 There are plans in The Netherlands to introduce a system of road pricing. It means car drivers are charged for the roads they use. Such a system could lead to better use of the available road capacity and reduction in traffic congestion. An automobile association wants to know what the attitude of the Dutch is toward road pricing. It conducts a survey in which a simple random sample of 1000 people is selected. Selected people are asked two questions:

- Are you in favor of road pricing?

- Do you have a car?

Unfortunately, not everybody wants to participate in the survey. Due to nonresponse, only a part of selected people answer the questions. The results are summarized below:

	In Favor of Road Pricing?	
Has a Car?	Yes	No
Yes	128	512
No	60	40

a. Compute the response percentage.
b. Using the available data, compute the percentage in favor of road pricing.
c. Using the available data, compute a lower bound and upper bound for the percentage in the complete sample in favor of road pricing.
d. From another source, it is known that 80% of the target population owns a car, and 20% does not have one. Use this additional information to apply weighting adjustment.

Compute a weight for car owners, and a weight for those without a car.

e. Make a table like the one above, but with weighted frequencies.
f. Compute a weighted estimate for the percentage in favor of road pricing.
g. Explain the difference between the weighted and unweighted estimate.

10.4 A transport company carries out a survey to determine how healthy its truck drivers are. From the population of all its drivers a simple random sample has been selected. Of course, there is nonresponse. Therefore, data on only 21 drivers become available. Each respondent has been asked if he has visited a doctor because of medical problems. Also, experience of the driver (little, much) and age (young, middle, old) have been recorded. The results are shown in the table:

No.	Age	Experience	Doctor Visits	No.	Age	Experience	Doctor Visits
1	Young	Much	2	12	Middle	Little	6
2	Young	Much	3	13	Middle	Little	6
3	Young	Much	4	14	Middle	Little	7
4	Young	Little	3	15	Old	Much	8
5	Young	Little	4	16	Old	Much	10
6	Young	Little	4	17	Old	Much	10
7	Young	Little	5	18	Old	Much	8
8	Middle	Much	5	19	Old	Little	8
9	Middle	Much	6	20	Old	Little	9
10	Middle	Much	7	21	Old	Little	10
11	Middle	Little	5				

EXERCISES

a. Estimate the average number of doctor visits, assuming that the response can be seen as a simple random sample.

b. Assume that the population distributions of experience and age are available for the population of all drivers of the company:

Experience	Percentage
Much	48%
Little	52%

Age	Percentage
Young	22%
Middle	30%
Old	48%

Establish whether or not the response is selective. Explain which of these two auxiliary variables should be preferred for computing adjustment weights.

c. For each auxiliary variable separately carry out weighting adjustment. Compute weights for each of the categories of the auxiliary variable.

d. Compute for both weighting adjustments a weighted estimate of the average number of doctor visits.

e. Compare the outcomes under (a) and (d). Explain differences and/or similarities.

CHAPTER 11

Online Surveys

11.1 THE POPULARITY OF ONLINE RESEARCH

Collecting data using a survey is a complex, costly, and time-consuming process. Traditionally, surveys were carried out using paper forms (PAPI). One of the problems of this mode of data collection was that data usually contained many errors. Therefore, extensive data editing was required to obtain data of acceptable quality. Data editing activities often consume a substantial part of the total survey budget. Chapter 7 described how rapid developments in information technology in the last few decades of the previous century have made it possible to use microcomputers for computer-assisted interviewing (CAI). This type of data collection has three major advantages: (1) It simplifies work of the interviewers because they do not have to pay attention any more to choosing the correct route through the questionnaire, (2) it improves the quality of the collected data because answers can be checked and corrected during the interview, and (3) it considerably reduces time needed to process the survey data. Thus, it improves the timeliness of survey results and reduces survey costs. More on the benefits of CAI can be found in Chapter 7 and Couper et al. (1998).

Computer-assisted interviewing comes in various modes. It started in the 1970s with *computer-assisted telephone interviewing* (CATI). More recent was *computer-assisted personal interviewing* (CAPI), that is, face-to-face interviewing in which interviewers use a laptop computer to ask the questions. CAPI emerged in the 1980s when lightweight laptop computers made face-to-face interviewing with a computer feasible. After more and more companies and households purchased their own computers, mail surveys could be replaced with their electronic analogue, and thus *computer-assisted self-interviewing* (CASI) emerged.

The rapid development of Internet in the last decade has led to *computer-assisted Web interviewing* (CAWI), a new type of computer-assisted interviewing. The questionnaire is designed as a Web site, which is accessed by respondents.

Applied Survey Methods: A Statistical Perspective, Jelke Bethlehem
Copyright © 2009 John Wiley & Sons, Inc.

These *online surveys*, also called *Web surveys*, are almost always self-administered: respondents visit the Web site and complete the questionnaire by answering the questions. Not surprisingly, survey organizations use, or consider using, online surveys. At first sight, they seem to have some attractive advantages:

- Now that so many people are connected to Internet, an online survey is a simple means to get access to a large group of potential respondents.
- Questionnaires can be distributed at very low costs. No interviewers are needed, and there are no mailing and printing costs involved.
- Surveys can be launched very quickly. Little time is lost between the moment the questionnaire is ready and the start of the fieldwork.
- Online surveys offer new, attractive possibilities, such as the use of multimedia (sound, pictures, animation, and movies).

Thus, online surveys seem to be a fast, cheap and attractive means of collecting large amounts of data. However, there are methodological problems, caused partly by the use of Internet for selecting respondents and partly by the use of the Web as a measuring instrument. If these problems are not seriously addressed, online surveys may result in low-quality data by which no proper inference can be made with respect to the target population of the survey.

This chapter discusses some of the methodological issues that are specific for online surveys. Particularly, attention is paid to the effects of undercoverage and self-selection. Some theory is developed, and it is shown what the effects of some correction techniques can be. Practical implications are explored using data from a fictitious population.

11.2 ERRORS IN ONLINE SURVEYS

In the process of carrying out a survey, a lot of things can happen that may have an impact on the quality of the survey outcomes. Chapter 8 presented a systematic overview of possible problems. Many of these problems can also occur in online surveys.

Undercoverage occurs when elements of the target population do not have a corresponding entry in the sampling frame. These elements can and will never be contacted. Undercoverage is a serious problem if Internet is used as a sampling frame and the target population contains people without Internet. These people can never be selected for the survey.

Selection errors can occur in an online survey when the sample is based on self-selection. The survey questionnaire is simply put on the Web. Respondents are those people who happen to have Internet access, visit the Web site, and decide to participate in the survey. The survey researcher is not in control of the selection process. Consequently, selection probabilities are unknown, and therefore unbiased estimation is not possible.

Nonresponse can also occur in online surveys. An online survey questionnaire is a self-administered questionnaire. Therefore, online surveys have a potential of high nonresponse rates. An additional source of nonresponse problems is technical problems of respondents having to interact with Internet (see, for example, Couper, 2000; Dillman and Bowker, 2001; Fricker and Schonlau, 2002; Heerwegh and Loosveldt, 2002). Slow modem speeds, unreliable connections, high connection costs, low-end browsers, and unclear navigation instructions may frustrate respondents.

Coverage and selection problems are discussed in more detail in the following subsections.

11.2.1 Coverage Problems

The collection of all elements that can be contacted through the sampling frame is called the *frame population*. Since the sample is selected from the frame population, conclusions drawn using the survey data will apply to the frame population, and not necessarily to the target population. Coverage problems can arise when the frame population differs from the target population.

Undercoverage occurs when elements in the target population do not appear in the frame population. These elements have zero probability of being selected in the sample. Undercoverage can be a serious problem for online surveys. If the target population consists of all people with an Internet connection, there is no problem. However, usually the target population is wider than that. Then, undercoverage occurs due to the fact that still many people do not have access to Internet. According to Eurostat (2007), the statistical office of the European Union, countries differ substantially in Internet coverage of households. Table 11.1 summarizes the extremes.

Internet access is very high in The Netherlands. More than four out of five households have an Internet connection. Internet coverage is also high in Scandinavian countries of Sweden and Denmark. Coverage is very low in the Balkan countries of Romania and Bulgaria. Approximately, only one out of five households has Internet access.

Table 11.1 Internet Access by Households in Europe in 2007

Country	Internet Access	Broadband Connection
Netherlands	83%	74%
Sweden	79%	67%
Denmark	78%	70%
...		
Greece	25%	7%
Romania	22%	8%
Bulgaria	19%	15%
EU	54%	42

Source: Eurostat (2007).

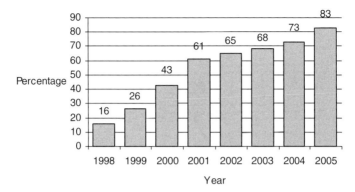

Figure 11.1 Percentage of persons having Internet (in The Netherlands).

Table 11.1 also contains information about the percentage of households with a broadband Internet connection. It is clear that still many Internet connections are based on slow modems. This may put restrictions on the questionnaires used. They may not be too long and too complicated, and prohibit advanced features such as the use of images, video, and animation. Slow questionnaire processing may cause respondents to break off the session, resulting in only partially completed questionnaires.

In The Netherlands, the percentage of persons having an Internet connection at home has increased from year to year (see Fig. 11.1). In 7 years, the number of Internet connections increased from 16 to 83%. Still, it is clear that not every household will have access to Internet in the near future.

An analysis of data on Internet access in The Netherlands in 2005 indicates that Internet access is unevenly distributed over the population. Figure 11.2 shows the distribution by gender. Apparently, more males than females have access to the Internet.

Figure 11.3 contains the percentage of Dutch people having Internet by age group (in 2005). The percentage of Internet access at home decreases with age. Particularly, the people of age 55 and older will be very much underrepresented when the Internet is used as a selection mechanism.

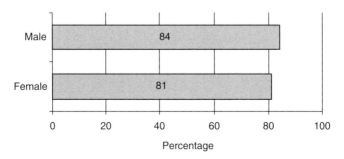

Figure 11.2 Having Internet by gender.

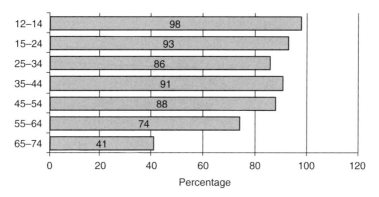

Figure 11.3 Having Internet by age.

Figure 11.4 shows the percentage of people in The Netherlands with an access to Internet by level of education (in 2005). It is clear that people with a higher level of education tend to have Internet access more frequently than people with a lower level of education.

According to De Haan and Van't Hof (2006), Internet access among nonnative young people is much lower in The Netherlands than among native young people: 91% of the young natives have access to Internet. This access is 80% for young people from Surinam and Antilles, 68% for young people from Turkey, and only 64% for young people from Morocco.

The results described above are in line with the findings of many authors in other countries (see Couper, 2000; Dillman and Bowker, 2001).

It is clear that the use of Internet as a sampling frame will cause problems, because specific groups are substantially underrepresented. Even if a proper probability sample is selected, the result will be a selective sample. Specific groups in the target population will not be able to fill in the (electronic) questionnaire form.

Note that there is some similarity with CATI surveys in which telephone directories are used as a sampling frame. Here, people without a telephone and people with an

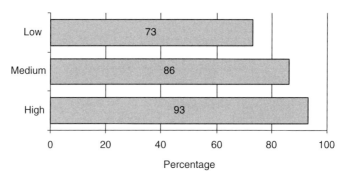

Figure 11.4 Having Internet by level of education.

unlisted number will be excluded from the survey. It is interesting to note that, for example, in The Netherlands between 60 and 70% of the people have listed fixed-line telephone number. This implies that by using a telephone directory as a sampling frame, one out of three households will never be selected. So Internet coverage is much higher than telephone coverage in The Netherlands (for listed fixed-line telephones). These numbers speak in favor of online surveys. However, the effects of undercoverage are also determined by the extent to which the undercovered part of the population differs from the covered part (with respect to the target variables of the survey) (see also Section 11.3.2).

11.2.2 Selection Problems

Horvitz and Thompson (1952) show in their seminal paper that unbiased estimates of population characteristics can be computed only if a real probability sample has been used, every element in the population has a nonzero probability of selection, and all probabilities are known to the researcher. Furthermore, only under these conditions can the accuracy of estimates be computed.

Many online surveys are not based on probability sampling. The survey questionnaire is simply put on the Web. Respondents are those people who happen to have Internet access, visit the Web site and decide to participate in the survey. The survey researcher is not in control of the selection process. Selection probabilities are unknown, and therefore neither can unbiased estimates be computed nor can the accuracy of estimates be determined. These surveys are called self-selection surveys.

The effects of self-selection can be illustrated by using an example related to the general elections in The Netherlands in 2003. Various organizations made attempts to use opinion polls to predict the outcome of these elections. The results of these polls are summarized in Table 11.2.

Table 11.2 Dutch Parliamentary Elections 2003: Outcomes and the Results of Various Opinion Surveys

	Election	Kennisnet	RTL4	SBS6	Nederland 1
Sample size		17,000	10,000	3,000	1,200
Seats in Parliament					
CDA (Christian democrats)	44	29	24	42	42
LPF (populist party)	8	18	12	6	7
VVD (liberals)	28	24	38	28	28
PvdA (social democrats)	42	13	41	45	43
SP (socialists)	9	22	10	11	9
GL (green party)	8	26	9	6	8
D66 (liberal democrats)	6	4	7	5	6
Other parties	5	14	9	7	7
Mean absolute difference		12.5	5.3	1.8	0.8

A typical example of a self-selection survey was the survey on the Dutch Web site *Kennisnet* (Knowledge net). This is a Web site for all those involved in education. More than 11,000 schools and other educational institutes use this Web site. The survey was an opinion poll for the general elections held on January 22, 2003. Everybody, including those not involved in education, could participate in the poll. Table 11.2 contains both the official results (seats in Parliament) of the election (column *Election*) and the results of this poll on the morning of the election day (column *Kennisnet*). The survey estimates were based on votes of approximately 17,000 people. No adjustment weighting was carried out. Although this was a large sample, it is clear that the survey results were no way near the true election results. The mean absolute difference (MAD) is equal to 12.5, which means that the estimated number of seats and the true number of seats differ on average by an amount of 12.5.

Another example of a self-selection Web survey was the election site of the Dutch television channel RTL4. It resembled to some extent the Kennisnet survey but was targeted at a much wider audience. Again, the survey researcher had no control at all over who was voting. There was some protection, by means of cookies, against voting more than once. However, this also had a drawback as only one member of the family could participate. Table 11.2 shows the survey results at noon on the day of the general elections (column *RTL4*). Figures were based on slightly over 10,000 votes. No weighting adjustment procedure was carried out. The results are better than that of the Kennisnet survey (the MAD decreased from 12.5 to 5.3). However, deviations between estimates and true figures are still substantial, particularly for the large parties. Note that even a large sample size of over 10,000 people did not help to get accurate estimates.

The Dutch commercial television channel SBS6 used an *access panel*. This is an Internet panel. Its members are regularly approached to complete a questionnaire on the Internet. Values of basic demographic variables were available for all panel members. A sample of size 3000 was selected. Selection was carried out such that the sample was representative with respect to the social-demographic and voting characteristics. Table 11.2 shows the results (column *SBS6*). The survey took place on the day before the general elections. Although attempts have been made to create a "representative" sample, the results differ still from the final result. The MAD has decreased to 1.8 but is still substantial.

A better prediction was obtained with a true probability sample. The table shows the results of a survey based on such a probability sample. It was carried out by the television channel *Nederland 1* in cooperation with the marketing agency *Interview-NSS*. A sample of size 1200 was selected by means of random digit dialing. The MAD was reduced to 0.8.

The conclusion from the analysis above is that a probability sample is a vital prerequisite for making proper inference about the target population of a survey. Even with a probability sample only of size 1200, better results can be obtained than with a nonprobability sample of size 10,000 or more.

A more recent comparison is presented in Table 11.3. *Politieke Barometer*, *Peil.nl* and *De Stemming* are opinion polls for the Dutch General Election of 2006. These polls are based on samples from online panels. To reduce a possible bias, adjustment

11.3 THE THEORETICAL FRAMEWORK

Table 11.3 Dutch Parliamentary Elections 2006: Outcomes and the Results of Various Opinion Surveys

	Election Result	Politieke Barometer	Peil.nl	De Stemming	DPES 2006
Sample size		1000	2500	2000	2600
Seats in Parliament					
CDA (Christian democrats)	41	41	42	41	41
PvdA (social democrats)	33	37	38	31	32
VVD (liberals)	22	23	22	21	22
SP (socialists)	25	23	23	32	26
GL (green party)	7	7	8	5	7
D66 (liberal democrats)	3	3	2	1	3
ChristenUnie (Christian)	6	6	6	8	6
SGP (Christian)	2	2	2	1	2
PvdD (animal party)	2	2	1	2	2
PvdV (conservative)	9	4	5	6	8
Other parties	0	2	1	2	1
Mean absolute difference		1.27	1.45	2.00	0.36

weighting has been carried out. *DPES* is the Dutch Parliamentary Election Study. The fieldwork was carried out by Statistics Netherlands. It used a true (two-stage) probability sample. Respondents were interviewed face-to-face (using CAPI). It is clear that the DPES outperformed the online polls.

Probability sampling has the additional advantage of providing protection against certain groups in the population attempting to manipulate the outcomes of the survey. This may typically play a role in opinion polls. Self-selection does not have this safeguard. An example of this effect could be observed in the election of the 2005 Book of the Year award (Dutch: NS Publieksprijs), a high-profile literary prize. The winning book was determined by means of a poll on a Web site. People could vote for one of the nominated books or mention another book of their choice. More than 90,000 people participated in the survey. The winner turned out to be the new interconfessional Bible translation launched by The Netherlands and Flanders Bible Societies. Although this book was not nominated, an overwhelming majority of respondents (72%) voted it. This was due to a campaign launched by (among others) Bible societies, a Christian broadcaster and Christian newspaper. Although this was all completely within the rules of the contest, the group of voters could clearly not be considered representative of the Dutch population.

11.3 THE THEORETICAL FRAMEWORK

11.3.1 The Internet Population

Let the target population of the survey consist of N identifiable elements, which are labeled $1, 2, \ldots, N$. Associated with each element k is a value Y_k of the target variable Y.

The aim of the Web survey is assumed to be an estimation of the population mean

$$\bar{Y} = \frac{1}{N}\sum_{k=1}^{N} Y_k \quad (11.1)$$

of the target variable Y.

The population U is divided into two subpopulations: U_I of elements having an access to Internet and U_{NI} of elements not having an access to the Internet. Associated with each element k is an indicator I_k, where $I_k = 1$ if element k has access to Internet (and thus is an element of subpopulation U_I), and $I_k = 0$ otherwise. The subpopulation U_I will be called the *Internet population* and U_{NI} is the *non-Internet population*. Let

$$N_I = \sum_{k=1}^{N} I_k \quad (11.2)$$

denote the size of subpopulation U_I. Likewise, N_{NI} denotes the size of the subpopulation U_{NI}, where $N_I + N_{NI} = N$.

The mean of the target variable for the elements in the Internet population is equal to

$$\bar{Y}_I = \frac{1}{N_I}\sum_{k=1}^{N} I_k Y_k. \quad (11.3)$$

Likewise, the mean of the target variable for the non-Internet population is denoted by

$$\bar{Y}_{NI} = \frac{1}{N_{NI}}\sum_{k=1}^{N} (1-I_k) Y_k. \quad (11.4)$$

11.3.2 A Random Sample from the Internet Population

The first situation to consider for an online survey is the more or less ideal case in which it is possible to select a random sample without replacement from the Internet population. This would require a sampling frame listing all elements with an access to Internet. No such list exists, but there are ways to get close to such a situation. One way to do this is to select a random sample from a larger sampling frame (e.g., a population or address register), approach the selected people in a classical way (by mail, telephone, or face-to-face), and filter out only those people having an access to Internet. Next, selected people are provided with an Internet address where they can fill in the questionnaire form. It is clear that initially such registers suffer from overcoverage, but with this approach every element in the Internet population has a positive and known probability of being selected.

A random sample selected without replacement from the Internet population is represented by a series

$$a_1, a_2, \ldots, a_N \quad (11.5)$$

of N indicators, where the kth indicator a_k assumes the value 1 if element k is selected, otherwise it assumes the value 0, for $k = 1, 2, \ldots, N$. Note that always $a_k = 0$ for elements k in the non-Internet population. The sample size is denoted by $n_\mathrm{I} = a_1 + a_2 + \cdots + a_N$.

The expected value $\pi_k = E(a_k)$ is the *first-order inclusion probability* of element k. Recall that Horvitz and Thompson (1952) have shown that always an unbiased estimator of the population mean can be defined if all elements in the population have known positive probability of being selected. The Horvitz–Thompson estimator for the mean of the Internet population is defined by

$$\bar{y}_{\mathrm{HT}} = \frac{1}{N_\mathrm{I}} \sum_{k=1}^{N} a_k I_k \frac{Y_k}{\pi_k}, \tag{11.6}$$

where by definition $Y_k/\pi_k = 0$ for all elements outside the Internet population. In case of a simple random sample from the Internet population, all first-order inclusion probabilities are equal to n/N_I. Therefore, expression (11.6) reduces to

$$\bar{y}_\mathrm{I} = \frac{1}{n} \sum_{k=1}^{N} a_k I_k Y_k. \tag{11.7}$$

This estimator is an unbiased estimator of the mean $\bar{Y}_\mathrm{I}$ of the Internet population but not necessarily of the mean $\bar{Y}$ of the target population. The bias is equal to

$$B(\bar{y}_{\mathrm{HT}}) = E(\bar{y}_{\mathrm{HT}}) - \bar{Y} = \bar{Y}_\mathrm{I} - \bar{Y} = \frac{N_{\mathrm{NI}}}{N}(\bar{Y}_\mathrm{I} - \bar{Y}_{\mathrm{NI}}). \tag{11.8}$$

The magnitude of this bias is determined by two factors. The first factor is the relative size N_{NI}/N of the subpopulation without Internet. The bias will increase as a larger proportion of the population does not have access to Internet. The second factor is the *contrast* $\bar{Y}_\mathrm{I} - \bar{Y}_{\mathrm{NI}}$ between the Internet population and the non-Internet population. The more the mean of the target variable differs for these two subpopulations, the larger the bias will be.

Presently, the size of the non-Internet population cannot be neglected in The Netherlands. Figure 11.1 shows that although the percentage of people without Internet is rapidly decreasing, it is still in the order of 17%.

Furthermore, there are substantial differences between these two subpopulations. The graphs in Section 11.2 show that specific groups are underrepresented in the Internet population; for example, the elderly, those with a low level of education, and ethnic minority groups. So, the conclusion is that generally a random sample from an Internet population will lead to biased estimates for the parameters of the target population.

11.3.3 Self-Selection from the Internet Population

For many online surveys no proper random sample is selected from the Internet population. These surveys rely on *self-selection* of respondents. Participation requires that respondents are first aware of the existence of a survey (they have to accidentally visit the Web site or they have to follow up a banner or an e-mail message). Second, they have to make the decision to fill in the questionnaire on the Internet. All this means that each element k in the Internet population has unknown probability ρ_k of participating in the survey, for $k = 1, 2, \ldots, N_I$. The responding elements can be denoted by a series

$$R_1, R_2, \ldots, R_N \tag{11.9}$$

of N indicators, where the kth indicator R_k assumes the value 1 if element k participates, and otherwise it assumes the value 0, for $k = 1, 2, \ldots, N$. Not that selection without replacement is assumed. The expected value $\rho_k = E(R_k)$ will be called the *response probability* of element k. For the sake of convenience also response probabilities are introduced for the elements in the non-Internet population. By definition, the values of all these probabilities are 0.

The realized sample size is equal to

$$n_S = \sum_{k=1}^{N} R_k. \tag{11.10}$$

A naive researcher assuming that every element in the Internet population has the same probability of being selected in the sample will use the sample mean

$$\bar{y}_S = \frac{1}{n_S} \sum_{k=1}^{N} R_k Y_k \tag{11.11}$$

as an estimator for the population mean. The expected value of this estimator is approximately equal to

$$E(\bar{y}_S) \approx \bar{Y}^* = \frac{1}{N_I \bar{\rho}} \sum_{k=1}^{N} \rho_k I_k Y_k, \tag{11.12}$$

where $\bar{\rho}$ is the mean of all response probabilities in the Internet population (see, for example, Bethlehem, 1988).

By using an approach similar to Cochran (1977, p. 31), it can be shown that the variance of the sample mean is approximately equal to

$$V(\bar{y}_S) \approx \frac{1}{(N_I \bar{\rho})^2} \sum_{k=1}^{N} \rho_k (1 - \rho_k)(Y_k - \bar{Y}^*)^2. \tag{11.13}$$

Note that this expression for the variance does not contain a sample size (because no fixed size sample is drawn) but the expected sample size $N_I \bar{\rho}$. Not surprisingly, the variance decreases as the expected sample size increases.

In general, the expected value of the sample mean is not equal to the population mean of the Internet population. The only situation in which the bias vanishes is that in which all response probabilities in the Internet population are equal. In terms of nonresponse correction theory, this comes down to missing completely at random (MCAR). See also Section 8.3 on imputation techniques for item nonresponse and Section 9.4 on the analysis of unit nonresponse.

Indeed, in case of equal selection probabilities, self-selection does not lead to an unrepresentative sample because all elements have the same selection probability.

THE THEORETICAL FRAMEWORK

Similar to Bethlehem (1988), it can be shown that the bias of the sample mean 11.11 can be written as

$$B(\bar{y}_S) = E(\bar{y}_S) - \bar{Y}_I \approx \bar{Y}^* - \bar{Y}_I = \frac{S_{\rho Y}}{\bar{\rho}} = \frac{R_{\rho Y} S_\rho S_Y}{\bar{\rho}}, \qquad (11.14)$$

in which

$$S_{\rho Y} = \frac{1}{N_I} \sum_{k=1}^{N} I_k (\rho_k - \bar{\rho})(Y_k - \bar{Y}) \qquad (11.15)$$

is the covariance between the values of target variable and the response probabilities in the Internet population, and $\bar{\rho}$ is the average response probability. Furthermore, $R_{\rho Y}$ is the correlation coefficient between target variable and the response behavior, S_ρ is the standard deviation of the response probabilities, and S_Y is the standard deviation of the target variable.

The bias of the sample mean (as an estimator of the mean of the Internet population) is determined by the following factors:

- The average response probability. The more likely people are to participate in the survey, the higher the average response probability will be, and thus the smaller the bias will be.
- The relationship between the target variable and response behavior. The higher the correlation between the values of the target variable and the response probabilities, the higher the bias will be.
- The variation in the response probabilities. The more these probabilities vary, the larger the bias will be.

Three situations can be distinguished in which this bias vanishes:

(1) All response probabilities are equal. Again, this is the case in which the self-selection process can be compared with a simple random sample.
(2) All values of the target variable are equal. This situation is very unlikely to occur. If this were the case, no survey would be necessary. One observation would be sufficient.
(3) There is no relationship between target variable and response behavior. It means participation does not depend on the value of the target variable.

Expression (11.14) for the bias of the estimator can be used to compute an upper bound for the bias. Given the mean response probability $\bar{\rho}$, there is a maximum value that the standard deviation S_ρ of the response probabilities cannot exceed

$$S_\rho \leq \sqrt{\bar{\rho}(1-\bar{\rho})}. \qquad (11.16)$$

This implies that in the worst case (S_ρ assumes its maximum value and the correlation coefficient $R_{\rho Y}$ is equal to either $+1$ or -1), the absolute value of the bias will be equal to

$$|B_{\max}| = S_Y \sqrt{\frac{1}{\bar{\rho}} - 1}. \qquad (11.17)$$

This worst-case value of the bias also applies to the situation in which a probability sample has been drawn and subsequently nonresponse occurs in the fieldwork. Therefore, expression (11.17) provides a means to compare potential biases in various survey situations.

For example, regular surveys of Statistics Netherlands have response rates of around 70%. This means the absolute maximum bias is equal to $0.65 \times S_Y$. One of the largest Web surveys in The Netherlands was *21minuten.nl*. This survey was supposed to provide answers to questions about important problems in the Dutch society. Within a period of 6 weeks in 2006 about 170,000 people completed the questionnaire (which took about 21 min). As everyone could participate in the survey, the target population was not defined properly. If it is assumed the target population consists of all Dutch citizens from the age of 18, the average response probability was 170,000/ 12,800,000 = 0.0133. Hence, the absolute maximum bias is equal to $8.61 \times S_Y$. The conclusion is that the bias of the large Web survey can be a factor 13 larger than the bias of the small probability survey.

In many cases, the objective of the survey is not to estimate the mean of the Internet population but the mean of the total population, the target population. In this case, the bias of the sample mean is equal to

$$B(\bar{y}_S) = E(\bar{y}_S) - \bar{Y} = E(\bar{y}_S) - \bar{Y}_I + \bar{Y}_I - \bar{Y} = \frac{N_{\mathrm{NI}}}{N}(\bar{Y}_I - \bar{Y}_{\mathrm{NI}}) + \frac{C(\rho, Y)}{\bar{\rho}}. \qquad (11.18)$$

The bias now consists of two terms: a bias caused by interviewing just the Internet population instead of the complete target population (undercoverage bias) and a bias caused by self-selection of respondents in the Internet population (self-selection bias). Theoretically, it is possible that these two biases compensate one another. If people without Internet resemble people with Internet who are less inclined to participate, the combined effects will produce a larger bias. Practical experiences suggest that this may often be the case. For example, suppose *Y* is a variable measuring the intensity of some activity on the Internet (surfing, playing online games). Then, a positive correlation between *Y* and response propensities is not unlikely. Also, the mean of *Y* for the Internet population will be positive whereas the mean of the non-Internet population will be 0. So, both bias terms have a positive value.

11.4 CORRECTION BY ADJUSTMENT WEIGHTING

Weighting adjustment is a family of techniques that attempt to improve the quality of survey estimates by using auxiliary information. *Auxiliary information* is defined as a set of variables that have been measured in the survey and for which information on their population distribution is available. By comparing the population distribution of an auxiliary variable with its sample distribution, it can be assessed whether or not the sample is representative for the population (with respect to this variable). If these distributions differ considerably, one must conclude that the sample is selective. To correct this, adjustment weights are computed. Weights are assigned to all records of observed elements. Estimates of population characteristics can now be obtained by using the weighted values instead of the unweighted values. Weighting adjustment used to correct surveys that are affected by nonresponse has been described in Chapter 10 (see also Bethlehem, 2002).

This section explores the possibility of reducing the bias of online survey estimates. The usefulness of adjustment weighting is described separately for undercoverage and self-selection. Section 11.4.1 shows how poststratification may reduce an undercoverage bias and Section 11.4.2 is about poststratification to reduce the self-selection bias.

11.4.1 Poststratification to Correct for Undercoverage

Poststratification is a well-known and often-used weighting adjustment method. It is typically used to correct the negative effects of nonresponse. It is now explored whether poststratification can also successfully reduce the bias caused by undercoverage.

To carry out poststratification, one or more qualitative auxiliary variables are needed. Here, only one such variable is considered. The situation for more variables is not essentially different. Suppose there is an auxiliary variable X having L categories. So it divides the target population into L strata. The strata are denoted by the subsets $U_1, U_2, \ldots, U_L$ of the population U. The number of target population elements in stratum U_h is denoted by N_h, for $h = 1, 2, \ldots, L$. The population size N is equal to $N = N_1 + N_2 + \cdots + N_L$. This is the population information assumed to be available.

Suppose a sample of size n_I is selected from the Internet population. If n_h denotes the number of sample elements in stratum h, then $n_I = n_1 + n_2 + \cdots + n_L$. The values of the n_h are the result of a random selection process, so they are random variables. Note that since the sample is selected from the Internet population, only elements in the substrata $U_I \cap U_h$ are observed (for $h = 1, 2, \ldots, L$).

Poststratification assigns identical adjustment weights to all elements in the same stratum. The weight w_k for an element k in stratum h is equal to

$$w_k = \frac{N_h/N}{n_h/n_I}. \tag{11.19}$$

The simple sample mean

$$\bar{y}_I = \frac{1}{n_I} \sum_{k=1}^{N} a_k I_k Y_k \tag{11.20}$$

is now replaced by the weighted sample mean

$$\bar{y}_{I,PS} = \frac{1}{n_I} \sum_{k=1}^{N} a_k w_k I_k Y_k. \tag{11.21}$$

Substituting the weights and working out this expression leads to the poststratification estimator

$$\bar{y}_{\text{I,PS}} = \frac{1}{N} \sum_{h=1}^{L} N_h \bar{y}_{\text{I}}^{(h)} = \sum_{h=1}^{L} W_h \bar{y}_{\text{I}}^{(h)}, \qquad (11.22)$$

where $\bar{y}_{\text{I}}^{(h)}$ is the sample mean in stratum h and $W_h = N_h/N$ is the relative size of stratum h. The expected value of this poststratification estimator is equal to

$$E(\bar{y}_{\text{I,PS}}) = \frac{1}{N} \sum_{h=1}^{L} N_h E(\bar{y}_{\text{I}}^{(h)}) = \sum_{h=1}^{L} W_h \bar{Y}_{\text{I}}^{(h)} = \tilde{Y}_{\text{I}}, \qquad (11.23)$$

where $\bar{Y}_{\text{I}}^{(h)}$ is the mean of the target variable in stratum h of the Internet population. Generally, this mean will not be equal to the mean $\bar{Y}^{(h)}$ of the target variable in stratum h of the target population. The bias of this estimator is equal to

$$B(\bar{y}_{\text{I,PS}}) = E(\bar{y}_{\text{I,PS}}) - \bar{Y} = \tilde{Y}_{\text{I}} - \bar{Y} = \sum_{h=1}^{L} W_h \left(\bar{Y}_{\text{I}}^{(h)} - \bar{Y}^{(h)} \right) = \sum_{h=1}^{L} W_h \frac{N_{\text{NI},h}}{N_h} \left(\bar{Y}_{\text{I}}^{(h)} - \bar{Y}_{\text{NI}}^{(h)} \right),$$

$$(11.24)$$

where $N_{\text{NI},h}$ is the number of elements in stratum h of the non-Internet population.

The bias will be small if there is (on average) no difference between elements with and without Internet within the strata. This is the case if there is a strong relationship between the target variable Y and the stratification variable X. The variation in the values of Y manifests itself between strata but not within strata. In other words, the strata are homogeneous with respect to the target variable. In nonresponse correction terminology, this situation comes down to *missing at random* (MAR).

It can be concluded that the application of poststratification will successfully reduce the bias of the estimator if proper auxiliary variables can be found. Such variables should satisfy three conditions:

- They have to be measured in the survey.
- Their population distribution $(N_1, N_2, \ldots, N_L)$ must be known.
- They must be strongly correlated with all target variables.

Unfortunately, such variables are not very often available or there is only a weak correlation.

The variance of the poststratification estimator is equal to

$$V(\bar{y}_{\text{PS}}) = \sum_{h=1}^{L} W_h^2 V(\bar{y}^{(h)}). \qquad (11.25)$$

Cochran (1977) shows that in the case of a simple random sampling from the complete population, this expression is equal to

$$V(\bar{y}_{\text{PS}}) = \frac{1-f}{n} \sum_{h=1}^{L} W_h S_h^2 + \frac{1}{n^2} \sum_{h=1}^{L} (1 - W_h) S_h^2, \qquad (11.26)$$

where $f = n/N$ and S_h^2 is the variance in stratum h. If the strata are homogeneous with respect to Y, the variance of estimator will be small.

In case of simple random sampling from the Internet population, the variance of the estimator 11.7 becomes

$$V(\bar{y}_{I,PS}) = \sum_{h=1}^{L} W_h^2 \left(\frac{1}{n_I W_{I,h}} + \frac{1 - W_{I,h}}{(n_I W_{I,h})^2} - \frac{1}{N_{I,h}} \right) S_{I,h}^2, \quad (11.27)$$

where $N_{I,h}$ is the size of stratum h in the Internet population, $W_{I,h} = N_{I,h}/N_I$ and $S_{I,h}^2$ is the variance in stratum h of the Internet population.

11.4.2 Poststratification to Correct for Self-Selection

It is now explored whether poststratification can also successfully reduce the bias caused by self-selection. Poststratification requires auxiliary variables. The population of these auxiliary variables must be known. For a probability sample in which nonresponse has occurred, it is also possible to use the distribution of the auxiliary variables in the complete sample instead of their population distribution. Such information can sometimes be retrieved from the sampling frame. This situation does not apply to self-selection samples as there is no sampling frame.

Suppose a self-selection sample is selected from the Internet population. The total sample size is denoted by n_S. If n_h denotes the number of respondents in stratum h, then $n_S = n_1 + n_2 + \cdots + n_L$. The values of the n_h are the result of a Poisson sampling process, so they are random variables.

Poststratification assigns identical adjustment weights to all elements in the same stratum. The weight w_k for a respondent k in stratum h is equal to

$$w_k = \frac{N_h/N}{n_h/n_S}. \quad (11.28)$$

The simple sample mean

$$\bar{y} = \frac{1}{n_S} \sum_{k=1}^{N} R_k Y_k \quad (11.29)$$

is now replaced by the weighted sample mean

$$\bar{y}_{PS} = \frac{1}{n_S} \sum_{k=1}^{N} w_k R_k Y_k. \quad (11.30)$$

Substituting the weights and working out this expression leads to the poststratification estimator

$$\bar{y}_{PS} = \frac{1}{N} \sum_{h=1}^{L} N_h \bar{y}_h = \sum_{h=1}^{L} W_h \bar{y}_h, \quad (11.31)$$

where $\bar{y}_h$ is the sample mean in stratum h and $W_h = N_h/N$ is the relative size of stratum h.

The expected value of this poststratification estimator is equal to

$$E(\bar{y}_{PS}) = \frac{1}{N}\sum_{h=1}^{L} N_h E(\bar{y}_h) = \sum_{h=1}^{L} W_h \bar{Y}_h^* = \tilde{Y}^*, \quad (11.32)$$

where

$$\bar{Y}_h^* = \frac{1}{N_{I,h}} \sum_{k=1}^{N_h} \frac{\rho_{k,h}}{\bar{\rho}_h} Y_{k,h} \quad (11.33)$$

is the weighted mean of the target variable in stratum h. The subscripts k, h denote the kth element in stratum h, and $\bar{\rho}_h$ is the average response probability in stratum h.

Expression 11.33 is the analogue of expression 11.12, but now computed for stratum h. Generally, this mean will not be equal to the mean of the target variable in stratum h of the target population. The bias of this estimator is equal to

$$B(\bar{y}_{PS}) = E(\bar{y}_{PS}) - \bar{Y} = \tilde{Y}^* - \bar{Y} = \sum_{h=1}^{L} W_h(\bar{Y}_h^* - \bar{Y}_h) = \sum_{h=1}^{L} W_h \frac{R_{\rho Y,h} S_{\rho,h} S_{Y,h}}{\bar{\rho}_h}, \quad (11.34)$$

where the subscript h indicates that the respective quantities are computed just for stratum h and not for the complete population.

This bias will be small if

- the response propensities are similar within strata;
- the values of the target variable are similar within strata;
- there is no correlation between response behavior and the target variable within strata.

These conditions can be realized if there is a strong relationship between the target variable Y and the stratification variable X. Then the variation in the values of Y manifests itself between strata but not within strata. In other words, the strata are homogeneous with respect to the target variable. Also, if the strata are homogeneous with respect to the response probabilities, the bias will be reduced. In nonresponse correction terminology, this situation comes down to missing at random (MAR).

It can be concluded that the application of poststratification will successfully reduce the bias of the estimator if proper auxiliary variables can be found. Such variables must have been measured in the survey, their population distribution must be known, and they must produce homogeneous strata. Unfortunately, such variables are rarely available.

In the case of a self-selection Web survey, the variance $V(\bar{y}_h)$ of the sample mean in a stratum is the analogue of variance 11.13 but restricted to observations in that stratum. Therefore, the variance of the poststratification estimator is approximately equal to

$$V(\bar{y}_{PS}) = \sum_{h=1}^{L} W_h^2 \frac{1}{(N_{I,h} \bar{\rho}_h)^2} \sum_{k \in U_h}^{N} \rho_k (1-\rho_k)(Y_k - \bar{Y}_h^*)^2. \quad (11.35)$$

This variance is small if the strata are homogeneous with respect to the target variable. So, a strong correlation between the target variable Y and the stratification variable X will reduce both the bias and the variance of the estimator.

11.5 CORRECTION USING A REFERENCE SURVEY

Poststratification can be an effective correction technique provided auxiliary variables that have a strong correlation with the target variables of the survey are available. If such variables are not available, it might be considered to conduct a *reference survey*. This reference survey is based on a small probability sample, where data collection takes place with a mode different from the Web, for example, computer-assisted personal interviewing, with laptops or computer-assisted telephone interviewing. The reference survey approach has been applied by several market research organizations (see Börsch-Supan et al., 2004; Duffy et al., 2005).

Under the assumption of no nonresponse, or ignorable nonresponse, this reference survey will produce unbiased estimates of quantities that have also been measured in the online survey. Unbiased estimates for the target variable can be computed, but due to the small sample size, these estimates will have a substantial variance. The question is now whether estimates can be improved by combining the large sample size of the online survey with the unbiasedness of the reference survey in improving estimates.

Section 11.5.1 explores the use of a reference survey to correct an undercoverage bias. Section 11.5.2 does the same for the self-selection bias.

11.5.1 Reducing the Undercoverage Bias with a Reference Survey

It is assumed that one qualitative auxiliary variable is observed both in the online survey and the reference survey, and that this variable has a strong correlation with the target variable of the survey. Then, a form of poststratification can be applied where the stratum means are estimated using online survey data and the stratum weights are estimated using the reference survey data. This leads to the poststratification estimator

$$\bar{y}_{I,\text{RS}} = \sum_{h=1}^{L} \frac{m_h}{m} \bar{y}_I^{(h)}, \qquad (11.36)$$

where $\bar{y}_I^{(h)}$ is the online survey-based estimate for the mean of stratum h of the Internet population (for $h = 1, 2, \ldots, L$) and m_h/m is the relative sample size in stratum h as estimated in the reference survey sample (for $h = 1, 2, \ldots, L$). Under the conditions described above, the quantity m_h/m is an unbiased estimate of $W_h = N_h/N$.

Let I denote the probability distribution for the online survey and let P be the probability distribution for the reference survey. Then, the expected value of the poststratification estimator is equal to

$$E(\bar{y}_{I,\text{RS}}) = E_I E_P(\bar{y}_{I,\text{RS}}|I) = E_I \left(\sum_{h=1}^{L} \frac{N_h}{N} \bar{y}_I^{(h)} \right) = \sum_{h=1}^{L} W_h \bar{Y}_I^{(h)} = \tilde{Y}_I, \qquad (11.37)$$

where $W_h = N_h/N$ is the relative size of stratum h in the target population. So, the expected value of this estimator is identical to that of the poststratification estimator 11.23. The bias of this estimator is equal to

$$B(\bar{y}_{I,RS}) = E(\bar{y}_{I,RS}) - \bar{Y} = \tilde{Y}_I - \bar{Y} = \sum_{h=1}^{L} W_h (\bar{Y}_I^{(h)} - \bar{Y}^{(h)}) = \sum_{h=1}^{L} W_h \frac{N_{NI,h}}{N_h} (\bar{Y}_I^{(h)} - \bar{Y}_{NI}^{(h)}). \tag{11.38}$$

If a strong relationship exists between the target variable and the auxiliary variable used for computing the weights, there is little or no variation of the target variable within the strata. This implies that if the stratum means for the Internet population and for the target population do not differ much, this results in a small bias. So, using a reference survey with proper auxiliary variables can substantially reduce the bias of online survey estimates.

Note that the expression for the bias of the reference survey estimator is equal to that of the poststratification estimator. An interesting aspect of the reference survey approach is that any variable can be used for adjustment weighting as long as it is measured in both surveys. For example, some market research organizations use "webographic" or "psychographic" variables that divide the population in "mentality groups." People in the same groups have more or less the same level of motivation and interest to participate in such surveys. Deployment of effective weighting variables resembles the MAR situation. This implies that within weighting strata there is no relationship between participating in an online survey and the target variables of the survey.

Bethlehem (2007) shows that if a reference survey is used, the variance of the poststratification estimator is equal to

$$V(\bar{y}_{I,RS}) = \frac{1}{m} \sum_{h=1}^{L} W_h (\bar{Y}_I^{(h)} - \tilde{Y}_I)^2 + \frac{1}{m} \sum_{h=1}^{L} W_h (1 - W_h) V(\bar{y}_I^{(h)}) + \sum_{h=1}^{L} W_h^2 V(\bar{y}_I^{(h)}). \tag{11.39}$$

The quantity $\bar{y}_I^{(h)}$ is measured in the online survey. Therefore, its variance $V(\bar{y}_I^{(h)})$ will be of the order $1/n_I$. This means that the first term in the variance of the poststratification estimator will be of the order $1/m$, the second term of order $1/mn_I$, and the third term of order $1/n_I$. Since n_I will generally be much larger than m in practical situations, the first term in the variance will dominate, that is, the (small) size of the reference survey will determine the accuracy of the estimates. So, the large number of observations in the online survey does not help to produce accurate estimates. One could say that the reference survey approach reduces the bias of estimates at the cost of a higher variance.

11.5.2 Reducing the Self-Selection Bias with a Reference Survey

This section explores the effects of using a reference survey to reduce the bias in a self-selection survey. Again, it is assumed that one qualitative auxiliary variable is observed both in the Web survey and the reference survey, and that this variable has a strong correlation with the target variable of the survey. A form of poststratification can be applied where the stratum means are estimated using Web survey data and the stratum weights are estimated using the reference survey data. This leads to the poststratification estimator

$$\bar{y}_{RS} = \sum_{h=1}^{L} \frac{m_h}{m} \bar{y}_h, \qquad (11.40)$$

where $\bar{y}_h$ is the Web survey-based estimate for the mean of stratum h of the target population (for $h = 1, 2, \ldots, L$) and m_h/m is the estimated relative sample size in stratum h using the reference survey (for $h = 1, 2, \ldots, L$). Under the conditions described above, the quantity m_h/m is an unbiased estimate of $W_h = N_h/N$.

Let I denote the probability distribution for the Web survey and let P be the probability distribution for the reference survey. Then the expected value of the poststratification estimator is equal to

$$E(\bar{y}_{RS}) = E_I E_P(\bar{y}_{RS}|m_1, m_2, \ldots, m_L) = E_I\left(\sum_{h=1}^{L} \frac{N_h}{N} \bar{y}_h\right) = \sum_{h=1}^{L} W_h \bar{Y}_h^* = \tilde{Y}^*. \qquad (11.41)$$

So, the expected value of this estimator is identical to that of the poststratification estimator 11.32. The bias of this estimator is equal to

$$B(\bar{y}_{RS}) = E(\bar{y}_{RS}) - \bar{Y} = \tilde{Y}^* - \bar{Y} = \sum_{h=1}^{L} W_h(\bar{Y}_h^* - \bar{Y}_h) = \sum_{h=1}^{L} W_h \frac{R_{\rho Y,h} S_{\rho,h} S_{Y,h}}{\bar{\rho}_h}. \qquad (11.42)$$

A strong relationship between the target variable and the auxiliary variable used for computing the weights means that there is little or no variation of the target variable within the strata. Consequently, the correlation between target variable and response behavior will be small, and the same applies to the standard deviation of the target variable. So, using a reference survey with the proper auxiliary variables can substantially reduce the bias of Web survey estimates.

Bethlehem (2008) shows that if a reference survey is used, the variance of the poststratification estimator is equal to

$$V(\bar{y}_{RS}) = \frac{1}{m}\sum_{h=1}^{L} W_h(\bar{Y}_h^* - \tilde{Y}^*)^2 + \frac{1}{m}\sum_{h=1}^{L} W_h(1 - W_h)V(\bar{y}_h) + \sum_{h=1}^{L} W_h^2 V(\bar{y}_h). \qquad (11.43)$$

The quantity $\bar{y}_h$ is measured in the online survey. Its variance $(\bar{y}_h)$ will be at most of the order $1/E(n_S) = 1/(N\bar{\rho})$. This means that the first term in the variance of the poststratification estimator will be of the order $1/m$, the second term will be of order $1/(mE(n_S))$, and the third term of order $1/E(n_S)$. Since $E(n_S)$ will generally be much larger than m in practical situations, the first term in the variance will dominate, that is, the (small) size of the reference survey will determine the accuracy of the estimates.

Moreover, since strata preferably are based on groups of people with the same psychographic characteristics, and target variables may very well be related to the psychographic variables, the stratum means $\bar{Y}_h^*$ may vary substantially. This also contributes to a large value of the first variance component.

The conclusion is that a large number of observations in the online survey do not help to produce accurate estimates. The reference survey approach may reduce the bias of estimates, but it does so at the cost of a higher variance.

The effectiveness of a survey design is sometimes also indicated by means of the effective sample size. This is the sample size of a simple random sample of elements that would produce an estimator with the same precision. Using a reference survey implies that the effective sample size is much lower than the size of the Web survey. See Section 11.9 for an example showing this effect.

11.6 SAMPLING THE NON-INTERNET POPULATION

The fundamental problem of online surveys is that persons without Internet are excluded from the survey. This problem could be solved by selecting a stratified sample. The target population is assumed to consist of two strata: the Internet population U_I of size N_I and the non-Internet population U_{NI} of size N_{NI}.

To be able to compute an unbiased estimate, a simple random sample must be selected from both strata. The online survey provides the data about the Internet stratum. If this is a random sample with equal probabilities, the sample mean

$$\bar{y}_I = \frac{1}{n} \sum_{k=1}^{N} a_k I_k Y_k \qquad (11.44)$$

is an unbiased estimator of the mean of the Internet population.

Now suppose a random sample (with equal probabilities) of size m is selected from the non-Internet stratum. Of course, there is no sampling frame for this population. This problem could be avoided by selecting a sample from the complete target population (a reference survey) and by using only people without Internet access. Selected people with Internet access can be added to the large online sample, but this will have no substantial effect on estimators. The sample mean of the non-Internet sample is denoted by

$$\bar{y}_{NI} = \frac{1}{m} \sum_{k=1}^{N} b_k (1-I_k) Y_k, \qquad (11.45)$$

where the indicator b_k denotes whether or not element k is selected in the reference survey, and

$$m = \sum_{k=1}^{N} b_k (1-I_k). \qquad (11.46)$$

The stratification estimator is now defined by

$$\bar{y}_{ST} = \frac{N_I}{N}\bar{y}_I + \frac{N_{NI}}{N}\bar{y}_{NI}. \qquad (11.47)$$

This is an unbiased estimator for the mean of the target population. Application of this estimator assumes the size N_I of the Internet population and the size N_{NI} of the non-Internet population to be known. The variance of the estimator is equal to

$$V(\bar{y}_{ST}) = \left(\frac{N_I}{N}\right)^2 V(\bar{y}_I) + \left(\frac{N_{NI}}{N}\right)^2 V(\bar{y}_{NI}). \qquad (11.48)$$

The variance of the sample mean in the Internet stratum is of order $1/n$ and the variance in the non-Internet stratum is of order $1/m$. Since m will be much smaller than n in practical situation, and the relative sizes of the Internet population and the non-Internet population do not differ that much, the second term will determine the magnitude of the variance. So the advantages of the large sample size of the online survey are for a great part lost by the bias correction.

Note that the sizes of the Internet and non-Internet population are usually unknown. In this case, they have to be estimated. This can, for example, be done using data from the reference survey.

11.7 PROPENSITY WEIGHTING

Propensity weighting is used by several market research organizations to correct a possible bias in their online survey (see Börsch-Supan et al., 2004; Duffy et al., 2005). The original idea behind propensity weighting goes back to Rosenbaum and Rubin (1983,1984). They developed a technique for comparing two populations. They attempted to make the two populations comparable by simultaneously controlling for all variables that were thought to explain the differences. Propensity weighting has already been described in Section 10.7 as a technique to reduce the nonresponse bias.

In the case of an online survey, there are also two populations: those who participate in the online survey and those who do not participate.

Propensity scores are obtained by modeling a variable that indicates whether or not someone participates in the survey. Usually, a logistic regression model is used where the indicator variable is the dependent variable and attitudinal variables are the explanatory variables. These attitudinal variables are assumed to explain why someone participates or not. Fitting the logistic regression model comes down to estimating the probability (propensity) of participation, given the values of the explanatory variables.

Application of propensity weighting assumes some kind of random process determining whether or not someone participates in the online survey. Each element k in the population has a certain, unknown probability ρ_k of participating, for $k = 1, 2, \ldots, N$. Let $R_1, R_2, \ldots, R_N$ denote indicator variables, where $R_k = 1$ if person k participates in the survey, and $R_k = 0$ otherwise. Consequently, $P(R_k = 1) = \rho_k$.

The propensity score $\rho(X)$ is the conditional probability that a person with observed characteristics X participates, that is,

$$\rho(X) = P(R = 1|X). \tag{11.49}$$

It is assumed that within the strata defined by the values of the observed characteristics X, all persons have the same participation propensity. This is the *missing at random* assumption. The propensity score is often modeled using a logit model:

$$\log\left(\frac{\rho(X_k)}{1-\rho(X_k)}\right) = \alpha + \beta'X_k. \tag{11.50}$$

The model is fitted using Maximum Likelihood estimation. Once propensity scores have been estimated, they are used to stratify the population. Each stratum consists of elements with (approximately) the same propensity scores. If indeed all elements within a stratum have the same response propensity, there will be no bias if just the elements in the Internet population are used for estimation purposes. Cochran (1968) claims that five strata are usually sufficient to remove a large part of the bias. The market research agency Harris Interactive was among the first to apply propensity score weighting in online surveys (see Terhanian et al., 2001).

To be able to apply propensity score weighting, two conditions have to be fulfilled. The first condition is that proper auxiliary variables must be available. These are variables that are capable of explaining whether or not someone participates in the online survey. Variables often used measure general attitudes and behavior. They are sometimes referred to as webographic or psychographic variables. Schonlau et al. (2004) mention as examples "Do you often feel alone?" and "On how many separate occasions did you watch news programs on TV during the past 30 days?"

The second condition for this type of adjustment weighting is that the population distribution of the webographic variables must be available. This is generally not the case. A possible solution to this problem is to carry out an additional reference survey. To allow unbiased estimation of the population distribution, the reference survey must be based on a true probability sample from the entire target population.

Such a reference survey can be small in terms of the number of questions asked. It can be limited to the webographic questions. Preferably, the sample size of the reference survey should be large to allow precise estimation. A small sample size results in large standard errors of estimates. This is similar to the situations described in Section 11.5.

Schonlau et al. (2004) describe the reference survey of Harris Interactive. This is a CATI survey, using random digit dialing. This reference survey is used to adjust several online surveys. Schonlau et al. (2003) stress that the success of this approach depends on two assumptions: (1) the webographic variables are capable of explaining the difference between the online survey respondents and the other persons in the target population and (2) the reference survey does not suffer from nonignorable nonresponse. In practical situations, it will not be easy to satisfy these conditions.

It should be noted that from a theoretical point of view propensity weighting should be sufficient to remove the bias. However, in practice the propensity score variable will

11.8 SIMULATING THE EFFECTS OF UNDERCOVERAGE

The possible consequences of undercoverage and the effectiveness of correction techniques are now illustrated using a simulation experiment. A fictitious population was constructed. For this population, voting intentions for the next general elections were simulated and analyzed.

The relationship between variables involved was such that it could resemble more or less a real-life situation. With respect to the Internet population, both missing at random (MAR) and not missing at random (NMAR) were introduced. The characteristics of estimators (before and after correction) were computed based on a large number of simulations.

First, the distribution of the estimator was determined in the ideal situation of a simple random sample from the target population. Then, it was explored how the characteristics of the estimator change if a simple random sample is selected just from the Internet population. Finally, the affects of weighting (poststratification and reference survey) were analyzed.

A fictitious population of 30,000 individuals was constructed. There were five variables:

- *Age in Three Categories.* Young (with probability 0.40), Middle aged (with probability 0.35), and Old (with probability 0.25).
- *Ethnic Origin in Two Categories.* Native (with probability 0.85) and Nonnative (with probability 0.15).
- *Having Access to Internet with Two Categories Yes and No.* The probability of having access to Internet depended on the two variables Age and Ethnic origin. For natives, the probabilities were 0.90 (for Young), 0.70 (for Middle aged), and 0.50 (for Old). So, Internet access decreases with age. For nonnatives, these probabilities were 0.20 (for Young), 0.10 (for Middle aged), and 0.00 (for Old). These probabilities reflect the much lower Internet access among nonnatives.
- *Vote for the National Elderly Party.* The probability to vote for this party depended on age. Probabilities were 0.00 (for Young), 0.40 (for Middle aged), and 0.60 (for Old).
- *Vote for the New Internet Party.* The probability to vote for this party depended on both age and having Internet. For people with Internet, the probabilities were 0.80 (for Young), 0.40 (for Middle aged), and 0.20 (for Old). For people without Internet, all probabilities were equal to 0.10. So, for people with Internet voting decreases with age. Voting probability is low for people without Internet (Fig. 11.5).

In the experiment, the variable NEP (National Elderly Party) suffered from missingness due to missing at random. There is a direct relationship between voting

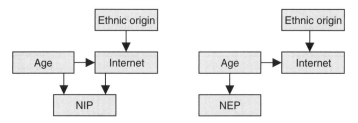

Figure 11.5 Relationships between variables.

for this party and age, and also there is a direct relationship between age and having Internet access. This will cause estimates to be biased. It should be possible to correct this bias by weighting using the variable age.

The variable NIP (New Internet Party) suffers from not missing at random. There exists (among other relationships) a direct relationship between voting for this party and having Internet access. Estimates will be biased, and there is no correction possible.

The distribution of estimators for the percentage of voters for both parties was determined in various situations by repeating the selection of the sample 800 times. In all cases, the sample size was $n = 2000$.

Figure 11.6 contains the results for the variable NEP (vote for the National Elderly Party). The upper-left graph shows the distribution of the estimator for simple random samples from the complete target population. The vertical line denotes the population value to be estimated (25.4%). The estimator has a symmetric distribution around this value. This clearly indicates that the estimator is unbiased.

The upper-right graph shows what happens if samples are not selected from the complete target population, but just from the Internet population. The shape of the distribution remains the same, but the distribution as a whole has shifted to the left. All values of the estimator are systematically too low. The expected value of the estimator is only 20.3%. The estimator is biased. The explanation of this bias is simple: relatively few elderly have Internet access. Therefore, they are underrepresented in samples selected from the Internet. These are typically people who will vote for the NEP.

The lower left graph in Fig. 11.6 shows the distribution of the estimator in case of poststratification by age. The bias is removed. This was possible because this is a case of missing at random.

Poststratification by age can be applied only if the distribution of age in the population is known. If this is not the case, one could consider conducting a small ($m = 100$) reference survey, in which this population distribution is estimated unbiased. The lower right graph in Fig. 11.6 shows what happens in this case. The bias is removed but at the cost of a substantial increase in variance.

Figure 11.7 shows the results for the variable NIP (vote for the New Internet Party). The upper left graph shows the distribution of the estimator for simple random samples from the complete target population. The vertical line denotes the population value to be estimated (39.5%). Since the estimator has a symmetric distribution around this value, it is clear that the estimator is unbiased.

SIMULATING THE EFFECTS OF SELF-SELECTION **301**

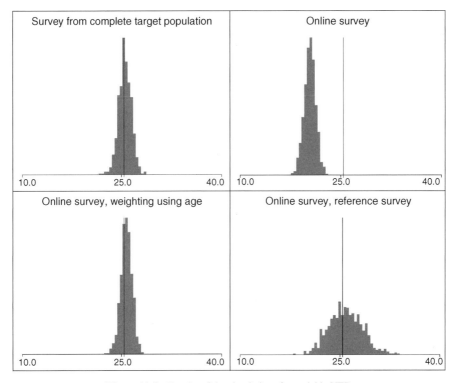

Figure 11.6 Results of the simulations for variable NEP.

The upper right graph shows what happens if samples are not selected from the complete target population, but just from the Internet population. The distribution has shifted to the right considerably. All values of the estimator are systematically too high. The expected value of the estimator is now 56.5%. The estimator is severally biased. The explanation of this bias is straightforward: voters for the NIP are overrepresented.

The lower left graph in Fig. 11.7 shows the effect of poststratification by age. Only a small part of the bias is removed. This is not surprising as there is a direct relationship between voting for the NIP and having Internet access. This is a case of not missing at random.

Also in this case, one can consider conducting a small reference survey if the population distribution of age is not available. The lower right graph in Fig. 11.7 shows what happens in this case. Only a small part of the bias is removed and at the same time there is a substantial increase in variance.

11.9 SIMULATING THE EFFECTS OF SELF-SELECTION

The possible consequences of self-selection and the effectiveness of correction techniques are also illustrated using a simulation experiment. A fictitious population

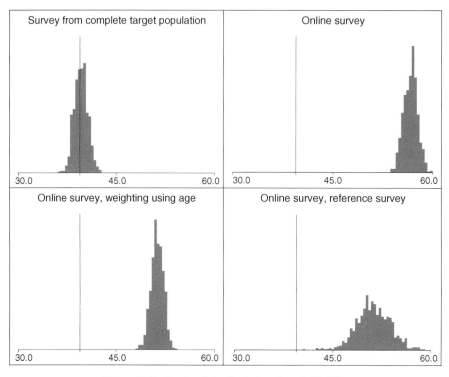

Figure 11.7 Results of the simulations for variable NIP.

was constructed. Again, voting intentions for the next general elections were simulated and analyzed. Relationships between variables involved were modeled somewhat stronger than they probably would be in a real-life situation. Effects are therefore more pronounced, making it clearer what the pitfalls are.

The characteristics of estimators (before and after correction) were computed based on a large number of simulations. First, the distribution of the estimator was determined in the ideal situation of a simple random sample from the target population. Then, it was explored how the characteristics of the estimator change if self-selection is applied. Finally, the effects of weighting (poststratification and reference survey) were analyzed.

A fictitious population of 100,000 individuals was constructed. There were five variables:

- The variable *Internet* indicates how active a person is on the Internet. There are two categories: very active users and more passive users. The population consists of 1% of active users and 99% of passive users. Active users have a response probability of 0.99 and passive users have a response probability of 0.01.
- The variable *Age* in three categories young, middle aged, and old. The active Internet users consist of 60% of young people, of 30% of middle-aged people, and of 10% of old people. The age distribution for passive Internet users is 40%

Figure 11.8 Relationships between variables.

young, 35% middle aged, and 25% old. So, typically younger people are more active Internet users.
- Vote for the NEP. The probability to vote for this party depends only on age. Probabilities are 0.00 (for Young), 0.30 (for Middle aged), and 0.60 (for Old).
- Vote for the NIP. The probability to vote for this party depends both on the age and on the use of Internet. The probabilities were 0.80 (for Young), 0.40 (for Middle aged), and 0.20 (for Old) for active Internet users. The probabilities are all equal to 0.10 for passive Internet users. So, for active users voting decreases with age. Voting probability is always low for passive users.

Figure 11.8 shows the relationships between the variables in a graphical way. The variable NEP suffers from missingness due to MAR. There is direct relationship between voting for this party and age, and also there is a direct relationship between age and propensity to participate in the survey. This will cause estimates to be biased. It should be possible to correct this bias by weighting using the variable age.

The variable NIP suffers from NMAR. There exists a direct relationship between voting for this party and response probability. Estimates will be biased, and there is no correction possible.

The distribution of estimators for the percentage of voters for both parties was determined in various situations by repeating the selection of the sample 500 times. The average response probability in the population is 0.01971. Therefore, the expected sample size in a self-selection survey is equal to 1971.

Figure 11.9 shows the results for the variable NEP (votes for National Elderly Party). The upper left graph shows the distribution of the estimator for simple random samples of size 1971 from the target population. The vertical line denotes the population value to be estimated (25.6%). The estimator has a symmetric distribution around this value. This clearly indicates that the estimator is unbiased.

The upper right graph shows what happens if samples are selected by means of self-selection. The shape of the distribution remains more or less the same, but the distribution as a whole has shifted to the left. All values of the estimator are systematically too low. The expected value of the estimator is only 20.5%. The estimator is biased. The explanation of this bias is simple: relatively few elderly are active Internet users. Therefore, they are underrepresented in the samples. These are typically people who will vote for the NEP.

The lower left graph in Fig. 11.9 shows the distribution of the estimator in case of poststratification by age. The bias is removed. Weighting works because this is a case of missing at random.

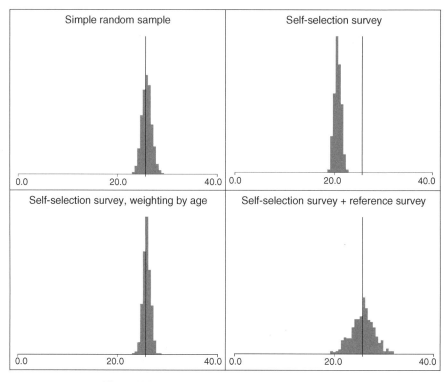

Figure 11.9 Results of the simulations for variable NEP.

Poststratification by age can be applied only if the distribution of age in the population is known. If this is not the case, one could consider conducting a small ($m = 100$) reference survey, in which this population distribution is estimated unbiased. The lower right graph in Fig. 11.9 shows what happens in this case. The bias is removed but at the cost of a substantial increase in variance. The variance is equal to that of a simple random sample of size of 290. So, the effective sample size is equal to 290. Apparently, an online survey of size 2000 is not more precise than a simple random sample of size 290 if a reference survey is used to correct the bias caused by self-selection.

Figure 11.10 shows the results for the variable NIP (vote for New Internet Party). The upper left graph shows the distribution of the estimator for simple random samples of size 1971 from the target population. The vertical line denotes the population value to be estimated (10.5%). Since the estimator has a symmetric distribution around this value, it is clear that the estimator is unbiased.

The upper right graph shows what happens if samples are selected by means of self-selection. The distribution has shifted to the right considerably. All values of the estimator are systematically too high. The expected value of the estimator is now 35.6%. The estimator is severally biased. The explanation of this bias is straightforward: voters for the NIP are overrepresented in the self-selection samples.

ABOUT THE USE OF ONLINE SURVEYS   305

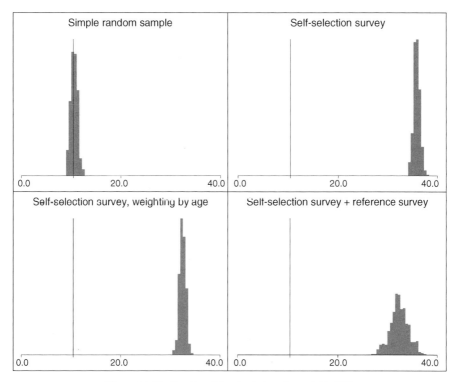

Figure 11.10   Results of the simulations for variable NIP.

The lower left graph in Fig. 11.10 shows the effect of poststratification by age. Only a small part of the bias is removed. Weighting is not successful. This is not surprising as there is a direct relationship between voting for the NIP and use of Internet. This is a case of NMAR.

Also in this case, one can consider conducting a small reference survey if the population distribution of age is not available. The lower right graph in Fig. 11.10 shows what happens in this case. Only a small part of the bias is removed, and at the same time there is a substantial increase in variance. The variance is equal to that of a simple random sample of size of 288. So, the effective sample size is 288. Apparently, an online survey of size 2000 is not more precise than a simple random sample of size 288. Moreover, the bias is not removed.

11.10   ABOUT THE USE OF ONLINE SURVEYS

This chapter discussed some of the methodological problems of online surveys. The underlying question is whether an online survey can be used as a data collection instrument for making valid inference about a target population. Costs and timeliness seem to be important arguments in favor of online survey. However, there are methodological challenges with respect to the properties of estimates.

Selecting a probability sample requires a sampling frame. The Internet is not an ideal sampling frame. It suffers from undercoverage. Certain groups in the population are underrepresented, for example, the elderly, low educated, and nonnatives. Therefore, estimates will often be biased and correction techniques are required to remove this bias. Unfortunately, correction techniques will be effective only if not having access to the Internet can be seen as missing at random.

It should be noted that other modes of data collection also have their coverage problems. For example, a CATI survey requires a sampling frame consisting of telephone numbers. Statistics Netherlands uses only fixed-line listed telephone numbers for this, as well as listed mobile numbers. Only between 60 and 70% of the people in The Netherlands have a listed phone number (see Cobben, 2004). This implies that only two out of three persons can be reached this way.

The undercoverage problem for CATI surveys will become even more severe over time. This is due to the popularity of mobile phones and the lack of lists of mobile phone numbers (see Kuusela, 2003). The situation is improving for surveys using the Internet as a sampling frame. In many countries, there is a rapid rise in households having Internet access. For example, the number of households with Internet is now over 80% in The Netherlands, and it keeps growing. So one might expect that online survey coverage problems will be less severe in the near future.

Unbiased estimators for population characteristics can be constructed only if all elements in the population have a known and positive probability of being selected. This is not always the case for online surveys. Market research agencies in The Netherlands have carried out an analysis of all their major online panels (see Vonk et al., 2006). It turned out that most of these panels are based on self-selection of respondents. The researchers concluded that panel members differ substantially from other people, and that therefore most of these panels cannot be considered representative for the population.

Can an online survey be an alternative for a CAPI or CATI survey? Coverage problems may be solved in the future, but there are also other aspects to consider. With respect to data collection, there is a substantial difference between CAPI and CATI on the one hand and online surveys on the other. Interviewers carry out the fieldwork in CAPI and CATI surveys. They are important in convincing people to participate in the survey, and they also can assist in completing the questionnaire. There are no interviewers in an online survey. It is a self-administered survey. Therefore, quality of collected data may be lower due to higher nonresponse rates and more errors in answering questions. According to De Leeuw and Collins (1997) response rates tend to be higher if interviewers are involved. However, response to sensitive questions is higher without interviewers. At present, little is known about the quality of the online survey data compared to CAPI or CATI survey data.

CAPI and CATI are both a form of computer-assisted interviewing. CAI has the advantage that error checking can be implemented. See also Chapter 7 about data collection. Answers to questions can be checked for consistency. Errors can be detected during the interview and therefore corrected during the interview itself. It has been shown (see Couper et al., 1998) that CAI can improve the quality of the collected data. The question is now whether error checking should be implemented in an online survey? What happens when respondents are confronted with error messages? Maybe they just correct their mistakes, but it may also happen that they will become annoyed and stop answering questions. There may be a trade-off here between nonresponse and data quality. Further research should make clear what the best approach is.

A reference survey is proposed as one way to remove the bias of estimates in an online survey. One of the advantages of a reference survey is that auxiliary variables can be used for weighting that are highly correlated with either target variables or participation probabilities. Therefore, correction will be effective. A disadvantage of a reference survey is that it results in large standard errors and therefore a small effective sample size. So a reference survey reduces the bias at the cost of a loss in precision. One attractive characteristic of an online survey is that it is rather easy to collect a large amount of data. If a reference survey is used, the large sample size of the online survey does not imply a high precision. So, one may wonder whether it is still worthwhile to carry out an online survey.

The reference survey only works well if it is a real probability sample without nonresponse, or with ignorable nonresponse (MCAR). This condition may be hard to satisfy in practical situations. Almost every survey suffers from nonresponse. If reference survey estimates are biased due to nonresponse, the online survey bias is replaced by a reference survey bias. This does not really help to solve the problem.

Reference surveys will be carried out in a mode other than CAWI. This means there may be mode effects that have an impact on estimates. Needless to say that a reference survey will dramatically increase survey costs.

If a reference survey is conducted, stratified estimation may be an option. The Internet population is one stratum and the non-Internet population is another stratum. In principle, this results in unbiased estimates. The drawback is that the complete questionnaire has to be used in the survey of the non-Internet population. If the reference survey is used just for weighting purposes, only relevant weighting variables need to be measured in both surveys. This reduces the reference survey in size and costs, and also the nonresponse may be less of a problem if a very short questionnaire is used.

One can say that an online survey based on self-selection and correction by means of a reference survey is not a reliable and cost-effective data collection instrument. This does not mean it is completely useless. When given a sound basis, for example, using probability sampling and more developed correction techniques, online surveys hold a promise for producing accurate and reliable information. This may make online survey an interesting and worthwhile topic for future research.

EXERCISES

11.1 Which of the statements below about an online survey is correct?
   a. An online survey always has a higher response rate than other types of surveys.
   b. Due to the large amount of respondents, estimates are always very close to the true values.

c. The quality of the results is often lower than those of CAPI or CATI surveys.

d. It is always possible to obtain unbiased estimates by using the Horvitz–Thompson estimator.

11.2 Why can a reference survey be useful to improve estimates based on data from an online survey?

a. A reference survey always provided unbiased estimates of population distributions of auxiliary variables.

b. The estimates after adjustment will be much more precise than the estimates before adjustment.

c. It is possible to use attitudinal variables for weighting. These variables suffer less from measurement errors than from factual variables.

d. The researcher can choose the most effective auxiliary variables for adjustment weighting.

11.3 A researcher wants to estimate the average number of hours per week the adult inhabitants of Samplonia spend on the Internet? He draws a simple random sample of Internet users. There is no nonresponse. The sample mean turns out to be 5 h.

a. Given that only three out of five inhabitants have access to Internet, compute an estimate of the bias of the sample mean.

b. Compute a better estimate for the average number of hours an inhabitant spends on the Internet.

11.4 A town council wants to know what percentage of the population is engaged in some form of voluntary work. Since there is only a limited budget available, it is decided to conduct an online survey. The target population consists of 1,000,000 persons. Only 70% of these persons have access to the Internet. It turns out that 10,000 persons participate in the survey. Of these respondents, 60% do some voluntary work.

a. Assuming that the 10,000 respondents are a simple random sample without replacement from the target population, compute the 95% confidence interval of the percentage of persons in the population doing voluntary work.

b. There is a strong suspicion that the survey estimates may be biased because only people with Internet access can participate. Therefore, a follow-up survey is conducted among people without Internet access. It turns out to be possible to draw a simple random sample of size 100 from this non-Internet population. The result is that 40% of the respondents in the follow-up survey do voluntary work.

Compute an improved estimate for the population percentage of people involved in voluntary work.

c. Compute a new 95% confidence interval of the percentage of persons in the population doing voluntary work.

d. Compare both confidence intervals and explain any differences.

11.5 A poll is conducted each year in The Netherlands to elect the best politician of the year. This poll is a self-selection Web survey. More than 21,000 people voted in 2006. Participants were also asked for which party they voted at the last general elections. Part of the results is summarized in the table below.

	Vote at Last Elections			
Politician	CDA	VVD	SP	Other
Jan-Peter Balkenende (CDA)	1980	254	38	218
Jan Marijnissen (SP)	135	97	3006	2080
Rita Verdonk (VVD)	385	1000	183	866
Other politicians	1427	1644	1540	6685

CDA is the party of the Christian democrats, VVD is the liberal party, and SP is the socialist party. Note that the category *Other* includes both other parties and people who did not vote.

a. Compute the percentages of votes for each politician. Determine the rank order of the three politicians.

b. Due to self-selection, the results will not properly reflect the situation in the population. Therefore, a weighting adjustment procedure is carried out. The CDA obtained 19.4% of the votes in the last general elections, the VVD got 10.7%, 12.1% voted for the SP, and 57.8% voted for another party or did not vote at all.

c. Compute adjustment weights for the three parties CDA, VVD, and SP and the category *Other*. Determine which parties are over- or underrepresented in the Web survey.

d. Compute a new table with weighted frequencies. Round the frequencies to integer numbers.

e. Compute weighted percentages of votes for the three politicians. Compare these percentages with those computed under (a). Explain the differences.

CHAPTER 12

Analysis and Publication

12.1 ABOUT DATA ANALYSIS

Statistics is a part of science that explains how to set up research, how to collect data, how to analyze these data, how to interpret the outcomes of the analysis, and how to publish the results of the analysis. The data are usually obtained by measuring or observing characteristics of people, objects, and phenomena. Survey research is a part of statistics in which data are collected by means of asking questions. The measuring instrument is the survey questionnaire.

Many statistical analysis techniques are available for analysis of the collected data. Most of these techniques assume a model stating that the data form an independent identically distributed random sample from some normal distribution. These assumptions are almost never satisfied in practical survey situations. More often, the *dirty data theorem* applies. It states that the data come from a dependent sample with unknown and unequal selection probabilities from a bizarre and unspecified distribution whereby some values are missing and many other values are subject to substantial measurement errors.

Analysts of survey data should take into account that their data maybe affected by measurement errors and nonresponse, that some values may not be observed but imputed, and that weights have to be used to compensate for a possible nonresponse bias. Many software packages for statistical data analysis assume the ideal model for the data, and have no possibilities to account for the effects of dirty data. Therefore, analysts should be very careful in their analysis. There are anecdotes about researchers discovering an interesting structure in the data, which in the end turned out to be the model used for imputing missing observations.

Survey data analysis will be carried out with some kind of statistical analysis package. A survey data file has to be prepared for this. Some general characteristics of such a survey data file are discussed here.

Applied Survey Methods: A Statistical Perspective, Jelke Bethlehem
Copyright © 2009 John Wiley & Sons, Inc.

ABOUT DATA ANALYSIS

Information is collected in a survey by means of asking questions. The questions correspond to the variables that have to be measured. The answers to the questions (i.e., the values of the variables) are recorded in the form of texts, numbers, and codes. Two types of variables are distinguished: qualitative variables and quantitative variables. It is important to also make this distinction in the analysis of the collected data.

Qualitative variables are variables that just divide the elements in groups. Their values are just labels. Examples of such variables are marital status, ethnic background, and region of the country. Most computations with these values are not meaningful. Only frequencies and percentages are relevant. Qualitative variables are usually measured by means of closed questions.

Quantitative variables measure a size, quantity, or value. Examples are the weight and length of a person, the profit of a company, and the number of students of a school. Computations with the values of these variables can be meaningful. Typical quantities are totals and averages. Quantitative variables are usually measured by means of numerical variables.

Many software packages for data analysis require the survey data to be in the format of a *survey data matrix*. This is a table in which each column denotes a variable and each row represents a record corresponding to an element. For reasons of efficiency, values of variables are stored in numeric format. This is obvious for quantitative variables. For qualitative variables, a code number is assigned to each category. These code numbers are stored instead of the labels of the categories. This requires much less storage space and moreover problems due to misspelling of labels are avoided. Table 12.1 shows this approach. It is part of the data matrix with data about Samplonia.

Table 12.1 shows the data. But just data are meaningless without a description of the data. The description is called *metadata*. So, there can be no data without metadata. Software packages for statistical analysis usually have ample facilities for documenting the data. It is important to do this properly and extensively, to avoid problems with the interpretation of the outcomes of the survey. Metadata are particularly important if a survey data set is reanalyzed long after the survey has been carried out. Table 12.2

Table 12.1 Part of the Data Matrix for Samplonia

Record	District	Province	Gender	Age	Employed	Income
1	5	2	1	65	2	0
2	6	2	1	36	2	0
3	7	2	2	73	2	0
4	6	2	1	6	2	0
5	3	1	2	33	1	158
6	1	1	2	82	2	0
7	2	1	1	2	2	0
8	1	1	1	32	1	525
9	5	2	2	66	2	0
10	3	1	2	2	2	0
...	...	...	...	...	...	...

Table 12.2 Metadata for Samplonia

Variable	Type	Description	Values
District	Qualitative	District of residence of the respondent	1 = Wheaton; 2 = Greenham; 3 = Newbay; 4 = Oakdale; 5 = Smokeley; 6 = Crowdon; 7 = Mudwater
Province	Qualitative	Province of residence of the respondent	1 = Agria; 2 = Induston
Gender	Qualitative	Gender of the respondent	1 = Male; 2 = Female
Age	Quantitative	Age if respondent (in years)	0 t/m 99
Employed	Qualitative	Respondent has a paid job for at least 12 h per week	1 = Yes; 2 = No
Income	Quantitative	Monthly net income of the respondent (in Samplonian dollars)	0 – 4500

contains a simple example of how the metadata of a survey about Samplonia could look like.

Many survey data sets suffer from missing values. These "holes" in the data matrix may be caused by item nonresponse or by errors in the values of variables that could not be corrected. Sometimes, analysis software uses special symbols or codes to indicate missing values. This facility sees to it that missing values are properly documented and also that missing values can be excluded from the analysis.

Traditionally, missing values were often denoted by filling the value field in the data matrix by a series of nines. For example, if the field for the variable "income" is four characters wide, a missing income would be represented by 9999. Of course, the analysis software must know that 9999 means a missing observation. If the software is not aware of this, something will go wrong in the computation of estimates. If 9999 is taken as a "real" income, estimates of the mean income will be systematically too high.

Some software packages allow distinguishing several types of missing values. For example, there could be special codes for "refusal" and "don't know."

Of course, it is possible to remove missing values from the survey data file. Imputation techniques can be used for this (see Chapter 8).

12.2 THE ANALYSIS OF DIRTY DATA

Many software packages for the analysis of survey data assume that the data can be seen as an independent identically distributed random sample from some normal distribution. Often this is not the case. This section describes three issues: sampling designs with unequal inclusion probabilities, weighting adjustment, and imputation.

12.2.1 Sampling Design Issues

The Horvitz–Thompson estimator has been introduced in Chapter 2. This estimator allows for unbiased estimation of population characteristics if all inclusion probabilities or selection probabilities are known and strictly positive. To be able to compute these estimates, inclusion or selection probabilities have to be included in the survey data file. Often *inclusion weights* (the inverse inclusion probabilities) are included.

Failure to include these probabilities in the survey data file, and to rely on standard estimation procedures implemented in software packages, may lead to biased estimates. An example illustrates this. Suppose a stratified sample of 50 persons is selected from the working population of Samplonia. There are two strata: the provinces of Agria and Induston. A simple random sample of 25 persons is selected without replacement from both strata. Since the province of Agria contains 121 persons, the inclusion probability here is $25/121 = 0.207$ for all elements. Induston contains 220 persons, which means the inclusion probability here is equal to $25/220 = 0.114$. So, the inclusion probabilities differ in both strata.

Figure 12.1 contains the results of an experiment in which 1000 samples were selected using this sampling design. The upper box plot contains the distribution of the estimator taking into account the inclusion probabilities. The vertical line corresponds to the population mean to be estimated. It is clear that the estimator is unbiased.

The lower box plot in Fig. 12.1 shows what happens if the inclusion probabilities are not taken into account and the simple sample mean is computed using some statistical analysis package. The distribution has shifted to the left. The estimator is clearly biased. Its values are systematically too low. The explanation is that people in high-income areas are underrepresented in the samples.

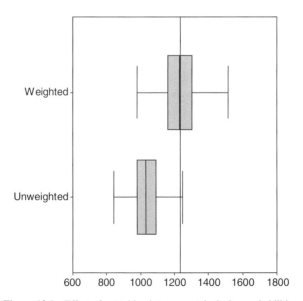

Figure 12.1 Effect of not taking into account inclusion probabilities.

Of course, it is possible to compute proper estimates with an analysis package. The trick is to include an extra variable in the survey data value containing the inclusion weight or similar quantity. One way to do this is to introduce a *correction weight*. It corrects for the wrong inclusion probability $n/N = 50/341$. For persons in the province of Agria, the correction weight is $(121/25)/(341/50) = 0.710$. This value is smaller than 1, because persons from Agria are overrepresented in the sample. For persons in the province of Induston, the correction weight is $(220/25)/(341/50) = 1.290$. This value is larger than 1 because persons from Induston are underrepresented in the sample. If a software package is instructed to use this correction weight in the computation of estimates, a weighted estimator is obtained that is identical to the Horvitz–Thompson estimator in case of stratification. Its distribution corresponds to the upper box plot.

12.2.2 Weighting Issues

It has already been said that many general software packages for statistical analysis can handle weights. However, it should be realized there are several types of weights. Each statistical package may interpret weights differently. Even weights can be interpreted differently within the same package. Here the following types of weights are considered:

- *Inclusion weights.* These weights are the inverse of the inclusion probabilities. Inclusion weights are determined by the sampling design. They must be known and nonzero to compute unbiased estimates (see Horvitz and Thompson, 1952).
- *Correction weights.* These weights are the result of applying some kind of weighting adjustment technique.
- *(Final) adjustment weights.* These weights combine inclusion weights and correction weights. When applied, they should provide unbiased estimates of population characteristics.
- *Frequency weights.* These weights are whole numbers indicating how many times a record occurs in a sample. It should be seen as a trick to reduce file size.

Problems may arise if weights are interpreted as frequencies weights while in fact they are inclusion weights. Suppose a sample of size n has been selected from a finite population of size N. The sample values of the target variable are denoted by $y_1, y_2, \ldots, y_n$. Let π_i be the inclusion probability of element i, for $i = 1, 2, \ldots, n$. Then, the inclusion weight for element i is equal to $1/\pi_i$. If these inclusion probabilities are used as frequency weights, the weighted sample mean is equal to

$$\bar{y}_W = \frac{\sum_{i=1}^{n} w_i y_i}{\sum_{i=1}^{n} w_i} = \frac{\sum_{i=1}^{n} (y_i/\pi_i)}{\sum_{i=1}^{n} (1/\pi_i)} \qquad (12.1)$$

According to the theory of Horvitz and Thompson (1952), the unbiased estimator is equal to

$$\bar{y}_{HT} = \frac{1}{N}\sum_{i=1}^{n} w_i y_i = \frac{1}{N}\sum_{i=1}^{n} \frac{y_i}{\pi_i} \qquad (12.2)$$

Generally, these two estimators are not the same. However, in the case of simple random sampling with equal probabilities ($\pi_i = n/N$), expression (12.1) reduces to (12.2).

Similar problems occur when computing estimates of variances. Many statistical packages assume the sample to be an independent random sample selected with equal probabilities. If the weights are interpreted as frequency weights, then the sample size is equal to

$$w_T = \sum_{i=1}^{n} w_i \qquad (12.3)$$

and the proper estimator for the variance of the sample mean is

$$v(\bar{y}_W) = \frac{\sum_{i=1}^{n} w_i(y_i - \bar{y}_W)^2}{w_T(w_T - 1)} \qquad (12.4)$$

Usually survey samples are selected without replacement, which means that the proper expression for the variance of the estimator is

$$v(\bar{y}_W) = \left(\frac{1}{w_T} - \frac{1}{N}\right) \frac{\sum_{i=1}^{n} w_i(y_i - \bar{y}_W)^2}{(w_T - 1)} \qquad (12.5)$$

If the finite population correction factor $f = w_T/N$ is small, expressions (12.4) and (12.5) are approximately the same.

The situation becomes more problematic if the weights w_i represent inclusion weights. In the simple case of an equal probability sample ($w_i = N/n$), expression (12.4) will be equal to

$$v(\bar{y}_W) = \frac{\sum_{i=1}^{n} (y_i - \bar{y}_W)^2}{n(N-1)}, \qquad (12.6)$$

which is a factor $(N-1)/(n-1)$ to small as a variance estimator.

In general, without replacement sampling designs a completely different expression should be used to estimate the variance of the estimator:

$$v(\bar{y}_W) = \sum_{i=1}^{n} \sum_{j=i+1}^{n} \frac{(\pi_i \pi_j - \pi_{ij})}{\pi_{ij}} \left(\frac{y_i}{\pi_i} - \frac{y_j}{\pi_j}\right)^2. \qquad (12.7)$$

Note that expression (12.7) involves second-order inclusion probabilities π_{ij}, which do not appear in expression (12.3).

The problems described above also occur in a more in depth analysis of the data. Many multivariate analysis techniques are based on the assumption of identically distributed independent samples. Due to complex sampling designs, adjustment weighting and imputation (see Section 12.3), estimates for the first and second-order moments of distributions are likely to be wrong.

12.2.3 Imputation Issues

Some imputation techniques affect the distribution of a variable. They tend to produce synthetic values that are close to the center of the original distribution. Hence, the imputed distribution is more "peaked." This may have undesirable consequences. Estimates of standard errors may turn out to be too small. Analysts using the imputed data (not knowing that the data set contains imputed values) may get the impression that their estimates are very precise while in reality this is not the case.

Possible effects of imputation are illustrated by analyzing one single imputation technique: imputation of the mean. This type of imputation does not affect the response mean of the variable: the mean $\bar{y}_{IMP}$ after imputation is equal to the mean $\bar{y}_R$ before imputation. As a result, the variance of the estimator also does not change.

Problems may arise when an unsuspecting analyst attempts to estimate the variance of an estimator, for example, for constructing a confidence interval. To keep things simple, it is assumed the available observations can be seen as a simple random sample without replacement, that is, missingness does not cause a bias. Then the variance after imputation is equal to

$$V(\bar{y}_{IMP}) = V(\bar{y}_R) = \frac{1-(m/N)}{m} S^2, \qquad (12.8)$$

in which $m \leq n$ is the number of "real" observations and S^2 is the population variance.

It is known from sampling theory that, in case of a simple random sample without replacement, the sample variance s^2 is an unbiased estimator of the population variance S^2. This also holds for the situation before imputation that the s^2 computed using the m available observations is an unbiased estimator of S^2.

What would happen if an analyst attempted to estimate S^2 using the complete data set, without knowing that some values have been imputed? He would compute the sample variance, and he would assume this is an unbiased estimator of the population variance. However, this is not the case. For the sample variance of the imputed data set, the following expression holds:

$$s^2_{IMP} = \frac{m-1}{n-1} s^2. \qquad (12.9)$$

Hence,

$$E(s^2_{IMP}) = \frac{m-1}{n-1} S^2. \qquad (12.10)$$

This is not an unbiased estimator of the population variance. The population variance is underestimated by a factor $(m-1)/(n-1)$. This creates the impression that estimators are very precise whereas in reality this is not the case. So, there is a substantial risk of drawing wrong conclusions from the data. This risk is larger as there are more imputed values.

Imputation also has an impact on the correlation between variables. Suppose the variable Y is imputed using imputation of the mean. And suppose the variable X is completely observed for the sample. In this case, it can be shown that the correlation after imputation is equal to

$$r_{\text{IMP},X,Y} = \sqrt{\frac{m-1}{n-1}} r_{XY} \qquad (12.11)$$

where r_{XY} is the correlation in the data set before imputation. So, the more observations are missing for Y, the smaller the correlation coefficient will be. Analysts not aware of their data set having been imputed will get the impression that relationships between variables are weaker than they are in reality. Also here, there is a risk of drawing wrong conclusions.

12.3 PREPARING A SURVEY REPORT

Analysis of the survey data will lead to a publication of the results. Form and contents of such a publication depend on the objective of the survey, the nature of the collected data and the intended audience. It is important that this audience understands what is said in the publication. Readers should be able to use the survey results to full advantage, to assess the reliability of the outcomes, and to be aware of their scope.

Furthermore, the publication should contain sufficient technical documentation about the survey. This documentation should enable survey researchers to understand how the survey was set up, how data were collected, what practical problems were encountered, what was done to correct problems, how accurate the results are, and so on.

The main purpose of a survey publication is communicating its results. Therefore, structure and style have to be such that this is accomplished as concisely and effectively as possible. This section presents some guidelines. Section 12.3.1 is about general issues. Section 12.3.2 concentrates on the general part of the publication. This part is intended to describe the results of the survey, usually for a nontechnical audience. Section 12.3.3 deals with the more technical survey documentation. Since the intended audience of the general part and the technical part can be very different, one might consider writing two separate publications.

12.3.1 General Issues

The audience for the general part of the publication can be very diverse. Therefore, it is important to use plain language. Technical jargon and mathematical formulas must be avoided. The language should be clear and concise. Short and simple sentences should be preferred.

The text should be written in a neutral and objective style. Informal language is not acceptable. Overfamiliar phrases like "*at the end of the day*" and "*in a nutshell*" should be avoided. The text should not contain personal opinions. For example, adjectives in "*painstaking data collection*" and "*careful analysis*," and "*surprising results*" must be left out.

It is usually advised to write the publications text in the passive voice (i.e., the third person). Use of "I," "we," and "you" must be avoided. So, "*the data were collected in a period of 4 weeks*" is better than "*We collected the data in a period of 4 weeks.*" However, sometimes the active voice should be preferred, for example, if the passive voice would mean hiding responsibility for specific activities. A text like "*Interviewers made errors during the fieldwork*" is more informative then "*errors were made during the fieldwork.*"

Most of the text should be written in the past tense. This particularly holds for the executive summary of the publication, the methodology section (the description of how the survey was carried out), and the section with the survey results. The introduction of the publication and the discussion of the results can be written in the present tense.

12.3.2 General Part

Most scientific reports have a common structure. This structure can also be used for the general part of the survey publication. It includes the following elements:

- Title
- Abstract
- Introduction
- Methodology
- Analysis
- Discussion
- References
- Appendices

The *title* must be short and precise. It should inform the reader about what has been investigated in the survey. Any unnecessary words (e.g., "*A study of . . .*") should be omitted.

The *abstract* is a self-contained summary of the whole report. It should therefore be written last and it is usually limited to just one paragraph of approximately 150 words. It must at least contain an outline of what has been investigated, the main results, and the conclusion.

The *introduction* consists of two parts. The first part describes the problem that has been addressed in the survey, and how it was addressed. The following topics must at least be included:

- Definition of the target population. This is the population to which the results refer.
- Major variables that have been measured.
- A nontechnical summary of the sampling frame and the sampling design.
- Description of the way the data were collected in the field.

- Initial sample size, the number of respondents, and the response rate.
- Indication of the accuracy of the results, for example, margins of error.

The second part of the introduction gives an overview of the main conclusions of the survey. It is just a list of conclusions, without any arguments or underpinning. Technical jargon and mathematical formulas must be avoided.

The introduction can also be seen as an executive summary of the survey report. It must be self-contained and readable for all those interested in the survey results.

The section on *methodology* must give a detailed description of every step in the survey process. The information should not be too technical so that a nonexpert can also get a good idea of how the survey was conducted. The following topics must be included in this section:

- A definition of the target population of the survey. What is exactly the population to which the survey results refer?
- The population characteristics that were estimated. What were the main survey questions? How were they translated into questions?
- A description of the questionnaire. How many questions did it contain? How long did it take to complete a form? Did respondents encounter any problems in answering the questions? The questionnaire itself could be included in an appendix of the report.
- A description of the sampling frame. What sampling frame was used? Was the frame up-to-date? How well covered this sampling frame the target population? Was there any undercoverage or overcoverage?
- A description of how the sample was selected from the sampling frame. What was the sampling design? What was the initial sample size?
- A description of the fieldwork. What mode of data collections were used? Were interviewers involved? If yes, how many and how were they trained?
- A description of data editing techniques. What kind of data editing took place? Where many errors detected? How were errors corrected?
- A description of the nonresponse. What was the number of respondents? What was the response rate? Was the nonresponse selective? What has been done to correct for this? Have any imputation techniques or adjustment weighting been carried out?
- A description of the accuracy of the survey results. What are the margins of error? Could anything be said about the magnitude of nonsampling errors?

The *analysis* section covers the results of the analysis of the survey data (after editing and nonresponse correction). This section could start with an exploratory analysis of all relevant survey variables separately. This can be done in the form of descriptive tables (frequency distributions, means, standard errors, and so on) or graphs.

The second part could explore relationships between variables. Also, here a choice can be made between numerical or graphical display of the results. In case specific patterns, structures or relationships are discovered, it must be made clear whether they are significant or could be attributed to sampling variation. If possible, artifacts must be distinguished from discovery of new knowledge that can be explained from the underlying subject-matter theory.

It should be realized that data cannot be used at the same time to formulate and to test a hypothesis about the target population. Hypothesis testing should always be based on new, independent data. It might be a good idea to split the survey data set randomly in two subsets: one for exporatory analysis, leading to the formulation of hypotheses, and the other for testing these hypotheses.

In case of hypothesis testing, always specify which tests were carried out and why they were chosen for this purpose. Also, mention significance levels or p-values.

There could be a third part with a more in-depth multivariate analysis of the survey data. It should be kept in mind that many multivariate analysis techniques require the data to be generated according to some models (e.g., an independent sample form a normal distribution). The "dirty data" produced by the survey may not always satisfy the underlying model assumptions.

The *discussion* section is an important part of the survey report. It places the survey results in the context of the relevant subject-matter area. It should enable the reader to understand the relevance of the results, also in relation to other research work in the area.

The discussion section starts with an overview of all main results of the analysis of the survey data. The next step is to interpret these results. Does it give new insight in the population? How do the results relate to other findings? Are the findings consistent with an underlying theory? Can this theory explain the findings?

It should be made clear what the implications are of the survey findings. They may suggest future research to obtain insight in specific topics. It may happen that all kinds of limitations were encountered during the survey process. Such limitations may restrict generalization of the findings. Were possible recommendations should be made to improve a possible future repetition of the survey.

The discussion may end with conclusions that summarize the most important elements of the discussion.

The *references* section contains a list of references to all literature mentioned in the survey report. It should include both references to subject-matter literature and statistical literature.

The *appendices* contain material that is relevant to the survey report, but that would disrupt its flow if it was contained within the main text. The appendices could contain the survey questionnaire and also the survey data (if the survey is not too large and there are no confidentiality problems). There could also be a glossary of terms, or other information that the reader may find useful. All appendices should be clearly labeled and referred to where appropriate in the main text, for example, "*See Appendix A for the complete questionnaire.*"

12.3.3 Technical Part

Already in 1948, a United Nations commission came with recommendations for the preparation of survey reports. Such a report should enable users of the survey data and the survey results to use the survey results to full advantage, to assess their reliability, and to utilize it in carrying out future surveys (see United Nations, 1964). These recommendations were updated in 1962.

The UN guidelines recommend making two reports, a general report and a technical report. The general report was already discussed in the previous subsection. This subsection is about the technical report. Such a report should be seen as the *survey documentation*.

The goal of the technical report is to provide complete, unambiguous information about all aspects of the survey. It should contain sufficient information to allow other researchers to assess the quality of the survey and the survey results. It also should contain sufficient information to carry out an exact copy of the survey at a future point in time.

A detailed description of the sampling frame should be given. It should be made clear whether the frame was constructed specifically for this survey. Particulars should be given of any known or suspected deficiencies, among which undercoverage, overcoverage errors.

The sampling design should be carefully specified, including details such as the type of sampling unit, sampling fractions, particulars of stratification, and so on. The procedure used in selecting sampling units should be described. If no random selection was applied, justification should be given for an alternative procedure.

It is desirable to give an account of the organization of the personnel employed in collecting, processing, and tabulating the primary data, together with information regarding their previous training and experience. Arrangements for training, inspection, and supervision of the staff should be explained. Also, a description should be given of applied data editing techniques. A brief mention of the equipment (for example, hardware and software) used is frequently of value to readers of the report. The statistical methods used for correcting item and unit nonresponse should be described.

If more elaborate statistical analysis techniques have been used than those for simple estimation of means and totals, these techniques should be explained, and the relevant formulas being reproduced where necessary. Where proper application of these techniques relies on specific conditions to be satisfied, this should be discussed.

A detailed account should be given of how the accuracy of the estimates is computed, taking into account the sampling design, and possible nonresponse correction techniques (like adjustment weighting). Where nonsampling errors are expected to have a substantial impact on the accuracy of the estimates, attempts should be made to compute at least some indication of the magnitude of these errors.

Every reasonable effort should be made to provide comparisons with other independent sources of information. Such comparisons should be reported along with the other results and the significant differences should be discussed. The objective of this is not to throw light on the sampling error since a well-designed survey provides adequate internal estimates of such errors, but rather to gain knowledge of biases, and other nonsampling errors. Where disagreement between sample survey results and other independent sources may be due, in whole or in part, to differences in concepts and definitions, this should be reported.

A sample survey can often supply the required information with greater speed and at lower cost than a complete enumeration. For this reason, information on the costs involved in sample surveys is of particular value for the development of sample surveys by other researchers. It is therefore recommended that fairly detailed information should be given on costs of a survey. Where possible, costs of different activities should be specified, like planning, fieldwork, supervision, processing, analysis, publication, and overhead costs.

The results of a survey often provide information that enables investigation of the efficiency of the sampling design in comparison to other sampling designs that might have been used in the survey. The results of any such investigations should be reported.

12.4 USE OF GRAPHS

12.4.1 Why Graphs?

Survey results can be communicated in various ways. One obvious way to do this would be to do it in plain text. If there is a lot of information or if the information is complex, readers may easily lose their way. More compact ways of presenting this type of information are tables and graphs. Particularly for statistically less educated users, graphs have a number of advantages over text and tables. These advantages were already summarized by Schmid (1983).

- A graph can provide a comprehensive picture. This makes it possible to obtain a more complete and better balanced understanding of the problem.
- Graphs can bring out facts and relationships that otherwise would remain hidden.
- Use of graphs saves time since the essential meaning of a large amount of data can be visualized at a glance.
- Relationships between variables as portrayed by graphs are more clearly grasped and more easily remembered.
- Well-designed graphs are more appealing and therefore more effective in creating the interest of the reader.

Graphs can be used in surveys in two ways. One way is to use them as tool for data analysis. Particularly, graphs can be very effective in an exploratory data analysis, to explore data sets, to obtain insight, and to detect unexpected patterns and structures. Users will have a background in statistics. Layout issues play a limited role here. Another use of graphs is use in survey publications. Most importantly, these graphs should be able to convey a message to statistically inexperienced readers. Therefore, the type of graph should be carefully selected. The visual display of the graph should be such that it reveals the message and not obscures it.

This section is devoted to the use of graphs in survey publications. The use of the *KISS principle* is promoted. KISS is an acronym for "Keep it simple, stupid." The principle states that simplicity should be a key design goal. Unnecessary complexity should be avoided. The KISS principle will be translated into a number of guidelines for the design of graphs. Furthermore, this section recommends which type of graph to use to display specific aspects of the data.

12.4.2 Some History of Graphs

It is not always simple to convey information in plain text, particularly if the message to be conveyed is complex. Therefore, it is not surprising that already, far back in history, attempts have been made to find different means. Probably the first graphs have been maps. Maps were already made thousands of years ago in China and Egypt. However, the idea to add statistical information did not appear until the seventeenth century. The first known graph of a time-series probably dates back to the tenth century (see Fig. 12.2). It shows the inclination of the orbits of the planets over time. For more information, see Tufte (1983).

It took 800 years before this type of graph was really used for statistical purposes. John Playfair (1759–1823) is seen by many as the inventor of the statistical graph. He published a book in 1786 that contains more than 40 graphs. Almost all of these graphs show time-series of economic variables. Figure 12.3 shows an example, the value of the trade between England and the East Indies.

Playfair (1786) has one other type of graph in his book. This is a bar chart. So, he can be seen as the inventor of the bar chart.

A classical graph is the map of the campaign of Napoleon in Russia in 1812. The French engineer Charles Joseph Minard made this graph in 1862. This graph is discussed in Tufte (1983) and Wainer (1997). It is reproduced in Fig. 12.4.

The graph is a combination of a map and a time-series. The map shows the route of Napoleon's army to Moscow and back. The size of the army is indicated by the width of the band. As it invades Russia, the army consists of 422,000 men. Only 100,000 men reach Moscow. The black band describes the retreat of the army. The graph dramatically shows the crossing of the Berezina river. Only 28,000 of the 50,000 reached the

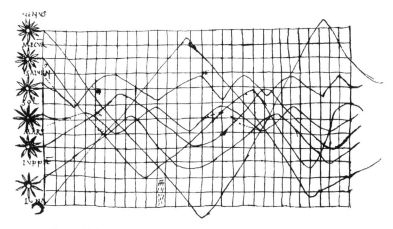

Figure 12.2 The first known time-series graph. *Source*: Tufte (1983).

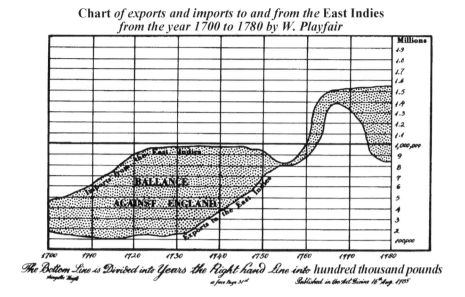

Figure 12.3 One of the first time-series of an economic variable. From Howard Wainer (1997), Visual Revelations. Reprinted by kind permission of Springer Science and Business Media.

other side of the river. The graph also shows daily temperature during the retreat. It was a very cold winter. This was one of the causes of the disaster of the Berezina crossing. In fact, this graph displays several variables simultaneously as a time-series: geographical position, size of the army, and temperature. Tufte (1983) suggests it may well be one of the best statistical graphs ever drawn. It tells with simple means and in a very clear way, a reasonably complex story.

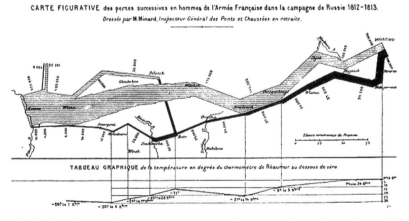

Figure 12.4 Minard's map of Napoleon's campaign in Russia. Reprinted by permission, Edward R. Tufte, The Visual Display of Quantitative Information (Cheshire, Connecticut, Graphics Press LLC, 1983, 2001).

12.4.3 Guidelines for Designing Graphs

A graph can be a powerful tool to convey a message contained in a survey data set, particularly for those without experience in statistics. Graphs can be more meaningful and more attractive than tables with numbers. Not surprisingly, graphs are often used in the popular media like newspapers and television. Use of graphs is, however, not without problems. Poorly designed graphs may convey the wrong message. There are ample examples of such graphs. Designers without much statistical expertise often make them. They pay much more attention to attractiveness of the graphic design than to its statistical content.

To avoid problems with graphs, a number of design principles should be followed. Some guidelines are proposed in this section that may help to produce proper graphs. Also, some examples are given of badly designed graphs.

Rule 1: Show the data. A graph should show the patterns, structures, and relationships that exist in the survey data set. It should do that in a clear way. It should be easy to see what the specific properties of the variables are. Graphs should be designed such that they support this principle. Every effort should be made to avoid graphs that obscure the message to be conveyed.

Graphs can be particularly powerful for displaying large amounts of data in one picture. Indeed, one picture can tell us more than a thousand words. Figure 12.5 shows an example of such a graph. Is shows the population density in Europe in 2004 (*source*: Eurostat).

The European Union uses the Nomenclature des Unités Territoriales Statistiques (NUTS) classification for dividing up to territory of its member countries, and other countries. NUTS is a hierarchical classification. It subdivides each country into three levels: NUTS 1, NUTS 2, and NUTS 3. Each classification is a subdivision of a previous level, respectively. NUTS 2 is used in the map. It divides Europe into 313 regions. The population density is shown for each region. This implies that the graph contains at least $3 \times 313 = 939$ numbers (geographical position of the region in longitude and latitude, and population density for 313 regions). Notwithstanding this large amount of numbers, the information in the graph is very readable. Not only global trends can be observed (high density in The Netherlands, Belgium, and the German Ruhr area) but also details like the relative high population density in the Stockholm area compared to the rest of Sweden.

Figure 12.5 is a typical example of a graph with a high density. Tufte (1983) proposed the *data density index* (DDI) as an indicator of the amount of data in a graph. It is defined as the number of data points per square inch. Research by Tufte (1983) showed that the DDI can assume values between 0 (graphs without data) to over 300.

The DDI of the population density map is around 50 (at this scale), which is reasonably high. Figure 12.6 shows an example of a graph with a very low DDI. It is a plot of the labor productivity of Japan versus that of the United States.

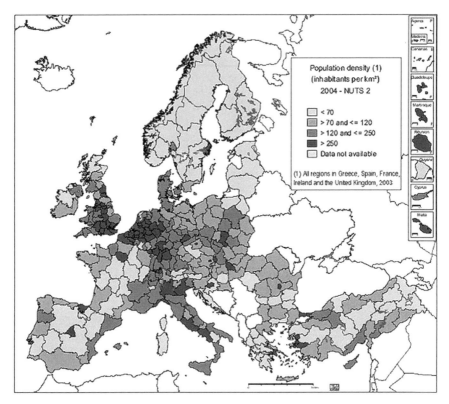

Figure 12.5 Population density in Europe in 2004. *Source:* Eurostat.

The graph on the left contains only three numbers: 44.0, 62.3, and 70.0. The DDI here is about 1.5, which is much lower than the DDI of the population density map in Fig. 12.5.

The labor productivity graph contains a lot of decoration that does not really helps to convey the statistical message. On the contrary, it obscures the message. It serves no other purpose than making the picture more attractive from an artistic point of view. This is what Tufte (1983) calls *chart junk*. It should be avoided. Tufte (1983) had introduced the data-ink ratio (DIR) as a measure of the amount of chart junk in a graph. It is defined as the ratio of the amount of ink used to draw to nonredundant parts of the graph (the real data) and the total amount of ink used. An ideal graph would have a DIR value of 1. Much smaller values of DIR are an indication that the graph contains too much chart junk. It will be clear that the DIR of the graph on the left in Fig. 12.6 is much smaller than the DIR of the graph on the right.

Rule 2: Do not mess around with the scales. The scales on the axes should help the reader interpreting the magnitude of the displayed phenomena correctly. Where the measurement scale of a variable has a clear interpretation of the value 0, the axis should start at this value, and not at an arbitrary larger

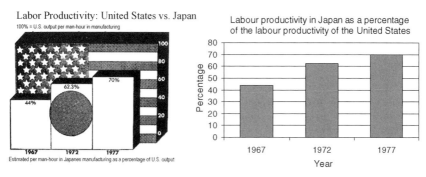

Figure 12.6 Labor productivity of the United States versus Japan. *Source:* Washington Post, 1978. From Howard Wainer (1997), Visual Revelations. Source: Washington Post, 1978. Reprinted by kind permission of Springer Science and Business Media.

value, as this could lead to a wrong interpretation of the graph. Figure 12.7 shows an example.

Both graphs show the average length of adult males in various parts of The Netherlands. Apparently, males are longer, on average, in the northern part of the country than in the southern part. The Y-axis in the graph on the left starts not at 0 but at 179. As a consequence, the differences between the regions are exaggerated. One almost gets the impression that men in the south are less than half as long as men in the north.

The graph on the right shows the same data, but now the scale at the Y-axis starts at the value 0. The difference between the regions turns out to be very small. This picture contains a more realistic message: the average age in the south is slightly smaller than in the north.

Figure 12.8 contains an example of a graph where two different Y-axes are used. The graph on the left shows the increase in average length (in centimeters) over the years for both men and women. It is reproduced version of a graph published by Statistics Netherlands (*Webmagazine*, January 17, 2008).

There seems to be a dramatic increase in length in 24 years and at first glance, the difference in length between men and women are substantial. Without looking at the vertical scale, one gets the impression that women are only half as long as men. By looking are the scale on the left, the difference seem to be a little

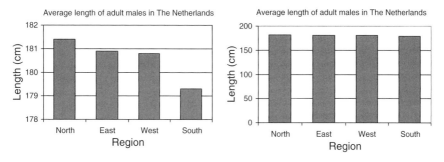

Figure 12.7 A graph with a scale not starting at zero.

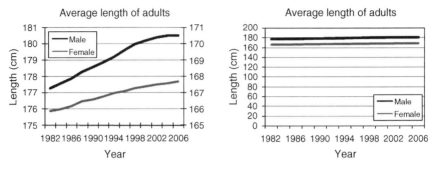

Figure 12.8 The increase in length of men and women over time.

over 1 cm in 1982. However, a closer look reveals that there is different scale for women. Its values for women are shown on the Y-axis on the right. So, the difference in length between men and women in 1982 is more than 11 cm. Another problem with this graph is that both Y-scales do not start a 0. The message conveyed by this graph can be confusing if not enough attention is paid to its details.

The graph on the right in Fig. 12.8 gives a more realistic picture. Now, the same scale for men and women is used and both scales start at zero. The changes over time are less profound.

A third example of the use of a wrong scale is also taken from Wainer (1997). The graph on the left in Fig. 12.9 displays the income of physicians from 1939 to 1976. The graph suggests a linear trend in the first part of the period. The yearly increase seems to slow down a little in the second part of the period.

A closer look at the scale of the X-axes reveals that the time gap between subsequent values is not the same everywhere. The first gap is 8 years, followed by periods of 4 years and at the end of the scale there is only a period of 1 year between subsequent values. The graph on the right in Fig. 12.9 shows the result in case of a proper regular scale for the X-axis. Now it becomes clear that the

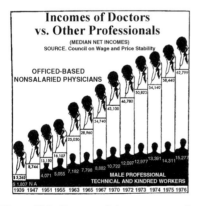
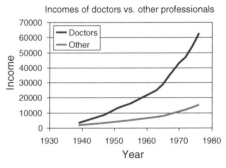

Figure 12.9 Incomes of doctors versus other professionals. *Source:* Washington Post, 1979. From Howard Wainer (1997), Visual Revelations. Reprinted by kind permission of Springer Science and Business Media.

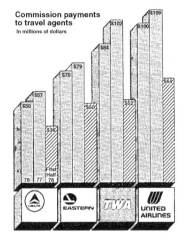 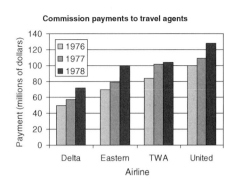

Figure 12.10 Commission payments to travel agents. *Source*: New York Times, 1978. From Howard Wainer (1997), Visual Revelations. Reprinted by kind permission of Springer Science and Business Media.

salaries of doctors increase much more than linear. So, the message is completely different.

Rule 3: Show the data in the proper context. The graph should promote presentation of the statistical information in the proper context so that the right conclusion is drawn by the user. Design and composition of the graph should be such that the correct message is conveyed. A misleading presentation must be avoided.

The graph on the left in Fig. 12.10 contains commission payments to travel agents by airlines. It seems to suggest that these payments have decreased dramatically in 1978. However, there is some small print in the graph explaining that the payments in 1979 only cover a period of 6 months and not the complete year. A more correct picture of the situation would be obtained if the commission payments for the whole year were estimated. This has been done in the graph on the right. This graph conveys a different, more correct, message: commission payments are still increasing.

Figure 12.11 shows another example of a misleading graph. The graph on the left shows the United States export to and import from China. The graph on the right does the same, but for Taiwan. At first sight, the impression is that there is not much difference between China and Taiwan with respect to trade.

However, a closer look would reveal that the Y-axes of both graphs are not the same. The scale for China runs from 0 to 3000 and the scale for Taiwan runs from 0 to 6000. To make the two graphs comparable, the graph for Taiwan should be twice as high.

Also, note that the shades have been interchanged in the graphs. Black corresponds in the left-hand graph to import and in the right-hand graph to export. This makes the comparison even more confusing. Finally, the scale of the X-axis on the left starts at the year 1972 while it starts on the right at the year 1970.

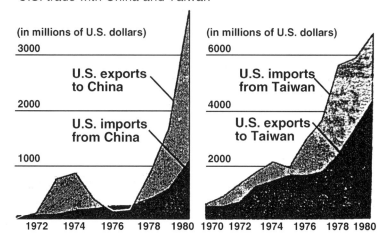

Figure 12.11 Trade of the United States with China and Taiwan. *Source:* New York Times, 1980. From Howard Wainer (1997), Visual Revelations. Reprinted by kind permission of Springer Science and Business Media.

Rule 4: Use the right metaphor. Graphs are used to visually display the magnitude of phenomena. There are many techniques to do this. Examples are bars of the proper lengths, or points on a scale. Whatever visual metaphor is used to represent the magnitude, it must be such that it enables correct interpretation. For example, it should retain the natural order of the values. If a value is twice as large as another value, the user should interpret the metaphor of the first as twice as large as the second metaphor. Unfortunately, this is not always the case. Particularly, graphs in popular printed media tent to violate this principle.

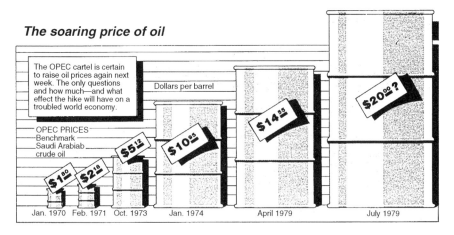

Figure 12.12 A bar chart using the wrong metaphor. Reprinted by permission of John Wiley & Sons, Inc.

Figure 12.12 shows a typical example of use of a wrong metaphor. The graph attempts to show the increase of oil prices in the years from 1970 to 1979. Schmid (1983) also discussed this graph. The price per barrel was $1.80 in 1970. In 1979, this price had increased to $20. Instead of bars, the graph uses oil barrels to indicate the oil price. The height of the oil barrels is taken proportional to the oil price. So, if the prices double, the oil barrel becomes twice as high. Something goes wrong here because the width of the barrel is also doubled. Consequently, the area of the picture of the oil barrel becomes four times larger. The visual impression of the value is that it becomes four times larger. A linear increase would therefore be displayed as a quadratic increase in the size of the metaphor. In this case, the reader gets the impression of a much faster increasing oil price.

Tufte (1983) introduced the *lie factor* for this type of graphs. It is defined as the value suggested by the graph divided by the true value. According to Fig. 12.12, oil prices have risen by a factor $20/1.8 = 11.1$ from 1970 to 1979. The areas of the oil barrels have increased by a factor 74.2 in the same period. So, the lie factor here is equal to $74.2/11.1 = 6.7$.

Note that there is another problem with this graph: the X-axis in not equally spaced over time. So, one also gets a wrong impression of the trend in oil prices in this respect.

Rule 5: Avoid three-dimensional graphs. Graphical designers instead of statisticians sometimes make graphs for popular media. They may find simple graphs boring and therefore attempt to make them more attractive, for example, by adding chart junk. Another way to do this is to add a three-dimensional perspective. Many statistical packages (e.g., Microsoft Excel) support this. However, three-dimensional graphs are not a good idea from a statistical point of view because they tend to make correct interpretation more difficult.

Figure 12.13 shows the distribution of the population of Samplonia over its districts. The three-dimensional shape of the graph on the left makes it very difficult to compare the size of sectors. For example, it is not clear whether Wheaton or Crowdon has more inhabitants. The three-dimensional perspective has been removed in the graph on the right. It is now easier to compare sectors although the situation is not ideal. It would be better to use a bar chart for this purpose.

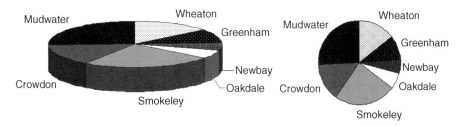

Figure 12.13 Pie charts of the population in the districts of Samplonia.

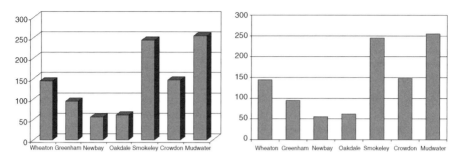

Figure 12.14 Bar charts of the population in the districts of Samplonia.

Figure 12.14 shows a bar chart of the population of Samplonia in a three-dimensional perspective. It is not easy to determine the length of the bars. This caused by the fact that there appears to be space between the bars and the background. For example, the graph seems to suggest that Mudwater has exactly 250 inhabitants, which is not correct. The three-dimensional perspective has been removed on the right. The design of the graph is much simpler, but it is also much easier to determine the lengths of the bars.

12.4.4 Types of Graphs

The available computer software offers ample possibilities of creating graphs. Generally, it is easy the produce all kinds of graphs. However, not every graph type is meaningful for every type of variable. Some graph types can only be used for qualitative variables and other types only for qualitative variables. Moreover, different graph types perform different functions. Some aim at displaying the distribution of variables and others at portraying relationships. Table 12.3 may be helpful in selecting the proper graph in a specific situation.

It is not only important to choose the proper type of graph but also attention should be paid to the graphic design of the graph. Chart junk should be avoided. Graphs should be simple and clear. Therefore, the KISS design principle already mentioned is advocated.

Table 12.3 Possible Graph Types

Variables	Distribution	Relationship
Quantitative	Histogram	Scatter plot
	Box plot	
Qualitative	Bar chart	Grouped bar chart
	Pie chart	Stacked bar chart
		Pie charts
Mixed		Box plots

It should be mentioned that there are more graph types than mentioned in Table 12.3. The graphs mentioned here are particularly useful in publications for nonstatisticians. Other graphs can be very meaningful in exploratory data analysis.

12.4.4.1 The Distribution of a Quantitative Variable

The *box plot* (or *box-and-whisker plot*) is a graphical summary of the distribution of the variable. See Fig. 12.15, for an example. The box represents the central part of the distribution. It stretches from the lower hinge (the first quartile) to the upper hinge (the third quartile). It contains the middle 50% of the values. The line in the box represents the median (the second quartile). Therefore, this is the dividing line between the lower half and the upper half of the distribution. It is the center of the distribution.

The *H-spread* is defined as the length of the box (the distance between the first and third quartile). The value of the *step* is equal to 1.5 times the H-spread. The *inner fences* are defined as values that are a distance equal to *step* outside the box (at both sides). The *lower adjacent value* is the smallest observation above the lower inner fence and the *upper adjacent value* is the largest observation below the upper inner fence. The whiskers run from the box to the adjacent values. Observations further away than the adjacent values are displayed as separate points. They should be seen as *outliers*. It indicates an element that substantially differs from other elements. An outlying value could also be caused by a measurement error.

A box plot gives a good impression of the location and spread of the observed values. Figure 12.15 shows that the incomes in Samplonia vary between 0 and 4500. The median income is 1000. The central half of the incomes are approximately in the range between 500 and 1500.

The box plot also provides some indication of the symmetry and skewness of the distribution. The distribution of incomes in Samplonia is clearly skewed to the right.

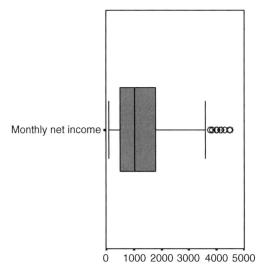

Figure 12.15 Box plot of income in the working population of Samplonia.

The plot shows some outlying values. As they are very close to a whisker, this may be the result of the skewness of the distribution instead of true outliers.

The *histogram* is another, maybe much more used, way to display the distribution of a quantitative variable. To that end the range of possible values is divided into a number of intervals. Then the number of observations in each interval is computed. For each interval, a column is drawn; the length of which is taken proportional to the number of observations. So, a *histogram* is graphical analogue of the frequency distribution.

Attention should be paid to the number of intervals used. In case of only a few intervals, a more global picture of the distribution will be obtained. Details may be hidden.

Figure 12.16 contains two histograms of the income distribution of the working population of Samplonia. In case of many intervals, the focus will be more on details and the global picture may be less clear. Sometimes, a rule of thumb is suggested to take the number of intervals equal to the square root of the number of observations, with a minimum of 5.

Both histograms in Fig. 12.16 are based on 341 observations. Applying the rule of thumb would mean 18 intervals. This number has been used in the *histogram* on the right. On the one hand, it shows the global shape of the (skewed) distribution, and on the other, it also shows some detail, like the relatively low number of incomes around 350.

Note that the columns have been drawn adjacent to each other without any space between them. This is in contrast to bar charts.

12.4.4.2 The Distribution of a Qualitative Variable

The only thing that can be done with a qualitative variable is to count the number of observations in each category. The resulting frequency distribution can be displayed as bar chart or a pie chart.

The bar chart consists of a number bars. Each bar represents a category and the length of the bar is taken proportional to the number of observations in that category.

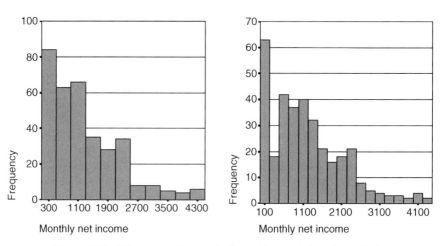

Figure 12.16 Histograms of income in the working population of Samplonia.

USE OF GRAPHS

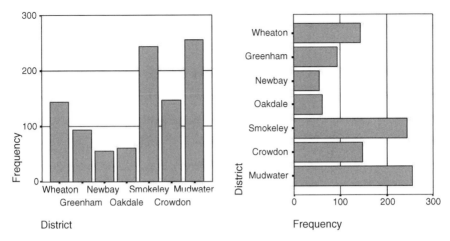

Figure 12.17 Bar charts of the population distribution in Samplonia.

The bars are drawn separate from each other, with some space between. Thus, the impression is avoided that the graph displays a quantitative variable. Figure 12.17 contains two examples of a bar chart. The same variable (District) is shown. So the bars represent the number of inhabitants in each district.

It is recommended using a bar chart with horizontal bars. This even more avoids confusion with a histogram. Moreover, there is ample space for labeling the bars. Note that there is no need for use of different colors. All bars can get the same color or shade. Use of different colors could be confusing. Intensive colors may create an impression that some categories are more important than other.

Usually, the categories of a qualitative variable have no natural order. So, no meaning can be attached to the order of the bars. If this is meaningful, one could decide to order the bars in increasing (or decreasing) order of magnitude. This also may enhance ease of interpretation.

Popular media often seem the prefer pie charts to bar charts. The pie chart consists of a circle divided in to sectors. Each sector represents a category. The angle of the sector (and thus its area) is taken proportional to its frequency. Figure 12.18 shows an example. It shows again the population distribution in Samplonia.

Maybe a pie chart has a less dull appearance, but its interpretation is more difficult. Particularly, comparison of the size sectors is not easy if they roughly have the same order of magnitude. Being aware of this problem, software often offers the possibility to include frequencies or percentages in the graph.

To be able to distinguish the sectors in the pie chart, different colors are shades have to be used. Selection of colors or shades should be done carefully. Their intensities should not differ so much that they suggest some sectors more important than others.

Some software packages offer the possibility to give bar charts or pie charts a three-dimensional look. Section 12.4.3 suggested avoiding such a three-dimensional look. It may increase the aesthetic value of a picture, but can seriously hamper correct interpretation.

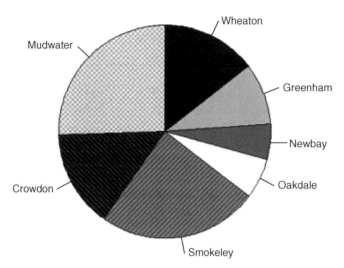

Figure 12.18 Pie chart of the population distribution in Samplonia.

12.4.4.3 The Relationship Between Quantitative Variables

The scatter plot is the obvious graphical tool to display the relationship between two quantitative variables. For each element i, the values x_i and y_i of two variables X and Y are seen as the coordinates of a point in two-dimensional space. Figure 12.19 shows an example. The incomes of working people is plotted against their ages.

Clear patterns in the cloud of points usually indicate some kind of relationship between the two variables. For example, it is easy to detect a linear relationship. This

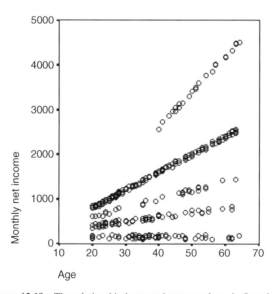

Figure 12.19 The relationship between income and age in Samplonia.

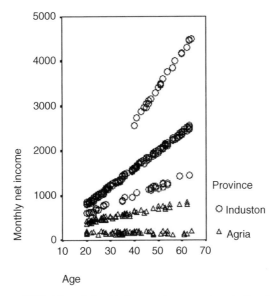

Figure 12.20 The relationship between income and age in Samplonia.

may be very helpful in explaining the behavior of one variable from another variable. Also, clustering of observations or outlying points will be clearly visible.

Figure 12.19 shows a number of clusters of points. Within each cluster, there seems to be a linear relationship between age and income. In several clusters, income increases with age but there are also cluster in which income seems to be independent of age.

It would be interesting to show what makes up all these different clusters. One way to do this is to introduce a third (qualitative) variable and to use different markers for different values of this variable. An example is shown in Fig. 12.20. There are two types of markers: circles for the province of Induston and triangles for Agria. It now becomes clear that incomes are higher in Induston than in Agria.

12.4.4.4 The Relationship Between Qualitative Variables

For showing the relationship between two qualitative variables the *clustered bar chart* and the *stacked bar chart* can be used.

The clustered bar chart consists of a number of simple bar charts of one variable. There is one for each category of the other variable. Figure 12.21 contains an example. It shows the age distribution for each district in Samplonia. Vertical bars have been used here, but, as was suggested earlier, it could have been better to use horizontal bars to avoid confusion with a histogram. Of course, it is possible to interchange to role of the two variables and to make bar charts of districts for each age category. This would show the data from a different perspective.

The clustered bar chart works well for showing some aspects like the absolute size of each age class. For example, it is clear that there are no old people in Newbay and no young people in Oakdale. Other aspects are more difficult to observe, like the total size

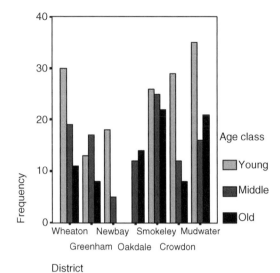

Figure 12.21 Clustered bar chart of the population distribution in Samplonia by district and age.

of each district or the relative contribution of each age class within each district. For example, it is hard to answer the question whether the percentage of young people is larger in Smokeley than in Mudwater.

Another way to show the relationship between two qualitative variables is to make a stacked bar chart. Within a category of one variable, the bars corresponding to the categories of the other variable are not drawn adjacent to each other, but stacked upon each other. Figure 12.22 shows two ways to do this.

The stacked bar chart on the left was obtained by stacking the bars of Fig. 12.21. It is now clear which district has the most inhabitants and which district the fewest. It is also possible to see which age class is relatively well represented in each district. For

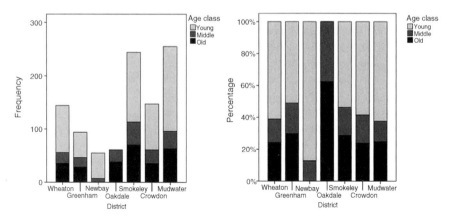

Figure 12.22 Stacked bar charts of the population distribution in Samplonia by district and age.

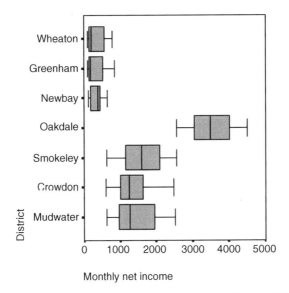

Figure 12.23 Box plots for the distribution of income in the districts of Samplonia.

example, one can observe that there are no elderly in Newbay and no young in Oakdale. However, it is still not easy to compare the age distributions of two districts. The stacked bar chart on the right in Fig. 12.22 may be better suited for this. Now, all bars have the same length (100%). There are no absolute numbers, just relative sizes. Age compositions within districts can be compared. For example, the percentage of elderly in Smokeley and Mudwater is larger than in Wheaton in Greenham.

12.4.4.5 The Relationship Between Mixed Variables

There are no specific graphic tools to show the relationship between a qualitative and a quantitative variable. Usually, use made of graphs for the distribution of a quantitative variable. These graphs are repeated within each category of the qualitative variable. The box plot is particularly suited for this. Figure 12.23 shows an example. The graph contains the income distribution in each district of Samplonia.

The graph clearly shows the substantial differences in the income distributions. Wheaton, Greenham, and Newbay are poor areas, and Oakdale is a very rich area. Of course, for each district separately, symmetry and possible outliers can be analyzed.

EXERCISES

12.1 If the sample is selected with unequal probabilities whereas an analyst assumes a simple random sample without replacement, then

  a. the estimator is biased and also the variance estimator is biased;

  b. the estimator is biased, but the variance estimator is unbiased;

  c. the estimator is unbiased, but the variance estimator is biased;

  d. the estimator is unbiased and also the variance estimator is unbiased.

12.2 A simple random sample of size 5 is selected without replacement is selected form a population of size 20. The sample values of the target variable are: 8, 9, 10, 11, and 12.

  a. Compute the estimate for variance of the sample mean.

  b. What value would have been obtained if the variance estimate was computed with a statistical package assuming the data to come from an independent sample selected with equal probabilities?

  c. Explain the difference of the estimates in (a) and (b).

12.3 The new political party "Social Democratic Harewood (SDH)" is taking part in the upcoming local elections in the town of Harewood. A local radio station carries out a poll to find out how popular the new party is. There are two neighborhoods in the town: Rhinegate and Millwood. A stratified sample has been selected. The sample size in each neighborhood was 500. All sampled persons were asked whether they would vote for the SDH or not. The table below summarizes all relevant information.

Neighborhood	Population Size	Sample Size	Percentage for SDH
Rhinegate	15,000	500	40
Millwood	5,000	500	20

  a. Compute an estimate of the percentage of voters in Harewoood that will vote for the SDH. Also, estimate the variance of the estimator and the 95% confidence interval.

  b. A lot of computer software for data analysis assumes the data to come from an independent equal probability sample. In this case, the proper estimator for the variance of sample percentage is equal to $p(100-p)/(n-1)$.

  Suppose such a computer program would have been used to analyze the Harewood poll data. What would be the estimate of the percentage of voters for the SDH? And what would be the estimated variance and the 95% confidence interval of the estimator?

  c. Compare the outcomes of (a) and (b). Explain the differences.

12.4 A survey report should at least contain the following three components:

  a. Underpinning and derivation of all formulas, summary of the problem and conclusions, and a detailed description of all steps in the analysis.

  b. Underpinning and derivation of all formulas, results in comprehensible language, and a detailed description of all steps in the analysis.

  c. Results in comprehensible language, summary of the problem, and conclusions and a detailed description of all steps in the analysis.

  d. Underpinning and derivation of all formulas, results in comprehensible language, summary of the problem, and conclusions.

EXERCISES

12.5 An executive summary of the survey results should at least satisfy the following three conditions:

    a. It contains an extensive description of the target population, it allows the commissioner of the survey to take policy decisions, and it does not contain arguments.

    b. It contains a concise overview of the conclusions, it allows the commissioner of the survey to take policy decisions, and it does not contain arguments.

    c. It contains a concise overview of the conclusions, it contains an extensive description of the target population, and it does not contain arguments.

    d. It contains a concise overview of the conclusions, it contains an extensive description of the target population, and it allows the commissioner of the survey to take policy decisions.

12.6 Which style should be preferred for the text of the survey report?

    a. The text should be written in the passive voice.

    b. The text should be written in comprehensible spoken language.

    c. The text should be written in the imperative voice.

    d. The text should be written in the active voice using "we," "you," or "I."

12.7 Describe at least two situations in which graphs with a three-dimensional perspective cause interpretation problems.

12.8 Describe at least six different ways to mislead readers of a statistical graph.

CHAPTER 13

Statistical Disclosure Control

13.1 INTRODUCTION

National statistical offices and other data collection agencies meet the increasing demand for releasing survey data files. These files contain for each respondent the scores on the variables measured in the survey. Because of this trend and an increasing public consciousness about the privacy of individuals, the problems involved in releasing survey data have become more serious over the years. Many national statistical offices, including Eurostat, the statistical office of the European Union, are confronted with these problems. For example, the situation in the United States was discussed by Cox et al. (1986), and CBS (1987) gives an account of a joint seminar of Sweden and The Netherlands on openness and protection of privacy.

This chapter explains why, at least in some countries, disclosure is a problem. The basic identification and disclosure problem is described in Section 13.2. In section 13.3 the concept of uniqueness is introduced. Uniqueness plays an important role in the identification of individuals, and the subsequent disclosure of information. Concentrating on the concept of identification, a basic, but probably impractical rule, for identification protection is formulated in this section. Various types of disclosure are distinguished in Section 13.4. In the analysis of disclosure risks, it is important to get some indication of the number of individuals who are unique in the population. Section 13.5 presents a model to estimate uniqueness and lays down two criteria for determining the disclosure risk. Many users of disseminated survey data sets are interested only in data relating to a particular subpopulation, for example, a specific region of the country. So the analysis of disclosure risks has to be extended to uniqueness in subpopulations, and Section 13.5 proposes a simple method to determine the critical size of such subpopulations. Section 13.5 also contains an example of the analysis of population uniqueness. A procedure that at least is able to cope with some types of disclosure risk is presented in Section 13.6.

Applied Survey Methods: A Statistical Perspective, Jelke Bethlehem
Copyright © 2009 John Wiley & Sons, Inc.

13.2 THE BASIC DISCLOSURE PROBLEM

The description of the basic disclosure problem is based on the fundamental assumption that the statistical agencies collect data from respondents for statistical purposes only and not for administrative purposes. The difference between statistics and administration is crucial: *statistics* deals with information on groups of individuals differentiated by some broad characteristics (income, social class, region, race, etc.), whereas *administration* deals with data of designated individuals. More on the difference between administrative and statistical use of data can be found in Begeer et al. (1986).

The *disclosure problem* relates to the possibility of identification of individuals in released statistical information (including publications on paper, tape, CD-ROM, Internet, etc.) and to the revelation of what these individuals consider to be sensitive information. Disclosure is a two-step process:

(1) *Identification of an Individual.* A one-to-one relationship can be established between a record in a released survey data file and a specific individual. For example, identification is very easy if the survey data file contains names and addresses of surveyed persons.
(2) *Disclosure of Sensitive Information.* This is information in the record of the identified individual that was not known beforehand and which this individual does not want to be known. This is the so-called *sensitive information*.

The definition of disclosure agrees to some extent with the definition of disclosure as suggested by Dalenius (1977) and the U.S. Department of Commerce (1978), which states that disclosure takes place if publication of statistical data makes it possible to determine characteristics of specified individuals more accurately than is possible without access to this statistical information.

Why is disclosure undesirable? First, it is undesirable for legal reasons. In countries like The Netherlands, for example, there is a law stating that firms should provide information to the national statistical office, while the office may not publish statistical information in such a way that information about separate individuals, firms, and institutions becomes available:

> ... Data, collected in accordance with this law, may not be disclosed in such a way that returns and information about a separate person, firm or institution can be observed, unless that person, the head of the firm, or the management of the institution has no objection.

Second, there is an ethical reason. When collecting data from individuals, the following statement is made by the Statistics Netherlands:

> The data requested from you and other persons by the Statistics Netherlands will be used exclusively for the preparation of statistical publications. From these publications no identifiable information concerning separate persons can be derived by others, including other government services. Statistics Netherlands takes great care to ensure that the information provided by you can never be used for other than statistical purposes.

The International Statistical Institute (ISI) Declaration on Professional Ethics, see ISI (1985), states that

> Statisticians should take appropriate measures to prevent their data from being published or otherwise released in a form that would allow any subject's identity to be disclosed or inferred.

Therefore, there is an ethical and legal obligation to avoid disclosure by any means.

Third, there is a very practical reason: if respondents do not trust statistical agencies, they will not respond. Nonresponse rates in household surveys in The Netherlands have increased over the last decade to a level of, say, 40%. Hence, confidence is of the utmost importance for the statistical office. The willingness of respondents to cooperate is a very important condition for the production of reliable statistical information.

Having stated that disclosure of data concerning individuals is unacceptable, the question arises to what extent statistical publications are to be protected to achieve this goal. Too heavy confidentiality protection of the data may violate another right: the *freedom of information*. It is the duty of every statistical agency to collect and disseminate statistical information. It is this dilemma, right of anonymity versus freedom of information, that is the core of the considerations about disclosure control of survey data.

The objective of *Statistical Disclosure Control* is to develop techniques that avoid identification of individuals. Often 100% protection is not possible. Therefore, disclosure control techniques aim at protecting survey data sets such that the identification and disclosure become very unlikely, and in fact can only be accomplished after disproportionately large efforts.

This chapter focuses on the disclosure problem in survey data files. Such files contain the individual values of survey variables that have been obtained in a survey. It should be noted that the disclosure problem can also occur in, for example, published statistical tables. For more information on this aspect of statistical disclosure control, see Hundepool et al. (2007) and Willenborg and De Waal (1996).

13.3 THE CONCEPT OF UNIQUENESS

A survey data file consists of records of values of the variables measured in the survey. The information in the records is considered to consist of two disjointed parts: *identifying* information on the one hand and *sensitive* information on the other.

Identifying information relates to those variables in the record (called *identification variables* or *key variables*) that allow one to identify a record, that is, to establish a one-to-one correspondence between the record and a specific individual. The well-known key variables are name and address, but household composition, age, race, gender, region of residence, occupation, and region of work can also help to identify individuals. All key variables are assumed to be qualitative variables.

Since identifying information is assumed to be known or accessible to others than the respondent (neighbors, relatives, friends, colleagues, etc.), this information is not considered to be sensitive in the sense of "information not to be revealed by statistical dissemination." Therefore, identifying and sensitive information are considered to be disjoint. However, in practice, a situation may arise where these types of information are not separable in this way. For example, in many confidentiality laws, no distinction is made between identifying information and other (sensitive) information. In many parts of the world, membership of an ethnic group may be both identifying and sensitive.

Sensitive information refers to the values of variables that belong to the private domain of the respondents, and hence to characteristics that they do not like to be revealed. No exact definition can be given of variables to be considered sensitive. Some general consensus exists about variables like sexual behavior and criminal past. For other variables, it may depend on the context and cultural background. A simple example is income, which in The Netherlands is considered to be sensitive whereas in Sweden it would sometimes be characterized as an identification variable.

Having established the distinction between identifying information and sensitive information, it is now possible to formulate the basic rule for disclosure control: a disseminated survey data set should be composed such that it is impossible for others to correctly link records to individuals by using the identifying information in the data set and prior knowledge.

A crucial element is the prior knowledge of the user of the data: if someone has no information whatsoever about a specific individual, identification and therefore disclosure is impossible. Hence, the risk of disclosure depends on the nature and amount of *a priori* available knowledge. Particularly, if the data are used by other government agencies that maintain comprehensive data files for administrative purposes such as tax collection, keeping disclosure risk at an acceptable level will pose severe problems.

Since protection against disclosure is very difficult, the basic rule implies that many survey data sets cannot be published. Therefore, in practice, this rule will have to be relaxed to continue the release of useful survey data sets.

To protect a survey data set against disclosure, it must be known how identification takes place in practice. Identification is made possible by *uniqueness*. To be able to define uniqueness, the key is introduced. The *key* denotes the set of variables to be used for identification purposes. Knowledge of the key constitutes the identifying information. The key will be taken to have K different actually occurring values. The score combinations of the key are denoted by $1, 2, \ldots, K$. If, for example, the key is composed of age (in 6 categories) and gender (in 2 categories), there are 12 different score combinations; so $K = 12$. The number of elements in the population with key value i is denoted by F_i ($i = 1, 2, \ldots, K$) and the corresponding number of elements in the sample is equal to f_i ($i = 1, 2, \ldots, K$). All F_i are strictly positive, but some of the f_i may be equal to zero.

The value of K need not necessarily be equal to the product of the numbers of categories of the key variables. If some combinations are impossible (i.e., there are so-called *structural zeros*), K will be less than the product of the categories. An example is a key consisting of age and marital status: the combination of being married and being younger than 10 years is impossible.

Let N be the size of the population. Then the probability that a person, selected at random from this population, has key value i is equal to $\pi_i = F_i/N$.

The *resolution* of a key is defined by

$$R = \frac{1}{\sum_{i=1}^{K} \pi_i^2}. \tag{13.1}$$

The resolution is equal to the reciprocal of the probability that two random elements, selected with replacement from the population, have the same key value. The resolution of a key gives some indication of the risk of identification. If, on the one hand, the resolution of the key is high, the probability of an accidental match is low. Therefore, there would be many persons who would differ on the set of key values. In this sense, there are many unique persons. So in many cases, it is possible to establish a one-to-one relationship between a specific person and a record in the data set. If, one the other hand, the resolution of the key is low, the probability of an accidental match is high. There will not be many persons with a unique set of values on the key variables. Hence, if a link is established between a specific person and a record in the data set, chances are high that this record contains data on a different person.

From the point of view of disclosure risk, high-resolution keys are dangerous. To get some feeling of which value of the resolution indicates dangerous keys, two extreme cases are considered. Disregarding the trivial case of $K = 1$, the risk of disclosure is least if the key assumes only two different values with equal probability in a large population. Since $\pi_i = 0.5$, the resolution is equal to $R = 2$, which is far less than N. The risk of disclosure is highest if every person is unique. This is the case if the key assumes as many values as there are elements in the population. Since $\pi_i = 1/K$ and $N = K$, the resolution is equal to $R = K = N$. So there are real disclosure problems if the resolution is of the same order as the population size. Note that if $\pi_i = 1/K$ and $K < N$, the resolution is equal to $R = K$.

An example of a harmless key is the key that consists only of the variable gender. Assuming that the probabilities of being male and female are the same, the resolution is equal to 2, which is generally much lower than the population size. The resolution will be much higher if more variables are included in the key. For example, the combination of age (in 17 categories), income (in 13 categories), and size of town (in 6 categories) produces a resolution of 500. For a specific population of households, consisting of father, mother, and two children, and a key consisting of ages of father and mother and the ages and sexes of both children, the resolution was found to be approximately equal to 500,000. Particularly in small regions, this is a dangerous key, as illustrated in Section 13.4.

Some individual is *unique in the population* if this person is the only one in the population with a particular set of scores on the key, that is, he/she has key value i with $F_i = 1$, for some i. Likewise, someone is *unique in the sample* if he/she is the only one in the sample with that set of scores on the key, that is, he/she has value i with $f_i = 1$.

Every unique person in the population will also be unique in the sample, if selected. However, uniqueness in the sample does not imply uniqueness in the population. Sample uniqueness may also occur if exactly one person out of several with the same key value is selected. It is clear that a statistical spy, interested in persons who are unique in the population and who have been selected in the sample, can concentrate on records with a unique key value in the sample.

Uniqueness in the population is vital for disclosure. Suppose some user of a data set knows that a specific person is unique in a well-defined population. Then there are two possibilities: either this person is in the sample or he is not. If he is in the sample, he will be identified and disclosed with certainty. If he is not in the sample, no harm can be done. Knowledge of population uniqueness should not be underestimated, in particular if the data set contains variables that make it possible to detect respondents living in a small area. For example, in many small areas, certain professions are unique (the doctor, the notary, the dentist). In such subpopulations, many persons are unique on a key consisting of only one identifier. It is thus clear that from the viewpoint of disclosure, geographical information is very dangerous identifying information.

13.4 DISCLOSURE SCENARIOS

It is important to know the prevalence of unique persons on a key of current identification variables. It should be realized that even if the categories of single identifiers are sufficiently filled, the combination of two such identifiers may still generate a large number of unique persons. Take, as an example, the two identifiers profession and region. Persons are certainly not unique if one variable at a time is considered. Although there are many dentists and many people live in small regions, often there is only one dentist in a small region. So, using only these two identifiers, it is possible to identify persons in a survey data set. And if the identifier gender is included, a female dentist may even be unique in a much larger area. So, in this example, gender is no longer a harmless key variable.

The danger of a high-resolution key is illustrated by means of an example based on figures for The Netherlands. The population in a certain region contained 83,799 households. Of these households, 23,485 were composed of father, mother, and two children. Suppose a key consists of the ages of father and mother and ages and sexes of the two children (all ages in years). On this key of 6 variables, 16,008 out of the 23,485 households turned out to be unique, which is about 68%! So, if a certain household with father, mother, and two children is known to be in a sample from this region, there is a high probability that this household can be identified.

High-resolution keys are dangerous, but that does not mean that low-resolution keys are always safe. In a Dutch health survey consisting of a sample of $n = 3500$ persons, about 250 persons (7%) were unique on a key, consisting of the variables age (17 categories), household income (13 categories), and size of the municipality (6 categories). In this case, K was equal to the product of the number of categories (1326) and the resolution was equal to 500.

Even on a low-resolution key, there will still be exceptional values, for example, a widow of 18. Disclosure of these "rare persons" happens often by accident. This type of disclosure will be called *disclosure by spontaneous recognition*. It is important to always check for these "rare persons" and do something about them, for example, remove them from the survey data file.

Disclosure with high-resolution keys can be accomplished by matching the data set with a register containing the key and also names and addresses. If a register contains a complete enumeration of the population or of a subpopulation (e.g., all inhabitants of a large town), nearly every record in the data set can be matched uniquely to a record in the register. This phenomenon will be called *disclosure by matching*. This type of disclosure can only be carried out by a specialized sleuth.

The danger of disclosure by matching was revealed more or less by accident by Statistics Netherlands in 1984 when, in the context of a project on real income changes, a successful exact matching procedure was carried out for statistical purposes on files with tax data (from the Internal Revenue Service). Subjects could be located and matched in files from several years, without using their exact names and addresses; see Van de Stadt et al. (1986). The danger of matching was also discovered by Paass and Wauschkuhn (1985) in a seminal study on exact and statistical matching. They showed that with information generally available to institutions such as police headquarters, credit organizations, health bureaus, a large proportion of the records in statistical data sets could be identified and disclosed.

Knowledge about uniqueness in the population is vital for a successful disclosure operation. In many cases, this type of information is limited. However, in case of complete enumeration of a population, uniqueness can easily be established from the data set. Someone who is unique in this data set is also unique in the population.

Another interesting case of additional knowledge is *response knowledge*, that is, knowledge that a person was interviewed for a particular survey. If the statistical spy knows that a specific individual has participated in a survey and, consequently, that his data must be in the data set, identification and disclosure is accomplished very easily if this individual is unique in the sample (not necessarily in the population!). Even knowledge of which primary sampling units were selected in a multistage survey increases the risk of disclosure substantially.

Identification can be established by a simple selection or elimination procedure. No advanced technology is needed. A computer and some generally available software (e.g., a statistical package like SPSS or STATA) are sufficient. Experiments have shown that records with specified key values for, say, 3–20 variables can be found in a file consisting of 10,000 records within a few minutes.

Response knowledge reduces population uniqueness to sample uniqueness. Population uniqueness is not always easy to verify, but a simple tabulation program is sufficient to determine sample uniqueness. Therefore, response knowledge significantly increases the dangers of disclosure.

A simple, but realistic, example of disclosure by response knowledge shows the danger of this scenario. The survey data set used contained all key variables from a health survey data file consisting of 3500 records. Now suppose it is known that colleague John is in the data set. John is 42 years old, has an academic degree, and works for the government. The disclosure attempt starts with all 3500 records. First, all records with an age outside the interval 40–44 are deleted. This leaves only 164 records. Next, excluding all records of persons without an academic degree reduces the number of remaining records to five. Finally, picking out only those people who are employed in government institutions results in just one record. This is John! And only three variables were required to identify him.

From the point of view of disclosure control, a sample is safer than a complete enumeration of the population. In large surveys, the sampling fraction could be 0.05. So, only 1 out of each 20 persons is in the sample and not everyone's private information can be revealed. Furthermore, a sample does not give information about uniqueness in the population. Is a sample therefore safe? No, certainly not. Since a small sample contains more unique persons than a large sample, the risk of disclosure by response knowledge even becomes larger as the sample size decreases.

Time may also be a factor affecting the disclosure risk. If the fieldwork of the survey was carried out a long time ago, all information necessary for identification must refer to that time. Since people are generally not very good in recalling events and facts from the past, disclosure based on old data sets may be more difficult than disclosure based on recent data sets.

13.5 MODELS FOR THE DISCLOSURE RISK

For estimating the number of population uniques using sample survey data, a simple model is proposed. The model is based on the assumption that the cell frequencies in the population are a realization of a superpopulation distribution. Let the population consist of N individuals, and suppose the key divides the population into K cells. Each cell i is assigned a superpopulation parameter $\pi_i > 0$ (a probability) and a random variable F_i denoting the population frequency in that cell. It is assumed that F_i has a Poisson distribution with expected value $\mu_i = N\pi_i$. Furthermore, let U_p denote the expected number of population uniques. Under these assumptions, U_p is equal to

$$U_p = \sum_{i=1}^{K} \mu_i e^{-\mu_i}. \tag{13.2}$$

The expected number of population uniques can be used as an approximation to the realized number of unique individuals under the superpopulation model. To estimate the number of uniques, all expected values $\mu_1, \mu_2, \ldots, \mu_K$ have to be estimated. Since the number of cells is usually very large, this can turn out to be a complex problem. To simplify calculations, a model is assumed that governs the generation of the superpopulation parameters $\mu_1, \mu_2, \ldots, \mu_K$. Two possible models are discussed here: the *Constant-Poisson model* and the *Poisson-Gamma model*.

The *Constant-Poisson model* assumes that all parameters $\mu_1, \mu_2, \ldots, \mu_K$ are equal. Consequently, all F_i have the same Poisson distribution with expected value μ. Since all probabilities have to sum to 1, it follows that

$$\pi_i = \frac{1}{K} \tag{13.3}$$

for $i = 1, 2, \ldots, K$. Using $\mu_i = \mu = N/K$, the expected number of population uniques is equal to

$$U_p = Ne^{-N/K}. \tag{13.4}$$

This is a nice and simple expression that can be computed quickly. Unfortunately, the Constant-Poisson model rarely holds in practical situations. It is too simple.

The idea of the *Poisson-Gamma model* is to allow variations in the π_i by considering them as realizations of Gamma(α, β) distributed random variables, G_i say. The first parameter α controls the magnitude of the π_i and the second parameter β controls the variation in the π_i. This distribution is used because it covers a wide range of possible distributions, and also arithmetic is rather simple. The usefulness of this model was investigated by Bethlehem et al. (1990), Skinner et al. (1990), and Greenberg and Zayatz (1992).

Although logically $\sum G_i = 1$, it is simply assumed that $\sum E(G_i) = 1$. Then $\alpha = 1/K\beta$, so there is only one unknown parameter left in the common distribution of the G_i's. This parameter (β) reflects the amount of dispersion of the superpopulation probabilities G_i around their common mean $1/K$.

The Poisson-Gamma model can now be summarized as

$$\begin{aligned} G_i &\sim \text{Gamma}(\alpha, \beta), \\ F_i &\sim \text{Poisson}(\mu_i = N\pi_i | \pi_i = G_i), \end{aligned} \tag{13.5}$$

for $i = 1, 2, \ldots, K$. An attractive property of this model is that the marginal distribution of each F_i is the negative-binomial distribution (Johnson and Kotz, 1969). Consequently, the expected value of F_i is

$$E(F_i) = \mu = N\alpha\beta = \frac{N}{K}, \tag{13.6}$$

and its variance is equal to

$$V(F_i) = \mu(1 + N\beta) = \frac{N}{K}(1 + N\beta). \tag{13.7}$$

Note that expressions (13.6) and (13.7) do not contain the parameter α. Due to the restriction $\alpha\beta = 1/K$, the choice of a value for β fixes the value of α.

Under the Constant-Poisson model, the variance of F_i is equal to N/K. Comparison with expression (13.7) shows that the Poisson-Gamma model allows more variation. The expected number of population uniques is under this model that is equal to

$$U_p = N(1 + N\beta)^{-(1+\alpha)}. \tag{13.8}$$

To estimate U_p, estimates of the parameters α and β of the Poisson-Gamma model are required. Expressions can be given for the maximum likelihood (ML) estimators, but the moment estimators can also be used. These can be found by equating the first and second sample moments to their expected values. An estimator b for β is obtained by solving

$$(1+nb)\frac{n}{K} = \frac{1}{K-1}\sum_{i=1}^{K}\left(f_i - \frac{n}{K}\right)^2. \qquad (13.9)$$

Then an estimator a for α is obtained from the equation

$$a = \frac{1}{bK}. \qquad (13.10)$$

Now the expected number of population uniques U_p can be estimated by

$$U_p = N(1+Nb)^{-(1+a)}. \qquad (13.11)$$

Only those records in the survey data file can be identified that are unique in the population. The expected number of population uniques in the survey data file is denoted by U_{ps}. Assuming equal selection probabilities, U_{ps} can be estimated by

$$U_{ps} = \frac{n}{N}U_p. \qquad (13.12)$$

Two criteria can be proposed for establishing the disclosure risk, based on the available information in the data file. The purpose of these criteria is to determine whether (additional) measures for disclosure protection should be taken. The first criterion is an absolute criterion of the form $U_p < C_a$, where C_a, the *absolute critical value*, is a constant, small enough to ensure that U_p is negligible. The second, relative and less stringent, criterion states that the proportion of possibly identifiable records ($U_{ps}/n = U_p/N$) must be smaller than some critical value C_r. This is a *relative critical value*. The motivation for a relative criterion is that it might be acceptable if just a few of the many sample elements are identifiable, because it will then be very unlikely that a specific record will be recognized as being unique.

Geographical variables in a survey data file may lead to even more severe disclosure control problems, particularly if such variables describe a detailed geographical classification. There is a dilemma here. On the one hand, researchers often want a detailed geographical classification for their analysis, and on the other, this may cause confidentiality of data to be at stake. This calls for a criterion that helps to determine which level of detail of a geographical variable is still acceptable in terms of disclosure risk.

One possibility for modeling uniqueness in geographical areas is to use a negative binomial distribution for each area separately, that is,

$$F_{ij} \sim \text{Negative binomial } (N_j, a_j, b_j), \qquad (13.13)$$

where F_{ij} is the frequency in cell i of area j. This model will, in general, give a better fit than a model that ignores the subpopulation structure by having only one α and β parameter and one population size N. Moreover, separate models for each area enable estimation of the number of unique elements U_{pj} in each area j. Hence, the number of unique elements in the entire population can be estimated by summing the U_{pj}.

However, if the number of areas is large, the computational effort can be considerable. A simpler model that requires only one α and one β to be estimated is obtained by assuming all α_j and all β_j to be equal, that is,

$$F_{ij} \sim \text{Negative binomial } (N_j, a, b). \tag{13.14}$$

This model can be used to choose which of the more or less refined several regional classifications to include in the data set. In such situations, some indication is required about which regional classification still satisfies the criterion for a "safe" data file, even for the smallest region in that classification. Assuming model (13.14) is a good enough approximation for this purpose; the relative criterion

$$\frac{U_p}{N} = (1 + N\beta)^{-(1+\alpha)} \tag{13.15}$$

can be seen as a function of N. Since α and β are positive, expression (13.15) is a monotonic decreasing function of N. Now the *critical population size* N_C is defined as the population size for which the relative criterion is just satisfied; so,

$$\frac{U_p(N_C)}{N_C} = C_r \tag{13.16}$$

and

$$N_C = \frac{\left(C_r^{-1/(1+\alpha)} - 1\right)}{\beta}. \tag{13.17}$$

The estimates a for α and b for β can be used to estimate the critical population size. And this estimate will indicate how refined the regional classification can be: the regional classification must be such that the smallest distinguished area has a population size larger than N_C.

The theory discussed in this section is illustrated with an example, using a survey data file containing data of 8399 individuals. There are four identification variables: household composition (H) in 24 categories, age (A) in 14 categories, marital status (M) in 2 categories, and gender (G) in 2 categories.

Four different keys were used. The first key consisted of variable H only. The second key $H \times A$ was obtained by crossing H and A. In the same way, the third and fourth keys were defined and denoted by $H \times A \times M$ and $H \times A \times M \times G$. For each of these keys, the contingency table containing the sample frequencies for all possible key values was formed. Since not all combinations are possible, the number of possible key values was smaller than the product of the categories of the variables involved. In contingency table terminology, structural zeros were excluded from the analysis but sampling zeros were not.

The performance of the Constant-Poisson model and the Poisson-Gamma model was analyzed by estimating the numbers of uniques in the sample (not necessarily also unique in the population). The estimated number of uniques in the sample could be compared with the corresponding observed number as a partial check of the model. The results are summarized in Table 13.1.

Table 13.1 Estimating the Number of Uniques and the Critical Population Size

Key	Number of Key Values	Number of Uniques in the Sample	Estimate for Constant-Poisson Model	Estimate for Poisson-Gamma Model	Critical Population Size
H	23	0	<0.01	0.1	743
$H \times A$	288	23	<0.01	21.6	17,422
$H \times A \times M$	554	50	0.002	37.9	32,206
$H \times A \times M \times G$	1108	108	4.3	80.2	63,624

The estimates based on the Constant-Poisson model differ substantially from the observed number of uniques. Clearly, this model does not fit in this example. Although the Poisson-Gamma model underestimates the number of sample uniques in all cases, the order of magnitude is roughly correct.

The critical population sizes were computed by using a relative criterion value of 0.1%, that is, the number of possibly identifiable records in any subpopulation must be smaller than 0.1%. At first sight this criterion value seems rather small, yet in a population of 14,000,000 (the Dutch population at that time), this would mean that 14,000 people were unique and therefore at the risk of disclosure. The results show that a data set containing any of the four keys (but no other key variables) can be released as long as they pertain to subpopulations with more than 63,624 inhabitants.

13.6 PRACTICAL DISCLOSURE PROTECTION

Experiences with the analysis of disclosure risks of real survey data files have led to a number of observations:

- In every survey data file containing 10 or more key variables, a large number of persons can be identified by matching this file with another file containing the key and names and addresses (disclosure by matching).
- Response knowledge nearly always leads to identification (disclosure by response knowledge), even on a low-resolution key.
- On a key consisting of only two or three identifiers, a considerable number of persons are unique in the sample, some of them being "rare persons," and therefore also unique in the population.

If someone is unique in the population, the question may arise: How high is the risk of identification? This risk depends on the amount of knowledge that is available to some user of the data. Furthermore, there are many respondents and many potential users, and the amount of available knowledge may vary substantially. This makes it difficult, but not impossible, to model additional knowledge and to quantify the probability that someone has knowledge of certain information; see, for example, Cassel (1976), Frank (1976, 1979), and Duncan and Lambert (1986, 1987). Sometimes statistical agencies take a different approach by asking the question: Does a respondent consider his private information in the survey data file safe? Hence, the risk of disclosure is considered not only from the legal, ethical, and practical viewpoints of the agency but also from the viewpoint of the single respondent who might have second thoughts about answering questions in a survey. A consequence of such a standpoint is that all respondents, who are either unique in the sample while it might be known that they are in the sample or who are unique in the population, have the right of protection. In particular, this right is appropriate for persons who are unique on a low-resolution key, that is, persons with exceptional characteristics.

In the literature, several techniques can be found that reduce the risk of disclosure. Spruill (1983), Paass (1985), Kim (1986), and McGuckin and Nguyen (1988) discuss *adding random noise* to the data. However, this works well only for quantitative variables and not so well for qualitative variables. For example, adding noise to the variable gender would turn males into females, and vice versa. For qualitative variables, noise may affect the structure and nature of the data too much. Furthermore, Paass (1985) has shown that adding noise to data does not significantly reduce the disclosure risk.

Another disclosure avoidance technique is *data swapping*, suggested by Dalenius and Reiss (1982). Data swapping transforms the data set into another data set by exchanging the values of variables. So the value of a variable in the record of a respondent is not his own value but the value of some other respondent. Data swapping affects the internal structure of the data, but knowing how much swapping has been done does allow one to correct the estimates of second-order moments.

A third technique to avoid disclosure is called *microaggregation* (Spruill, 1983; Cox et al., 1986). The individual data are not published, but aggregated data are. In the case of quantitative variables, it is often sufficient to publish means, variances, and covariances only. With these aggregates, many multivariate analyses techniques, for example, regression analysis, can be carried out (McGuckin and Nguyen, 1988). For qualitative variables, microaggregation means publishing two-dimensional, or higher dimensional, tables. To satisfy the needs of all users and to make possible all kinds of analysis techniques for this type of data, for example, loglinear analysis, the released data set should contain the frequency counts for the crossing of all variables, and this will come down to the individual data.

A final technique to be mentioned here is the reduction of the resolution of the key. This reduction can be obtained by removing identification variables from the data set or by collapsing categories of identification variables. This often means a reduction to at most 10 identifiers. Removal of so much vital information can make disseminated data sets useless for scientific research.

The risk of disclosure by response knowledge can be reduced by not publishing the survey data file immediately after finishing the fieldwork. Still, by allowing a number of years to pass between fieldwork and publication, one may wonder whether the respondent will feel comfortable if he knows that response knowledge will reveal his private data almost with certainty.

The problem of disclosure by spontaneous recognition of rare persons can be tackled. This risk will be diminished if population uniqueness is removed from all low-dimensional tables of key variables.

The following procedure is proposed to remove uniqueness in low-dimensional tables. A disclosure analysis always starts with establishing the key variables. A file is created that contains the values of only these key variables. Next, it must be decided what the criterion should pertain to. If the criterion pertains to population uniques, the Poisson-Gamma model can be used to estimate the number of uniques. However, it is also possible to apply the criterion to the sample uniques. On the one hand, this is much simpler and straightforward, but on the other it is a conservative criterion: it causes more protection measures to be undertaken than really necessary (sample uniques need not be population uniques). On the file with key variables, an analysis is carried out that consists of four steps:

Step 1: Univariate Scan. Check the univariate frequency distributions and locate variables with small frequencies that do not satisfy the criterion.

Step 2: Collapse/Remove. If a variable does not satisfy the criterion, it can be removed entirely from the data set, but often a better approach is to collapse categories or to recode a bad (rare) category as "unknown" or "otherwise."

Step 3: Bivariate Scan. Check the bivariate distributions frequency distributions and locate the tables that do not satisfy the criterion.

Step 4: Collapse/Remove/Recode. If a bivariate table does not satisfy the criterion, something has to be done about at least one of the two variables concerned. The choice may depend on the behavior of the variables in other tables. Variables causing problems can be removed entirely or some categories may be collapsed. If the problems are caused by only a few records, the relevant scores in these records may be set to "unknown," thus minimizing the loss of information.

Of course, the analysis can be extended to trivariate tables, but if the number of key variables is substantial, this will be very time consuming.

If the data set contains some kind of regional classification, it is recommended to perform the analysis for each region separately. In fact, this is a trivariate analysis in which one variable is always equal to region.

It should be noted that this procedure will not protect the data set against disclosure by matching, and hardly will it protect against disclosure by response knowledge. Specifying more stringent criterion values will produce data sets that might to some extent be protected against these two types of disclosure, but the subsequent loss of information will generally be unacceptably large.

It turns out that disclosure of sensitive information in survey data files is often possible, and difficult to prevent, unless the information in the data set is severely reduced. Disclosure of "rare persons" can be prevented by taking care of the uniques in two- or three-dimensional tables. The risk of disclosure by response knowledge can be limited by advising the respondents not to tell anyone else that they were in a survey. Furthermore, delaying the release of the survey data may help. The third type of disclosure, disclosure by matching, requires considerable resources in terms of methodology, computing power, and manpower. Therefore, if survey data files are released under the conditions that the data may be used for statistical purposes only and that no matching procedures would be carried out at the individual level, any huge effort to identify and to disclose clearly shows malicious intent. In view of the duty of statistical agencies to disseminate statistical information, disclosure protection for this kind of malpractice could and should be taken care of by legal arrangements and not by restrictions on the data to be released.

EXERCISES

13.1 How is the number of key values K defined?
  a. The sum of the numbers of categories for all key variables.
  b. The product of the numbers of categories of all key variables.
  c. The outcome under (a) minus the number of impossible key combinations.
  d. The outcome under (b) minus the number of impossible key combinations.

13.2 What happens to the value of the resolution R if the number of records in a survey data set is made four times large by adding three copies of the data set to the data set.
  a. The value of the resolution will be four times as large.
  b. The value of the resolution will be two times as large.
  c. The value of the resolution does not change.
  d. The value of the resolution will be half as large.

13.3 A survey data set relates to a province consisting of three districts. The population sizes in the districts are 40,000, 20,000, and 10,000. A variable age has been measured in three categories. The age distribution is the same in each district: 30% young, 40% middle aged, and 30% elderly. Also gender has been recorded. It is known that within each combination of district and age, the number of males is equal to the number of females.
  a. Compute the resolution of the key consisting of district, age, and gender.
  b. What would have been the resolution of this key if the number of people for each combination of district, age, and gender was exactly the same?

13.4 A simple key just splits the population into two categories. The number of persons in the first category is F and the number in the second category is $N-F$. Assuming that the value of F can vary, compute the minimum and maximum values of the resolution. For which values of F are these extreme values obtained?

EXERCISES   357

13.5 Which of the following statements is correct?

   a. The Constant-Poisson model fits better in practice than the Poisson-Gamma model because it allows more variation in the frequencies of the key values.

   b. The Constant-Poisson model fits worse in practice than the Poisson-Gamma model because it allows less variation in the frequencies of the key values.

   c. In most practical applications, the fit of the Constant-Poisson model is as good as the fit of the Poisson-Gamma model.

   d. The Constant-Poisson model fits better in practice than the Poisson-Gamma model because it contains less parameters.

13.6 A population consists of 100,800 persons. There are 5 key variables: gender (2 categories), region (12 categories), composition of the household (6 categories), age (10 categories), and education (7 categories). Suppose, the number of persons F_i with key value i has a Poisson distribution. Also suppose that the expected value of all F_i is the same.

   a. Compute the expected number of key values i with $F_i > 0$.

   b. Compute the expected number of key values i with $F_i = 1$.

13.7 A research agency intends to disseminate a survey data file, but wants to keep the disclosure risk to a minimum. The file contains the data of a sample from a population of 7,000,000 employed persons. The survey agency considers making available one of the following two files:

   - A file with a detailed regional classification, but with less detailed other variables. The identification variables are municipality (600 categories), gender (2 categories), age (10 categories), level of education (7 categories), and function type (12 categories).

   - A file with a less detailed regional classification, but with more detailed other variables. The identification variables are province (12 categories), gender (2 categories), age (20 categories), level of education (7 categories), function type (12 categories), marital status (2 categories), and composition of the household (13 categories).

   The research agency applies a relative criterion value of 0.001 for disclosure control for all its survey data files.

   a. Using the Constant-Poisson model, determine whether or not both survey data files satisfy this criterion, and thus whether or not they can be published. It can be assumed that there are no structural zeros among all possible combinations of the categories of the key variables.

   b. There is different survey data set that relates to the same target population. This file does not contain a regional classification variable. The number of different key values of the available key variables is equal to 1,000,000. An analysis shows that the Poisson-Gamma model fits well. The estimate for the parameter α is equal to 0.00005 and the estimate for the parameter β is equal to 0.02.

Estimate the number of uniques in the total population. Can this file be published if the relative criterion of 0.001 is applied?

c. Compute the critical population size. What conclusion can be drawn from the result of this computation?

13.8 The town council of a large city has carried out a survey among its inhabitants. The total population size is $N = 600,000$ and the sample size of the survey was $n = 10,000$. The town council intends to make the survey data file available to other organizations. Before making a decision, a disclosure analysis is carried out. There are 5 key variables in the file: gender (2 categories), marital status (4 categories), age (20 categories), neighborhood (40 categories), and occupation (15 categories).

a. Compute the number of different key values K. If it is assumed that every key value appears with the same frequency in the table, compute the resolution R.

b. Using the Constant-Poisson model, estimate the number of population uniques. Can this file be disseminated under a relative criterion of 0.001?

Using the sample data, it can be shown that

$$\frac{1}{K-1} \sum_{i=1}^{K} \left(f_i - \frac{n}{K}\right)^2 = 2.1875.$$

c. Compute estimates a and b for the parameters α and β of the Poisson-Gamma model. Next, estimate the number of population unique. Can this survey data set be published under a relative criterion of 0.001?

d. Using the estimates a and b, compute the critical population size. What conclusion can be drawn with respect to publishing data at the level of a neighborhood?

References

AAPOR (2000), *Standard Definitions: Final Dispositions of Case Codes and Outcome Rates for Surveys*. American Association for Public Opinion Research, Ann Arbor, MI.

Begeer, W., De Vries, W.F.M. & Dukker, H.D. (1986), Statistics and administration. *Netherlands Official Statistics: Quarterly Journal of the Central Bureau of Statistics* 1, 7–17.

Bethlehem, J.G. (1988), Reduction of nonresponse bias through regression estimation. *Journal of Official Statistics* 4, 251–260.

Bethlehem, J.G. (1996), *Bascula for Weighting Sample Survey Data: Reference Manual*. Statistical Informatics Department, Statistics Netherlands, Voorburg/Heerlen, The Netherlands.

Bethlehem, J.G. (1997), Integrated control systems for survey processing. In: Lyberg, L., Biemer, P., Collins, M., De Leeuw, E., Dippo, C., Schwarz, N. & Trewin, D. (Eds.), *Survey Measurement and Process Control*. Wiley, New York, pp. 371–392.

Bethlehem, J.G. (1999), Cross-sectional research. In: Adèr, H.J. & Mellenbergh, G.J. (Eds.), *Research Methodology in the Social, Behavioural & Life Sciences*. Sage Publications, London, 110–142.

Bethlehem, J.G. (2002), Weighting nonresponse adjustments based on auxiliary information. In: Groves, R.M., Dillman, D.A., Eltinge, J.L. & Little, R.J.A. (Eds.), *Survey Nonresponse*. Wiley, New York.

Bethlehem, J.G. (2007), Reducing the bias of web survey based estimates. Discussion Paper 07001, Statistics Netherlands, Voorburg/Heerlen, The Netherlands.

Bethlehem, J.G. (2008), How accurate are self-selection web surveys? Discussion Paper 08014, Statistics Netherlands, The Hague/Heerlen, The Netherlands.

Bethlehem, J.G. & Hofman, L.P.M.B. (2006), Blaise—alive and kicking for 20 years. *Proceedings of the 10th Blaise, Users Meeting, Statistics Netherlands, Voorburg/Heerlen, The Netherlands*, pp. 61–88.

Bethlehem, J.G. & Hundepool, A.H. (2004), TADEQ: a tool for the documentation and analysis of electronic questionnaires. *Journal of Official Statistics* 20, 233–264.

Bethlehem, J.G. & Keller, W.J. (1987), Linear weighting of sample survey data. *Journal of Official Statistics* 3(2), 141–154.

Bethlehem, J.G., Keller, W.J. & Pannekoek, J. (1990), Disclosure control of microdata. *Journal of the American Statistical Association* 85, 38–45.

Bethlehem, J.G. & Kersten, H.M.P. (1985), On the treatment of non-response in sample surveys. *Journal of Official Statistics* 1, 287–300.

Bethlehem, J.G. & Kersten, H.M.P. (1986), *Werken met non-respons*. Statistische Onderzoekingen M30. Netherlands Central Bureau of Statistics, Voorburg, The Netherlands.

Bethlehem, J.G. & Van der Pol, F. (1998), The future of data editing. In: Couper, M.D., Baker, R.P., Bethlehem, J.G., Clark, C.Z.F., Martin, J., Nicholls, W.L. & O'Reilly, J.M. (Eds), *Computer Assisted Survey Information Collection*. John Wiley & Sons, New York, pp. 201–222.

Biemer, P.P. & Lyberg, L.E. (2003), *Introduction to Survey Quality*. Wiley, Hoboken, NJ.

Börsch-Supan, A., Elsner, D., Faßbender, H., Kiefer, R., McFadden, D. & Winter, J. (2004), Correcting the participation bias in an online survey. Report, University of Munich, Germany.

Boucher, L. (1991), Micro-editing for the Annual Survey of Manufactures: What is the value added? *Proceedings of the Bureau of the Census Annual Research Conference*, pp. 765–781.

Bowley, A.L. (1906), Address to the Economic Science and Statistics Section of the British Association for the Advancement of Science. *Journal of the Royal Statistical Society* 69, 548–557.

Bradburn, N.M., Sudman, S. & Wansink, B. (2004), *Asking Questions: The Definitive Guide to Questionnaire Design—For Market Research, Political Polls, and Social and Health Questionnaires*. Jossey-Bass, San Francisco, CA.

Cassel, C.M. (1976), Probability based disclosures. In: Dalenius, T. & Klevmarken, A. (Eds.), *Personal Integrity and the Need for Data in the Social Sciences*. Swedish Council for the Social Sciences, Stockholm, Sweden, pp. 189–193.

Cassel, C.M., Särndal, C.E. & Wretman, J.H. (1983), Some uses of statistical models in connection with the nonresponse problem. In: Maddow, W.G. & Olkin, I. (Eds.), *Incomplete Data in Sample Surveys. Vol. 3: Proceedings of the Symposium*. Academic Press, New York.

CBS (1987), *Proceedings of the Seminar on Openness and Protection of Privacy in the Information Society*.

Chaudhuri, A. & Vos, J.W.E. (1988), *Unified Theory and Strategies of Survey Sampling*. North-Holland, Amsterdam, The Netherlands.

Cobben, F. (2004), Nonresponse correction techniques in household surveys at Statistics Netherlands: a CAPI–CATI comparison. Technical Report, Methods and Informatics Department, Statistical Netherlands, Voorburg, The Netherlands.

Cobben, F. & Bethlehem, J.G. (2005), Adjusting undercoverage and non-response bias in telephone surveys. Discussion Paper 05006, Statistics Netherlands, Voorburg/Heerlen, The Netherlands.

Cochran, W.G. (1953), *Sampling Techniques*. Wiley, New York.

Cochran, W.G. (1968), The effectiveness of adjustment by subclassification in removing bias in observational studies. *Biometrics* 24, 205–213.

Cochran, W.G. (1977), *Sampling Techniques*, 3rd edition. Wiley, New York.

Converse, J.M. & Presser, S. (1986), *Survey Questions: Handcrafting the Standardized Questionnaire*. Sage University Paper Series on Quantitative Applications in the Social Sciences, 07-063. Sage Publications, Beverly Hills, CA.

Couper, M.P. (2000), Web surveys: a review of issues and approaches. *Public Opinion Quarterly* 64, 464–494.

Couper, M.P., Baker, R.P., Bethlehem, J.G., Clark, C.Z.F., Martin, J., Nicholls, W.L. II & O'Reilly, J.M. (Eds.) (1998), *Computer Assisted Survey Information Collection*. Wiley, New York.

Couper, M.P. & Nicholls, W.L. (1998), The history and development of computer assisted survey information collection methods. In: Couper, M.P., Baker, R.P., Bethlehem, J., Clark, C.Z.F., Martin, J., Nicholls, W.L. & O'Reilly, J. (Eds.), *Computer Assisted Survey Information Collection*. Wiley, New York.

Cox, L.H., McDonald, S. & Nelson, D. (1986), Confidentiality issues at the United States Bureau of the Census. *Journal of Official Statistics* 2, 135–160.

Dalenius, T. (1977), Towards a methodology for statistical disclosure control. *Statistisk Tidskrift* 5, 429–444.

Dalenius, T. & Hodges, J.L. (1959), Minimum variance stratification. *Journal of the American Statistical Association* 54, 88–101.

Dalenius, T. & Reiss, S.P. (1982), Data-swapping: a technique for disclosure control. *Journal of Statistical Planning and Inference* 6, 73–85.

De Bie, S.E., Stoop, I.A.L. & de Vries, K.L.M. (1989), *CAI Software: An Evaluation of Software for Computer Assisted Interviewing*. VOI Uitgeverij, Amsterdam, The Netherlands.

DeFuentes-Merillas, L., Koeter, M.W.J., Bethlehem, J.G., Schippers, G.M. & Van Den Brink, W. (1998), Are scratchcards addictive? The prevalence of pathological scratchcard gambling among adult scratchcard buyers in the Netherlands. *Addiction* 98, 725–731.

De Haan, J. & Van't Hof, C. (2006), *Jaarboek ICT en samenleving, de digitale generatie*. Sociaal en Cultureel Planbureau, The Hague, The Netherlands.

Dehjia, R. and Wahba, S. (1999), Causal effects in non-experimental studies: re-evluating the evaluation of training programs. *Journal of the American Statistical Association* 94, 1053–1062.

De Leeuw, E.D. (2005), To mix or not to mix data collection modes in surveys. *Journal of Official Statistics* 21(2), 233–255.

De Leeuw, E.D. & Collins, M. (1997), Data collection methods and survey quality. In: Lyberg, L., Biemer, P., Collins, M., De Leeuw, E., Dippo, C., Schwarz, N. & Trewin, D. (Eds.), *Survey Measurement and Process Control*. Wiley, New York, pp. 199–220.

Deming, W.E. (1950), *Some Theory of Sampling*. Wiley, New York.

Deming, W.E. (1986), *Out of the Crisis*. Massachusetts Institute of Technology, Cambridge, MA.

Deming, W.E. & Stephan, F.F. (1940), On a least squares adjustment of a sampled frequency table when the expected marginal tables are known. *The Annals of Mathematical Statistics* 11, 427–444.

De Ree, S.J.M. (1978), Hutspot, een nieuw gerecht van oude ingrediënten. In: CBS (Ed.), *Denken en Meten*. Staatsuitgeverij, The Hague, The Netherlands.

Desrosières, A. (1998), *The Politics of Large Numbers: A History of Statistical Reasoning*. Harvard University Press, Cambridge, MA.

Deville, J.C. & Särndal, C.E. (1992), Calibration estimators in survey sampling. *Journal of the American Statistical Association* 87, 376–382.

Deville, J.C., Särndal, C.E. & Sautory, O. (1993), Generalized raking procedures in survey sampling. *Journal of the American Statistical Association* 88, 1013–1020.

Dijkstra, E.W. (1968), Go to considered harmful. *Communications of the ACM* 11(3), 147–148 (Letter).

Dillman, D.A. & Bowker, D. (2001), The web questionnaire challenge to survey methodologists. In: Reips, U.D. & Bosnjak, M. (Eds.), *Dimensions of Internet Science*. Pabst Science Publishers, Lengerich, Germany.

Dillman, D.A., Tortora, R.D. & Bowker, D. (1998), Principles for construction web surveys. Technical Report 98-50, Social and Economic Sciences Research Center, Washington State University, Pullman, WA.

Dillman, D.A. (2007), *Mail and Internet Surveys, The Tailored Design Method*. Second Edition. Wiley, Hoboken, NJ.

Duffy, B., Smith, K., Terhanian, G. & Bremer, J. (2005), Comparing data from online and face-to-face surveys. *International Journal of Market Research* 47, 615–639.

Duncan, G.T. & Lambert, D. (1986), Disclosure-limited data dissemination. *Journal of the American Statistical Association* 81, 10–28.

Duncan, G.T. & Lambert, D. (1987), The risk of disclosure for microdata. *Proceedings of the 3rd Annual Research Conference of the Bureau of the Census*, pp. 263–278.

Engström, P. (1995), A study on using selective editing in the Swedish surveys on wages and employment in industry. Room Paper No. 11, ECE Working Session on Statistical Data Editing, Athens, Greece.

Eurostat (2007), *More than 40% of households have broadband internet access*. Eurostat News Release 166/2007. Eurostat, Luxembourg.

Everaers, P. & Van Der Laan, P. (2001), The Dutch Virtual Census. *E-Proceedings of the 53rd Session of the International Statistical Institute, Seoul, Korea*.

Fellegi, I.P. & Holt, D. (1976), A systematic approach to automatic edit and imputation. *Journal of the American Statistical Association* 71, 17–35.

Frank, O. (1976), Individual disclosures from frequency tables. In: Dalenius, T. & Klevmarken, A. (Eds.), *Personal Integrity and the Need for Data in the Social Sciences*. Swedish Council for the Social Sciences, Stockholm, Sweden, pp. 175–187.

Frank, O. (1979), Inferring individual information from released statistics. *Bulletin of the International Statistical Institute, Proceedings of the 42nd Session (Book 3), Manila, Philippines*.

Fricker, R. & Schonlau, M. (2002), Advantages and disadvantages of Internet research surveys: evidence from the literature. *Field Methods* 15, 347–367.

Gelman, A. and Carlin, J.B. (2000), Poststratification and Weighting Adjustments. In: Groves, R.M., Dillman, D.A., Eltinge, J.L., Little, R.J.A. (Eds), *Survey Nonresponse*, John Wiley & Sons, New York.

Granquist, L. (1990), A review of some macro-editing methods for rationalizing the editing process, *Proceedings of the Statistics Canada Symposium*, pp. 225–234.

Graunt, J. (1662), *Natural and Political Observations Upon the Bills of Mortality*. Martyn, London.

Greenberg, B.V. & Zayatz, L.V. (1992), Strategies for measuring risk in public use microdata files. *Statistica Neerlandica* 46, 33–48.

Hansen, M.H. & Hurvitz, W.N. (1943), On the theory of sampling from a finite population. *Annals of Mathematical Statistics* 14, 333–362.

Hansen, M.H. & Hurwitz, W.H. (1946), The problem of nonresponse in sample surveys. *Journal of the American Statistical Association* 41, 517–529.

Hansen, M.H., Hurvitz, W.N. & Madow, W.G. (1953), *Survey Sampling Methods and Theory*. Wiley, New York.

Heerwegh, D. & Loosveldt, G. (2002), An evaluation of the effect of response formats on data quality in web surveys. Paper presented at the *International Conference on Improving Surveys, Copenhagen, Denmark*.

Hemelrijk, J. (1968), Back to the Laplace definition. *Statistica Neerlandica* 22, 13–21.

Hidiroglou, M.A., & Berthelot, J.-M. (1986), Statistical editing and imputation for periodic business surveys. *Survey Methodology* 12, 73–83.

Horvitz, D.G. & Thompson, D.J. (1952), A generalization of sampling without replacement from a finite universe. *Journal of the American Statistical Association* 47, 663–685.

Huang, E.T. & Fuller, W.A. (1978), Nonnegative regression estimation for sample survey data. *Proceedings of the Social Statistics Section, American Statistical Association*, pp. 300–305.

Hundepool, A., Domnigo-Ferrer, J., Franconi, L., Giessing, S., Lenz, R., Longhurst, J., Schulte Nordholt, E. & De Wolf, P.P. (2007), *Handbook on Statistical Disclosure Control*. Centre of Excellence for Statistical Disclosure Control, http://neon.vb.cbs.nl/casc.

ISI (1985), Declaration on professional ethics. *Bulletin of the International Statistical Institute, Proceedings of the 45th Session (Book 5, Appendix V)*.

Jabine, T.P. (1985), Flow charts: a tool for developing and understanding survey questionnaires. *Journal of Official Statistics* 1, 189–207.

Johnson, N.L. & Kotz, S. (1969), *Discrete Distributions*. Wiley, New York.

Kalsbeek, W.D. (1980), A conceptual review of survey error due to nonresponse. *Proceedings of the Section on Survey Research Methods, American Statistical Association*, pp. 131–136.

Kalton, G., Collins, M. & Brook, L. (1978), Experiments in wording opinion questions. *Applied Statistics* 27, 149–161.

Kalton, G. and Maligalig, D.S. (1991), A Comparison of Methods of Weighting Adjustment for Nonresponse. *Proceedings of the 1991 Annual Research Conference*, U.S. Bureau of the Census, Washington, pp. 409–428.

Kalton, G. & Schuman, H. (1982), The effect of the question on survey responses: a review. *Journal of the Royal Statistical Society* 145, 42–57.

Katz, I.R., Stinson, L.L. & Conrad, F.G. (1999), Questionnaire designer versus instrument authors: bottlenecks in the development of computer-administered questionnaires. Memo, University of Michigan, Michigan, USA.

Kersten, H.M.P. & Bethlehem, J.G. (1984), Exploring and reducing the nonresponse bias by asking the basic question. *Statistical Journal of the U.N. Economic Commission for Europe* 2, 369–380.

Kiaer, A.N. (1895), Observations et expériences concernant des dénombrements représentatives. *Bulletin of the International Statistical Institute, Proceedings of the 9th Session (Book 2)*, pp. 176–183.

Kiaer, A.N. (1997), *The Representative Method of Statistical Surveys* (English translation). Statistics Norway, Oslo, Norway (Reprinted from *Den repräsentative*

undersökelsesmetode. Christiania Videnskabsselskabets Skrifter. II. Historiskfilosofiske klasse, No. 4, 1897).
Kim, J. (1986), A method for limiting disclosure in microdata based on random noise and transformation. *Proceedings of the Section on Survey Research, American Statistical Association*, pp. 370–374.
Kish, L. (1967), *Survey Sampling*. John Wiley & Sons, New York.
Krosnick, J.A. (1991), Response strategies for coping with the cognitive demands of attitude measures in surveys. *Applied Cognitive Psychology* 5, 213–236.
Kruskal, W. & Mosteller, F. (1979a), Representative sampling. I. Non-scientific literature. *International Statistical Review* 47, 13–24.
Kruskal, W. & Mosteller, F. (1979b), Representative sampling. II. Scientific literature excluding statistics. *International Statistical Review* 47, 111–123.
Kruskal, W. & Mosteller, F. (1979c), Representative sampling. III. Current statistical literature. *International Statistical Review* 47, 245–265.
Kuusela, V. (2003), Mobile phones and telephone survey methods. In: Banks, R., Currall, J., Francis, J., Gerrard, L., Kahn, R., Macer, T., Rigg, M., Ross, E., Taylor, S. & Westlake, A. (Eds.), ASC *2003—The Impact of New Technology on the Survey Process. Proceedings of the 4th ASC International Conference*. Association for Survey Computing (ASC), Chesham Bucks, UK, pp. 317–327.
Kuusela, V., Vehovar, V. & Callegaro, M. (2006), Mobile phones—influence on telephone surveys. Paper presented at the *2nd International Conference on Telephone Survey Methodology, Florida, USA*.
Lahiri, D.B. (1951), A method of sample selection providing unbiased ratio estimates. *Bulletin of the International Statistical Institute, Proceedings of the 33rd Session (Book 2)*, pp. 133–140.
Laplace, P.S. (1812), *Théorie Analytique des Probabilités. Oevres Complètes*, Vol. 7. Gauthier-Villar, Paris, France.
Latouche, M. & Berthelot, J.-M. (1992), Use of a score function to prioritize and limit recontacts in business surveys. *Journal of Official Statistics* 8, 389–400.
Lindell, K. (1997), Impact of Editing on the Salary Statistics for Employees in County Council. *Statistical Data Editing*, Volume 2, UN Economic Commission for Europe, Geneva, pp. 2–7.
Lindström, H., Wretman, J., Forsman, G. & Cassel, C. (1979), *Standard Methods for Non-Response Treatment in Statistical Estimation*. National Central Bureau of Statistics, Stockholm, Sweden.
Linehard, J.H. (2003), *The Engines of Our Ingenuity: An Engineer Looks at Technology and Culture*. Oxford University Press, Oxford, UK.
Little, R.J.A. (1993), Post-stratification: A Modeler's Perspective. *Journal of the American Statistical Association* 88, 1001–1012.
Little, R.J.A. & Rubin, D.B. (1987), *Statistical Analysis with Missing Data*. Wiley, New York.
Little, R.J.A. & Smith, P.J. (1987), Editing and imputation for quantitative survey data. *Journal of the American Statistical Association* 82, 58–68.
Little, R.J.A. & Wu, M.M. (1991), Models for contingency tables with known margins when target and sampled populations differ. *Journal of the American Statistical Association* 86, 87–95.

Lodge, M., Steenbergen, M.R. & Brau, S. (1995), The responsive voter: campaign information and the dynamics of candidate evaluation. *American Political Science Review* 89, 309–326.

Lynn, P., Beerten, R., Laiho, J. & Martin, J. (2002), Towards standardisation of survey outcome categories and response rate calculations. *Research in Official Statistics* 1, 63–86.

Madow, W.G. & Madow, L.H. (1944), On the theory of systematic sampling. *Annals of Mathematical Statistics* 15, 1–24.

McGuckin, R.H. & Nguyen, S.V. (1988), Use of 'surrogate files' to conduct economic studies with longitudinal microdata. *Proceedings of the 4th Annual Research Conference of the Bureau of the Census*, pp. 193–211.

Neyman, J. (1934), On the two different aspects of the representative method: the method of stratified sampling and the method of purposive selection. *Journal of the Royal Statistical Society* 97, 558–606.

Nicholls, W.L. & Groves, R.M. (1986), The status of computer assisted telephone interviewing. *Journal of Official Statistics* 2, 93–134.

Nieuwenbroek, N.J. (1993), An integrated method for weighting characteristics of persons and households using the linear regression estimator. Report 8445-93-M1-1, Statistics Netherlands, Voorburg, The Netherlands.

Paass, G. (1985), Disclosure risk and disclosure avoidance for microdata. Working Paper, Institute for Applied Information Technique, Gesellschaft fur Mathematik und Datenverarbeitung, Sankt Augustin, Germany.

Paass, G. & Wauschkuhn, U. (1985), Datenzugang, Datenschutz und Anonymisierung— Analysepotential und Identifzierbarkeit von anonymisierten Individualdaten. *Berichte der Gesellschaft für Mathematik und Datenverarbeitung*, No. 148. Oldenbourg, München, Germany.

Pierzchala, M. (1990), A Review of the State of the Art in Automated Data Editing and Imputation. *Journal of Official Statistics* 6, 355–377.

Playfair, J. (1786), The Commercial and Political Atlas. London, UK.

Quetelet, L.A.J. (1835), Sur l'Homme et le Développement de ses Facultés, Essai de Physique Sociale. Paris, France.

Quetelet, L.A.J. (1846), Lettre à S.A.R. le Duc Régant de Saxe Coburg et Gotha sur la Théorie des Probabilités, Appliquée aux Sciences Morales at Politiques. Brussels, Belgium.

Raj, D. (1968), *Sampling Theory*. McGraw-Hill, New York.

Renssen, R.H., Nieuwenbroek, N.J. & Slootbeek, G.T. (1997), Variance module in Bascula 3.0: theoretical background. Research Paper 9712, Department for Statistical Methods, Statistics Netherlands, Voorburg, The Netherlands.

Rosenbaum, P.R. & Rubin, D.B. (1983), The central role of the propensity score in observational studies for causal effects. *Biometrika* 70, 41–55.

Rosenbaum, P.R. & Rubin, D.B. (1984), Reducing bias in observational studies using subclassification on the propensity score. *Journal of the American Statistical Association* 79, 516–524.

Rubin, D.B. (1987), *Multiple Imputation for Nonresponse in Surveys*. Wiley, New York.

Saris, W.E. (1997), The public opinion about the EU can easily be swayed in different directions. *Acta Politica* 32, 406–436.

Saris, W.E. (1998), Ten years of interviewing without interviewers: the Telepanel. In: Couper, M.P., Baker, R.P., Bethlehem, J.G., Clark, C.Z.F., Martin, J., Nicholls, W.L. II & O'Reilly,

J.M. (Eds.), *Computer Assisted Survey Information Collection*. Wiley, New York, pp. 409–430.

Schafer, J.L. (1997), *Analysis of Incomplete Multivariate Data*. Chapman & Hall, London, UK.

Schmid, C.F. (1983), *Statistical Graphics: Design Principles and Practices*. Wiley, New York.

Schonlau, M., Fricker, R.D. & Elliott, M.N. (2003), *Conducting Research Surveys via E-mail and the Web*. Rand Corporation, Santa Monica, CA.

Schonlau, M., Zapert, K., Payne Simon, L., Haynes Sanstad, K., Marcus, S., Adams, J., Kan, H., Turber, R. & Berry, S. (2004), A comparison between responses from propensity-weighted web survey and an identical RDD survey. *Social Science Computer Review* 22, 128–138.

Schouten, B. (2007), A follow-up of nonresponse in the Dutch Labour Force Survey. Discussion Paper 07004, Statistics Netherlands, Voorburg/Heerlen, The Netherlands.

Schwarz, N., Knäuper, B., Oyserman, D. & Stich, C. (2008), The psychology of asking questions. In: de Leeuw, E.D., Hox, J.J. & Dillman, D.A. (Eds.), *International Handbook of Survey Methodology*. Lawrence Erlbaum Associates, New York, pp. 18–34.

Sikkel, D. (1983), Geheugeneffecten bij het rapporteren van huisartsencontacten. *Statistisch Magazine, Netherlands Central Bureau of Statistics* 3(4), 61–64.

Sirken, M. (1972), *Designing Forms for Demographic Surveys*. Laboratory for Population Statistics Manual Series, No. 3. University of North Carolina, Chapel Hill, NC.

Skinner, C.J. (1991), On the efficiency of raking ratio estimation for multiple frame surveys. *Journal of the American Statistical Association* 86, 779–784.

Skinner, C.J., Marsh, C., Openshaw, S. & Wymer, C. (1990), Disclosure avoidance in Great Britain. *Proceedings of the 1990 Annual Research Conference of the Bureau of the Census*, pp. 131–143.

Spruill, N. (1983), The confidentiality and analytic usefulness of masked business microdata. *Proceedings of the Section on Survey Research, American Statistical Association*, pp. 602–607.

Statistics Netherlands (1998), *Statistical Yearbook of The Netherlands 1998*. Statistics Netherlands, Voorburg/Heerlen, The Netherlands.

Statistics Netherlands (2002), *Blaise: A Survey Processing System—Developers Guide*. Methods and Informatics Department, Statistics Netherlands, Heerlen, The Netherlands.

Stoop, I.A.L. (2005), *The Hunt for the Last Respondent: Nonresponse in Sample Surveys*. Social and Cultural Planning Agency, The Hague, The Netherlands.

Terhanian, G., Smith, R., Bremer, J. & Thomas, R.K. (2001), Exploiting analytical advances: minimizing the biases associated with Internet-based surveys of non-random samples. *ARF/ESOMAR: Worldwide Online Measurement*, Vol. 248. ESOMAR Publication Services, pp. 247–272.

Tiemeijer, W.L. (2008), *Wat 93,7 Procent van de Nederlanders moet weten over Opiniepeilingen*. Aksant, Amsterdam, The Netherlands.

Tippet, L.C. (1927), *Random Sampling Numbers*. Tracts for Computers, No. 15. Harvard University Press, Cambridge, MA.

Tremblay, V. (1986), Practical Criteria for Definition of Weighting Classes. *Survey Methodology* 12, 85–97.

Tufte, E.R. (1983), *The Visual Display of Quantitative Information*. Graphics Press, Cheshire, CT.

Tukey, J.W. (1977), *Exploratory Data Analysis*. Addison-Wesley, London.

United Nations (1964), Recommendations for the preparation of sample survey reports. Provisional Report, Sales No. 64.XVII.4, United Nations, New York.

United Nations (1994), *Statistical Data Editing, Vol. 1: Methods and Techniques*. Statistical Standards and Studies, No. 44. United Nations Statistical Commission and Economic Commission for Europe, Geneva, Switzerland.

U.S. Department of Commerce (1978), Report on statistical disclosure and disclosure-avoidance techniques. Statistical Policy Working Paper 2, U.S. Government Printing Office, Washington, DC.

Utts, J.M. (1999), *Seeing Through Statistics*. Duxbury Press, Belmont, CA.

Van de Pol, F. and Molenaar, W. (1995), Selective and automatic editing with CADI-applications, In Kuusela, V. (ed.), *Essays on Blaise 1995, Proceedings of the third International Blaise Users's Conference*, Heslinki: Statistics Finland, pp. 159–168.

Van de Stadt, H., Ten Cate, A., Hundepool, A.J. & Keller, W.J. (1986), *Koopkracht in kaart gebracht: een statistiek van de inkomensdynamiek*. Statistische Onderzoekingen M28. Netherlands Central Bureau of Statistics, Voorburg, The Netherlands.

Vonk, T., Van Ossenbruggen, R. & Willems, P. (2006), The effects of panel recruitment and management on research results: a study among 19 online panels. *Panel Research 2006, ESOMAR World Research*, Vol. 317. ESOMAR Publication Services, pp. 79–99.

Voogt, R. (2004), I'm not interested. Nonresponse bias, response bias and stimulus effects in election research. Ph.D. thesis, University of Amsterdam, The Netherlands.

Wainer, H. (1997), *Visual Revelations*. Copernicus/Springer-Verlag, New York.

Willenborg, L. & De Waal, T. (1996), *Statistical Disclosure Control in Practice*. Lecture Notes in Statistics, 11, Springer-Verlag, New York.

Yates, F. (1949), *Sampling Methods for Censuses and Surveys*. Charles Griffin & Co., London, UK.

Zaller, J.R. (1992), *The Nature and Origins of Mass Opinion*. Cambridge University Press, Cambridge, UK.

Index

Absolute critical value: 351
Access panel: 282
Adjusted population variance: 19
Adjustment weight: 249, 314
Administration: 343
Aggregation method: 205
Allocation of the sample: 105
Allocation:
  Neyman, 105
  optimal, 105
  proportional, 106
Ambiguous question: 46
Analogy principle: 67
Analysis:
  exploratory, 4
  inductive, 4
Asymptotically design unbiased (ADU): 140
Authoring language: 164
Automatic editing: 200
Auxiliary information: 249, 288
Auxiliary variable: 17, 134, 227, 249

Bar chart: 334
  clustered, 337
  stacked, 337, 338
Basic question approach: 239
Bethlehem: 5
Bias of an estimator: 36, 220, 222, 225, 285, 287
Bias:
  relative, 224

Big element: 92, 121
Blaise: 158, 164, 167
Box plot: 333, 339
Box-and-whisker plot: 333

CADE, computer assisted data input, 197
CADI, computer assisted data input, 197
CAI, computer assisted interviewing, 3, 155, 156, 276
Calibration: 250, 263
CAPI, computer assisted personal interviewing, 3, 156, 276, 293, 306
CASAQ, computer assisted self-administered questionnaire, 158
CASES: 164
CASI, computer assisted self-interviewing, 3, 158, 276
CATI, computer assisted personal interviewing, 3, 156, 276, 280, 293, 306
CAWI, computer assisted web interviewing, 3, 159, 276
Census: 1
Central limit theorem: 6, 37
Chart junk: 326
Check: 163, 172
Check box: 54
Check:
  hard, 199
  soft, 199
Check-all-that-apply question: 54

Checking:
  dynamic, 167
  static, 167
Closed question: 51
Cluster: 108
Cluster effect: 113, 116
Cluster sampling: 108
Clustered bar chart: 337
Coefficient of variation: 141
Collapsing strata: 253, 269
Complete enumeration: 1
Completeness error: 182
Computation: 163, 172
Computer assisted data entry (CADE): 197
Computer assisted data input (CADI): 197
Computer Assisted Interviewing (CAI): 3, 155, 156, 276
Computer Assisted Personal Interviewing (CAPI): 3, 156, 276, 293, 306
Computer Assisted Self Administered Questionnaire (CASAQ): 158
Computer Assisted Self Interviewing (CASI): 3, 158, 276
Computer Assisted Telephone Interviewing (CATI): 3, 156, 276, 280, 293, 306
Computer Assisted Web Interviewing (CAWI): 3, 159, 276
Concurrent approach: 160
Confidence interval: 8, 37, 223
Confidence level: 37
Consistency error: 3, 166, 182
Constant-Poisson Model: 349
Contrast: 220, 285
Correction weight: 250, 314
Critical population size: 352
Critical stream: 203
Critical value:
  absolute, 351
  relative, 351
Cumulative scheme: 84, 85
Cumulative-square-root-f rule: 105

Data analysis: 310, 312
Data collection: 3, 153
Data collection:
  mixed-mode, 160
  single mode, 160
Data density index (DDI): 325
Data editing: 3, 181, 195

Data Entry Program (DEP): 168
Data ink ratio (DIR): 326
Data matrix: 311
Data model: 168
Data swapping: 354
Date question: 55
DDI, data density index, 325
Deductive imputation: 185
Deletion:
  list-wise, 183
  pair-wise, 183
Descriptive model: 135
Dichotomous variable: 18
DIR, data ink ratio, 326
Direct estimator: 138
Dirty data: 312
Dirty Data Theorem: 310
Disclosure: 343
Disclosure by matching: 348
Disclosure by response knowledge: 348
Disclosure by spontaneous recognition: 347
Disclosure control: 4
  statistical, 344
Disclosure problem: 343, 344
Disclosure risk: 349
Distribution method: 205
Domain error: 182
Domesday book: 5
Don't know: 53
Donor imputation: 186
Donor table: 125
Double question: 49
Double-barreled question: 49
Dummy variable: 18, 256, 257
Dynamic checking: 167
Dynamic error reporting: 167
Dynamic question text: 173
Dynamic routing: 166

Editing:
  automatic, 200
  macro-, 182, 200, 204
  micro-, 182
  over-, 199
  selective, 200, 203
Effective sample size: 113
Efficiency of an estimator: 137
Electronic questionnaire: 60

Error reporting:
  dynamic, 167
  static, 167
Error: 178
  completeness, 182
  consistency, 3, 166, 182
  domain, 182
  estimation, 180
  measurement, 181
  nonobservation, 181
  nonresponse, 181
  nonsampling, 181
  observation, 181
  overcoverage, 181
  processing, 181
  range, 3, 166
  routing, 3, 183
  sampling, 180
  selection, 181, 277
  skip pattern, 183
  sources of, 178
  total, 178
  undercoverage, 181
Estimate: 34
Estimation error: 180
Estimator: 34, 35, 67, 71, 78, 93, 103, 104, 110, 111, 115, 116, 118, 119, 127, 134
  direct, 138
  generalized regression, 134, 254, 255
  Horvitz-Thompson, 38, 83, 86
  poststratification, 134, 148, 250, 251, 256, 289, 291
  ratio, 134, 140
  regression, 134, 144
  unbiased, 9, 35
Expected value: 35
Exploratory analysis: 4

Face-to-face interviewing: 154
Factual question: 43, 44
Familiar wording: 45
Felligi-Holt methodology: 201
Fieldwork: 153
Filter question: 53
Filter-table: 125
Finite population correction: 68
First-order inclusion expectation: 29
First-order inclusion probability: 30, 67, 77, 90, 285

First-order selection probability: 31, 84
Fixed response model: 218, 237
Follow-up survey: 237
Frame population: 278
Frequency weight: 314

Generalized regression estimator: 134, 254, 255
Geographical variable: 351
Graph: 322
  three-dimensional, 331

Hard check: 199
Hierarchical questionnaire: 165
Histogram: 334
Homogeneous: 104
Horvitz-Thompson estimator: 38, 83, 86
Hot-deck imputation: 186
Hypothetical question: 50

Identification: 343
Identification variable: 344
Identifying information: 344
Imputation: 3, 4, 184, 316
  deductive, 185
  donor, 186
  hot-deck, 186
  multiple, 185, 193
  nearest neighbor, 187
  random, 186, 192
  random within groups, 186
  ratio, 187
  regression, 187
  single, 185
Imputation of the mean: 186, 189, 316
Imputation of the mean within groups: 186
Inclusion expectation: 29
  first-order, 29
  second-order, 29
Inclusion probability:
  first-order, 30, 67, 77, 90, 285
  second-order, 30
Inclusion weight: 249, 313, 314
Inductive analysis: 4
Integrated survey processing: 175
Internet population: 284
Item nonresponse: 209
Iterative proportional fitting: 260

Key: 345
    resolution of, 346
Key variable: 344
KISS principle: 322

L'homme moyenne: 7, 23
Lahiri scheme: 84, 85
Large weights: 264
Layout instruction: 172
Leading question: 47
Learning effect: 55
Lie factor: 331
Linear weighting: 250, 263
Linear weighting including propensity
    scores: 268
Linear weighting with adjusted inclusion
    probabilities: 267
List-wise deletion: 183
Long question text: 46

Macroediting: 182, 200, 204
Mail interviewing: 153
Margin of error: 72
Mean square error of an estimator: 36, 179
Measurement error: 181
Memory error: 47
Memory-based model: 44
Metadata: 167, 311
Metadata file: 168
Microaggregation: 354
Microediting: 182
Missing at random (MAR): 185, 194, 226,
    290, 292, 299, 303
Missing completely at random (MCAR): 185,
    193, 225, 286
Missing value: 312
Mixed-mode data collection: 160
Mobile phone: 155, 306
Mode effect: 162
Model parameters: 135
Modes of data collection: 153
Monograph study: 7
Multi-stage sample: 8
Multiple imputation: 185, 193
Multiplicative weighting: 250, 260, 263

Nearest neighbor imputation: 187
Negative question: 50
Negative weights: 264

Neyman allocation: 105
Non-internet population: 284, 296
Noncontact: 212
Noncritical stream: 203
Nonfactual question: 43, 44
Nonobservation error: 181
Nonresponse: 209, 278
    item, 209
    selective, 210
    unit, 209
nonresponse bias: 220, 222
nonresponse error: 181
Nonresponse stratum: 218
Nonsampling error: 181
Normal distribution: 8
Not missing at random (NMAR): 185, 194,
    226, 299
Not-able: 212
Numerical question: 54

Observation error: 181
OK index: 204
Online processing model: 44
Online survey: 159, 277
Open question: 50
Opinion poll: 9
Opinion question: 43
Optimal allocation: 105
Outlier: 200, 333
Overcoverage: 21, 212
Overcoverage error: 181
Overediting: 199

Pair-wise deletion: 183
Partial investigation: 6
Pie chart: 334, 335
Poisson-Gamma Model: 349, 350
POLS: 229, 268
Population:
    internet, 284
    non-internet, 284, 296
    two-dimensional, 122
Population mean: 18, 67
Population parameter: 17
Population percentage: 18, 71
Population total: 18
Population uniques: 350, 351
Population variance: 19
    adjusted, 19

INDEX

Poststratification estimator: 134, 148, 250, 251, 256, 289, 291
Precise: 36, 178
Primary unit: 113
Privacy: 342
Probability sample: 28, 306
Probability sampling: 2, 281, 283
Processing error: 181
Propensity score: 266, 297
Propensity score stratification: 267
Propensity score weighting: 266, 297
Proportional allocation: 106
Pseudorandom number: 27
Pseudorandomizer: 27
Purposive selection: 8

Qualitative variable: 311
Quantitative variable: 311
Question: 163
   ambiguous, 46
   check-all-that-apply, 54
   closed, 51
   date, 55
   double, 49
   double-barreled, 49
   factual, 43, 44
   filter, 53
   hypothetical, 50
   leading, 47
   negative, 50
   nonfactual, 43, 44
   numerical, 54
   open, 50
   opinion, 43
   recall, 46
   sensitive, 49
Question order: 55
Question text: 45
   dynamic, 173
   long, 46
Questionnaire: 43
   electronic, 60
   hierarchical, 165
Questionnaire fatigue: 59
Questionnaire testing: 58
Quipu: 5
Quipucamayoc: 5
Quota sampling: 9

Radio button: 52
Raking: 260
Random digital dialing (RDD): 155
Random imputation: 186, 192
Random imputation within groups: 186
Random noise: 354
Random number: 8, 26
Random response model: 218, 221, 266
Random sampling: 8
Randomizer: 25
Range error: 3, 166
Ratio estimator: 134, 140
Ratio imputation: 187
Recall question: 46
Reference date: 21
Reference survey: 293, 294
Refusal: 212
Regression estimator: 134, 144
Regression imputation: 187
Relative bias: 224
Relative critical value: 351
Representative: 20, 23, 24, 249
Representative Method: 7, 8, 100
Representative sample: 100
Representativeness: 7
Residual: 135
Residual sum of squares: 135
Resolution of the key: 346
Response: 212
Response burden: 1
Response probability: 221, 286
Response rate: 212, 213
   unweighted, 214
   weighted, 214
Response stratum: 218
Route instruction: 57, 163, 172
Routing: 56
   dynamic, 166
   static, 166
Routing error: 3, 183

Sample: 1, 28
   representative, 100
   self-weighting, 106, 117, 122
Sample mean: 34
Sample selection scheme: 31
Sample size: 28, 72
   effective, 113
Sample value: 33

Sample variance: 34, 68
Sampling: 22
  cluster, 108
  simple random, 65
  stratified, 100
  systematic, 75
  systematic with unequal probabilities, 89
  two-dimensional, 122
  two-stage, 113
  unequal probability, 82
Sampling design: 28
Sampling error: 180
Sampling fraction: 67
Sampling frame: 20
Sampling in space and time: 123
Sampling strategy: 35
Sampling with replacement: 31
Sampling without replacement: 30
Samplonia: 11
Satisficing: 53
Scatter plot: 336
Second-order inclusion expectation: 29
Second-order inclusion probability: 30
Second-order selection probability: 32
Secondary unit: 113
Selection error: 181, 277
Selection probability:
  first-order, 31, 84
  second-order, 32
Selective editing: 200, 203
Selective nonresponse: 210
Self-selection: 281, 285, 301
Self-selection survey: 281
Self-weighting sample: 106, 117, 122
Sensitive information: 344, 345
Sensitive question: 49
Sequential approach: 161
Showcard: 153
Simple random sampling: 65
SimSam: 13
Single-imputation: 185
Single-mode data collection: 160
Skip pattern error: 183
Soft check: 199
Sources of error: 178
Stacked bar chart: 337, 338
Standard error of an estimator: 36
Starting point: 76, 90

Static checking: 167
Static error reporting: 167
Static routing: 166
Statistic: 34
Statistical disclosure control: 344
Statistics: 343
Step length: 75, 90
Stratification variable: 104
Stratification:
  propensity score, 267
Stratified sampling: 100
Stratum: 100, 108, 250
Straw poll: 9
Structural zero: 345
Subquestionnaire: 174
Survey: 1
Survey data matrix: 311
Survey design: 2, 15
Survey documentation: 321
Survey objectives: 15
Survey report: 317
Survey:
  online, 159, 277
  reference, 293, 294
  self-selection, 281
  web, 159, 277
Synthetic value: 184
Systematic sampling: 75
Systematic sampling with unequal probabilities: 89

TADEQ project: 62
Tailored Design Method: 154
Target population: 2, 16
Target variable: 16, 17
Telepanel: 159
Telephone interviewing: 154
Telescoping: 47
Three-dimensional graph: 331
Total error: 178
True value: 7
Two-dimensional population: 122
Two-dimensional sampling: 122
Two-stage sampling: 113

Unbiased: 178
Unbiased estimator: 9, 35
Undercoverage: 21, 277, 278, 299, 306

Undercoverage error: 181
Unequal probability sampling: 82
Unique in the population: 346
Unique in the sample: 346
Uniqueness: 342, 345
Unit nonresponse: 209
Unweighted response rate: 214

Validity of the question: 58
Variable:
   geographical, 351
   identification, 344
   key, 344
   qualitative, 311
   quantitative, 311
   stratification, 104
Variance of an estimator: 36, 78. 87, 94, 103,104, 110, 112, 116, 118, 119, 128

Web survey: 159, 277
Weight coefficient: 260
Weight factor: 260
Weight:
   adjustment, 249, 314
   correction, 250, 314
   frequency, 314
   inclusion, 249, 313, 314
Weighted response rate: 214
Weighting adjustment: 4, 236, 249, 288
Weighting:
   linear, 250, 263
   multiplicative, 250, 260, 263
   propensity score, 266, 297
Weights:
   large, 264
   negative, 264
With replacement sampling, 31
Without replacement sampling, 30

WILEY SERIES IN SURVEY METHODOLOGY
Established in Part by WALTER A. SHEWHART AND SAMUEL S. WILKS

Editors: *Robert M. Groves, Graham Kalton, J. N. K. Rao, Norbert Schwarz, Christopher Skinner*

The *Wiley Series in Survey Methodology* covers topics of current research and practical interests in survey methodology and sampling. While the emphasis is on application, theoretical discussion is encouraged when it supports a broader understanding of the subject matter.

The authors are leading academics and researchers in survey methodology and sampling. The readership includes professionals in, and students of, the fields of applied statistics, biostatistics, public policy, and government and corporate enterprises.

ALWIN · Margins of Error: A Study of Reliability in Survey Measurement
BETHLEHEM · Applied Survey Methods: A Statistical Perspective
*BIEMER, GROVES, LYBERG, MATHIOWETZ, and SUDMAN · Measurement Errors in Surveys
BIEMER and LYBERG · Introduction to Survey Quality
BRADBURN, SUDMAN, and WANSINK ·Asking Questions: The Definitive Guide to Questionnaire Design—For Market Research, Political Polls, and Social Health Questionnaires, *Revised Edition*
BRAVERMAN and SLATER · Advances in Survey Research: New Directions for Evaluation, No. 70
CHAMBERS and SKINNER (editors · Analysis of Survey Data
COCHRAN · Sampling Techniques, *Third Edition*
CONRAD and SCHOBER · Envisioning the Survey Interview of the Future
COUPER, BAKER, BETHLEHEM, CLARK, MARTIN, NICHOLLS, and O'REILLY (editors) · Computer Assisted Survey Information Collection
COX, BINDER, CHINNAPPA, CHRISTIANSON, COLLEDGE, and KOTT (editors) · Business Survey Methods
*DEMING · Sample Design in Business Research
DILLMAN · Mail and Internet Surveys: The Tailored Design Method
GROVES and COUPER · Nonresponse in Household Interview Surveys
GROVES · Survey Errors and Survey Costs
GROVES, DILLMAN, ELTINGE, and LITTLE · Survey Nonresponse
GROVES, BIEMER, LYBERG, MASSEY, NICHOLLS, and WAKSBERG · Telephone Survey Methodology
GROVES, FOWLER, COUPER, LEPKOWSKI, SINGER, and TOURANGEAU · Survey Methodology, *Second Edition*
*HANSEN, HURWITZ, and MADOW · Sample Survey Methods and Theory, Volume 1: Methods and Applications
*HANSEN, HURWITZ, and MADOW · Sample Survey Methods and Theory, Volume II: Theory
HARKNESS, VAN DE VIJVER, and MOHLER · Cross-Cultural Survey Methods
KALTON and HEERINGA · Leslie Kish Selected Papers
KISH · Statistical Design for Research
*KISH · Survey Sampling
KORN and GRAUBARD · Analysis of Health Surveys
LEPKOWSKI, TUCKER, BRICK, DE LEEUW, JAPEC, LAVRAKAS, LINK, and SANGSTER (editors) · Advances in Telephone Survey Methodology
LESSLER and KALSBEEK · Nonsampling Error in Surveys

*Now available in a lower priced paperback edition in the Wiley Classics Library.

LEVY and LEMESHOW · Sampling of Populations: Methods and Applications, *Fourth Edition*
LYBERG, BIEMER, COLLINS, de LEEUW, DIPPO, SCHWARZ, TREWIN (editors) · Survey Measurement and Process Quality
MAYNARD, HOUTKOOP-STEENSTRA, SCHAEFFER, VAN DER ZOUWEN · Standardization and Tacit Knowledge: Interaction and Practice in the Survey Interview
PORTER (editor) · Overcoming Survey Research Problems: New Directions for Institutional Research, No. 121
PRESSER, ROTHGEB, COUPER, LESSLER, MARTIN, MARTIN, and SINGER (editors) · Methods for Testing and Evaluating Survey Questionnaires
RAO · Small Area Estimation
REA and PARKER · Designing and Conducting Survey Research: A Comprehensive Guide, *Third Edition*
SARIS and GALLHOFER · Design, Evaluation, and Analysis of Questionnaires for Survey Research
SÄRNDAL and LUNDSTRÖM · Estimation in Surveys with Nonresponse
SCHWARZ and SUDMAN (editors) · Answering Questions: Methodology for Determining Cognitive and Communicative Processes in Survey Research
SIRKEN, HERRMANN, SCHECHTER, SCHWARZ, TANUR, and TOURANGEAU (editors) · Cognition and Survey Research
SUDMAN, BRADBURN, and SCHWARZ · Thinking about Answers: The Application of Cognitive Processes to Survey Methodology
UMBACH (editor) · Survey Research Emerging Issues: New Directions for Institutional Research No. 127
VALLIANT, DORFMAN, and ROYALL · Finite Population Sampling and Inference: A Prediction Approach